Food Texture and Viscosity: Concept and Measurement

FOOD SCIENCE AND TECHNOLOGY

A SERIES OF MONOGRAPHS

A complete list of the books in this series appears at the end of the volume.

Food Texture and Viscosity:
Concept and Measurement

Malcolm C. Bourne
Department of Food Science and Technology
New York State Agricultural Experiment Station
and
Institute of Food Science
Cornell University
Geneva, New York

ACADEMIC PRESS
A Subsidiary of Harcourt Brace Jovanovich, Publishers
New York London
Paris San Diego San Francisco São Paulo Sydney Tokyo Toronto

ACADEMIC PRESS, INC.
111 Fifth Avenue, New York, New York 10003

United Kingdom Edition published by
ACADEMIC PRESS, INC. (LONDON) LTD.
24/28 Oval Road, London NW1 7DX

Library of Congress Cataloging in Publication Data

Bourne, Malcolm C.
Food texture and viscosity.

(Food science and technology)
Includes bibliographical references and index.
1. Food texture. 2. Viscosity. 3. Food--Analysis.
I. Title. II. Series.
TX531.B685 664'.07 82-6711
ISBN 0-12-119060-9 AACR2

PRINTED IN THE UNITED STATES OF AMERICA

82 83 84 85 9 8 7 6 5 4 3 2 1

To my beloved wife, Elizabeth

Contents

3. Principles of Objective Texture Measurement

4. Practice of Objective Texture Measurement

5. Viscosity and Consistency

6. Sensory Methods of Texture and Viscosity Measurement

7. Selection of a Suitable Test Procedure

Preface

This book is intended for those who want to know more about the texture and viscosity of food and how these properties are measured. It draws together literature from many sources including journals in chemistry, dentistry, engineering, food science, food technology, physics, psychology, and rheology. The *Journal of Texture Studies* and scientific and trade journals dedicated to special commodity groups, books, proceedings, and commercial literature have also been utilized.

The treatment is descriptive and analytical but not mathematical. Equations are given only when they illuminate the discussion and then in only the simplest form. Their derivations, however, are not given; this is not a mathematics textbook.

Chapter 1 defines texture terms, discusses the importance of textural properties of foods, points out the present status of food texture and viscosity measurements in the food industry, and gives a brief history of early developments in the field. Chapter 2 describes physical interactions between the human body and food—a necessary background for the ensuing chapters. Chapter 3 discusses the principles of objective methods of texture measurements (including ideas that have yet to evolve into commercially available instruments) and provides a foundation for the following chapter. Chapter 4 describes commercially available instruments and their use. Chapter 5 discusses the various types of viscous flow, followed by a brief description of commercial viscometers. Chapter 6 describes sensory methods for measuring texture and viscosity. Chapter 7 outlines a system for selecting a suitable instrument and method with the minimum of time cost. The Appendix lists names and addresses of suppliers of instruments for those who are interested in purchasing equipment. I have no vested interest in any corporation that sells texture-measuring instruments and have endeavored to be unbiased in describing commercial instruments and to make the list as complete as possible.

Many people will read this book selectively. The practicing food technologist and quality controller might want to concentrate on Chapter 4 and the latter part of Chapter 5, where instruments are described. The professor and college student might spend more time on Chapter 3 and the first part of Chapter 5, which describe principles. The sensory specialist will find Chapter 6 the most interesting. The researcher wanting to establish a texture laboratory will find Chapter 7 most useful.

I have expressed my own opinions, attitudes, and interpretations in this volume. For example, my personal conviction that empirical tests have been responsible for most of the successes in food texture measurements is reflected in the extended discussion of empirical methodology and the brief discussion of fundamental tests. Even if subsequent reports show the guidance to be wrong at times, I hope most readers will find useful the methods and yardsticks offered.

I have received help from many sources in the preparation of this book . A number of individuals and organizations provided figures or compiled tables; their contributions are noted where the figure or table appears. I particularly appreciate the assistance of C. Cohen, J. M. deMan, W. F. Shipe, A. S. Szczesniak, and P. W. Voisey, each of whom critically read one or more chapters in the draft stage and made numerous suggestions for improvement. S. Comstock has faithfully assisted with experiments in my laboratory for many years. R. Bowers made a major contribution by typing the manuscript. I sincerely thank all of them for their contributions.

CHAPTER 1

Texture, Viscosity, and Food

Importance of Textural Properties

The four principal quality factors in foods are the following:

1. *Appearance,* comprising color, shape, size, gloss, etc., is based on optical properties and a visual manifestation of size and shape.
2. *Flavor,* comprising taste (perceived on the tongue) and odor (perceived in the olfactory center in the nose), is the response of receptors in the oral cavity to chemical stimuli.
3. *Texture* is the response of the tactile senses to physical stimuli that result from contact between some part of the body and the food.
4. *Nutrition.*

Other factors, such as cost, convenience, and packaging, are also important but are not considered quality factors of foods. Of the above listed the first three are termed "sensory acceptability factors" because they are perceived by the senses directly. Nutrition is a quality factor that is not an acceptability factor because it is not perceived by the senses.

The sensory acceptability factors of foods are extremely important because people obtain great enjoyment from eating their food and, furthermore, the enjoyment of food is a sensory pleasure that we can appreciate from the cradle to the grave. One prominent food scientist has said, and with good reason, "To most people, eating is a very personal, sensual, highly enjoyable experience; enjoyment here and now, with little worry about long-term consequences" (Clausi, 1973).

The importance of texture in the overall acceptability of foods varies widely, depending upon the type of food. We could arbitrarily break it into three groups:

1. *Critical:* Those foods in which texture is the dominant quality characteristic; for example, meat, potato chips, and celery.

2. *Important:* Those foods in which texture makes a significant but not a dominant contribution to the overall quality, contributing, more or less equally, with flavor and appearance; for example, most fruits, vegetables, bread, and candy fall into this category.

3. *Minor:* Those foods in which texture makes a negligible contribution to the overall quality; examples are most beverages and thin soups.

The importance of texture in foods was indirectly pointed out by Schiffman (1973), who fed 29 different foods to people who had been blindfolded and asked them to identify the foods based only on flavor. The samples had been pureed by blending and straining in order to eliminate textural clues. Some of the data from Schiffman's work are shown in Table 1. It is remarkable to discover how poorly many foods are identified when their texture and color are concealed and flavor is the only attribute that can be used to identify the food. Young adults of normal weight were able to identify correctly only 40.7% of the foods used in the study. It is surprising to find, for example, that only 4% of the respondents could

TABLE 1

PERCENTAGE OF CORRECT IDENTIFICATION OF PUREED FOODS[a]

Food	Normal weight (young)	Obese (young)	Normal weight (aged)
Apple	81	87	55
Strawberry	78	62	33
Fish	78	81	59
Lemon	52	25	24
Carrot	51	44	7
Banana	41	69	24
Beef	41	50	27
Rice	22	12	15
Potato	19	69	38
Green pepper	19	25	11
Pork	15	6	7
Cucumber	8	0	0
Lamb	4	6	—
Cabbage	4	0	7
Mean for 29 foods	40.7	50.0	30.4

[a]From Schiffman (1973), with permission from author.

identify cabbage correctly by flavor only, 15% for pork, 41% for beef, and 51% for carrots.

Szczesniak and Kleyn (1963) gave a word association test to 100 people to determine their degree of texture consciousness and the terms they used to describe texture. Seventy-eight descriptive words were used by the participants. These authors concluded that texture is a discernible characteristic, but that it is more evident in some foods than others. Foods that elicited the highest number of texture responses either were bland in flavor or possessed the characteristics of crunchiness or crispness.

Yoshikawa *et al.* (1970a,b,c) conducted tests in Japan that were similar to those conducted by Szczesniak's group in the United States. They asked 140 female college students to describe the texture of 97 foods and collected 406 different words that describe textural characteristics of foods. These studies showed the importance of textural properties as a factor in food quality and the great variety of textures found in food. The 10 most frequently used words in these two studies are listed in Table 2. It is interesting to notice that 7 of these 10 words are common to both lists, although the Japanese culture and food habits are substantially different from those of North America. It is also interesting to note that the Japanese used 406 descriptive words as compared to 78 words in the United States.

Perhaps the richer textural vocabulary of the Japanese is due partly to the greater variety of textures presented in Japanese cuisine, making them more

TABLE 2
MOST FREQUENTLY USED TEXTURE WORDS[a]

United States[b]	Japan[c]
Crisp	Hard
Dry	Soft
Juicy	Juicy
Soft	Chewy
Creamy	Greasy
Crunchy	Viscous
Chewy	Slippery
Smooth	Creamy
Stringy	Crisp
Hard	Crunchy
78 words	406 words

[a]In descending order of frequency.
[b]Szczesniak and Kleyn (1963).
[c]Yoshikawa *et al.* (1970a).

sensitive to subtle nuances in textures, and partly to the picturesque Japanese language which uses many onomatopoeic words. For example, Yoshikawa *et al.* (1970a) assign to each of the following expressions the meaning of some form of crispness: *kori-kori, pari-pari, saku-saku, pori-pori, gusha-gusha, kucha-kucha,* and *shaki-shaki.*

In a second study (Szczesniak, 1971), a word association test was given to 150 respondents and the results were similar to the first study. This test again showed that texture is a discernible characteristic of foods and the awareness of it generally equivalent to that of flavor. This study also found that women and people in the higher economic brackets showed a higher level of awareness of the textural properties of foods than did the general population.

Szczesniak and Kahn (1971) conducted in-depth interviews with homemakers and found that texture awareness in the United States is often apparent at a subconscious level and that it is taken more or less for granted; however, when the textural aspects did not come up to expectations, there was a sharp increase in the awareness of the texture and criticism of the textural deficiencies. The authors state that

> If the texture of a food is the way people have learned to expect it to be, and if it is psychologically and physiologically acceptable, then it will scarcely be noticed. If, however, the texture is not as it is expected to be . . . it becomes a focal point for criticism and rejection of the food. Care must be taken not to underestimate the importance of texture just because it is taken for granted when all is as it should be.

Szczesniak and Kahn (1971) also reported that time of day exerted a strong influence on textural awareness and flavor. At breakfast, most people prefer a restricted range of familiar textures that lubricate the mouth, remove the dryness of sleep, and can be swallowed without difficulty. New or unfamiliar textures, and textures that are difficult to chew, are not wanted at breakfast.

People are willing to accept a wider range of textures at the midday meal just so long as it is quick and easy to prepare and not messy to eat. After all, this is a practical meal with a limited time for preparation and consumption.

Texture is most appreciated and enjoyed at the evening meal. This is the time for relaxation, which comes after the day's work and, for most people, is the largest meal of the day when several courses are served and a wide range of textures is expected and relished. The appetizer (nondemanding textures and flavors that stimulate the flow of saliva) is perceived as a preparation for the main meal which follows, and this in turn features a great variety of textures, including some items that require considerable energy to chew. No texture seems to be completely inappropriate for the main course so long as there are several contrasting textures.

The dessert features textures that require low energy for mastication and restore the mouth to a relaxed and pleasant feeling. This is the time for ''fun'' foods that are easy to manipulate and leave a nice feeling in the mouth. Soft, smooth, creamy, or spongy textures are desired. Hard, chewy textures are not wanted at the conclusion of the meal (Szczesniak and Kahn, 1971).

In yet another report, Szczesniak (1972) studied the attitudes of children and teenagers to food texture and found it to be an important aspect of their liking or disliking of specific foods. The young child prefers simple soft textures that can be managed within the limited development of the structures of the mouth. The child extends its range of relished textures as its teeth, jaws, and powers of coordination develop. This study also showed that teenagers have a high degree of texture awareness that sometimes surpasses that of adults, suggesting that perhaps the next generation of adult consumers may be more sophisticated and demanding in terms of textural qualities of tbe foods that they purchase.

In a survey of consumer attitudes toward product quality conducted by the A. C. Nielsen Co. in 1973, complaints about product quality were recorded (Anonymous, 1973). The results are shown in Table 3. Complaints about a broken or crumbled product (a texture defect) headed the list at 51% of respondents. The second item (product freshness) is frequently measured by textural properties such as firmness. These data indicate that there is room for considerable improvement in textural properties of foods that are presently marketed.

The importance of texture, relative to other quality factors of foods, may be affected by culture. For example, in a study of food patterns of the United States and Caribbean blacks, Jerome (1975) stated: "For Afro-Americans of southern rural origin, the element of primary importance associated with food patterns is *texture;* flavor assumes secondary importance."

Another indication of the importance of texture in food is the large size of the dental industry in developed countries. This is due primarily to the fact that people do not want to be deprived of the gratifying sensations that arise from eating their food. From the nutritional standpoint it is possible to have a completely adequate diet in the form of fluid foods that require no mastication, but few people are content to live on such a diet. As their tooth function deteriorates

TABLE 3

CONSUMER COMPLAINTS ABOUT PRODUCT QUALITY[a]

Type of complaint	Total respondents (%)
Broken or crumbled product	51
Product freshness	47
Contaminated product	28
Incorrect carbonation	23
Bulged can	16
Other	9

[a]From Anonymous (1973).

with age, they undergo the inconvenience and cost of dental care that restores tooth function and enables them to continue to enjoy the textural sensations that arise from masticating their food.

The deeply ingrained need to chew on things is also found among infants. Growing infants are provided with teething rings and similar objects in order to give them something to satisfy their need for biting and chewing. If the baby is not given something on which it can chew, it will usually satisfy its need to chew on items such as the post of its crib, father's best slipper, or the expensive toy given to it by the doting grandmother.

There is an enormous range in textural characteristics of foods: the chewiness of meat, the softness of marshmallows, the crispness of celery and potato chips, the juiciness of fresh fruits, the smoothness of ice cream, the soft toughness of bread, the flakiness of fish, the crumbliness of cake, the melting of jelly, the viscosity of thick soup, the fluidity of milk, and many others. This great range of types of rheological and textural properties found in foods arises from the human demand for variety in the nature of their food.

An historical example of this human need for variety in food is found in the Old Testament. When the children of Israel made their historic 40-year march from Egypt to Palestine across the great desert, God provided their food in the form of manna, which fell nightly in sufficient quantity to feed daily this migrating nation. Manna was a delicious food to eat; it was known as ''Bread from Heaven,'' and is described as being ''crisp and sweet as honey.'' We know it provided all the essential nutrients because the people were free from illness during this long period of time. Despite the high quality and excellent sensory characteristics of manna, people became tired of eating it every day and demanded a change. The record says

> and the children of Israel also wept again, and said, Who shall give us flesh to eat? We remember the fish, which we did eat in Egypt freely; the cucumbers, and the melons, and the leeks, and the onions, and the garlick. But now our soul is dried away: There is nothing at all, beside this manna, before our eyes (Num. 11:4–6).

On another occasion the children of Israel complained about manna, saying ''Our soul loatheth this worthless bread'' (Num. 21:5).

The people of the twentieth century are just as insistent in demanding a variety of textures and flavors in their food as were the children of Israel many centuries ago. A large part of the effort of the food industry of our day is directed toward providing both high quality and a wide variety of textures and viscosities in the foods that are provided to the public.

The Status of Food Texture Measurements

Muller (1969b) reported upon a survey of food quality measurements made by the food processing industry in the United Kingdom. One hundred twenty-five companies responded to the survey and reported upon 228 different food products, with the following results:

154 Products (67.5%) used some kind of chemical quality control test.
125 (54.8%) Used some kind of rheological test.
49 (21.5%) Used neither rheological nor chemical tests.

Of those using rheological tests, 9 out of the 125 (7%) were not satisfied with the tests and instruments they were using. One aspect of this survey that shows the need for more knowledge and better techniques in this area comes from responses on the 103 products in which no rheological tests were used. Here the comments of the manufacturers were as follows:

Used a rheological test and gave it up: 9 products (8.7%).
Would use a rheological test if a good one could be found: 48 products (46.6%).
Not worth bothering about a rheological test: 21 products (20.4%).
No comments: 26 products (25.2%).

Another survey taken on food texture measurements in Canada (Cumming *et al.*, 1971) gave the results shown in Table 4. This group found in their survey of the food processing industry that the most widely used instruments in quality control were rotational viscometers and penetrometers and that 48% of the respondents do not use texture evaluation instruments. It is to be expected that few rheological tests would be made on beverages since beverages are predominantly water and there is usually little change in the viscosity of the water when it is formulated into most beverages.

TABLE 4
USE OF FOOD TEXTURE MEASUREMENTS IN CANADA[a]

Category	No. of replies	Using a texture measuring device (%)	Avg. no. of instruments per organization
Meat	14	35.7	2.4
Fish	10	20.0	1.0
Canner/freezer	14	78.6	3.4
Dairy	12	41.7	2.4
Confectionery	11	72.7	2.9
Baking	12	50.0	1.6
Beverage	10	0	0
Fats and oils	9	77.8	1.9
Multiproduct	17	76.5	2.9
Government	8	87.5	—
Universities	2	100.0	—

[a]From Cumming *et al.* (1971).

One concludes from these surveys that the status of food texture measurements in the food industry is far from satisfactory. Although adequate techniques exist to measure the texture of some foods, there are many foods for which texture measurements are unsatisfactory or nonexistent. There is much more to be learned about texture of foods and how to measure these textures.

Definition of Texture

This is a difficult term to define since it means different things to various people. The dictionary definition of texture is of little help because it relates mainly to textiles and the act or art of weaving and, in general, to "the disposition or manner of union of particles or smaller constitutent parts of a body or substance, the fine structure." The dictionary definition that comes closest to the needs of the food technologist states that texture is "the manner of structure, interrelation of parts, structural quality." Webster's dictionary gives examples of texture for textiles and fibers, weaving, artistic compositions, music, poetry, petrography (the study of rocks), texture of a bone or plant, but does not even mention foods. In view of this lack of coverage in the dictionary, food technologists have endeavored to produce their own definition of what is meant by texture. These definitions fall into two groups.

The first group comprise what might be called "commodity-oriented" definitions in which the term texture is applied to a particular quality attribute of a given type of food. For example, in ice cream grading, texture means the smoothness of the ice cream but does not include other factors such as hardness and melting properties; in bread grading, texture means uniformity of the crumb and even distribution in size of the gas bubbles but does not include the softness or toughness of the bread. Ball *et al.* (1957) give two definitions for texture of meat. The first, which they call a sight definition, is "texture of meat is the macroscopic appearance of meat tissues from the standpoint of smoothness or fineness of grain." The second, which they call a feel definition, is "the texture of cooked meat is the feel of smoothness or fineness of muscle tissue in the mouth." It is interesting to note that neither of these definitions includes the property of toughness, which most people consider of great importance in the quality of meat.

Davis (1937) defines texture of cheese as

> that which is evident to the eye, excluding color. . . . Texture varies in meaning in different localities, but is frequently taken to include both closeness (absence of cracks) and shortness or brittleness (easy breaking of a plug).

Davis also defines "body" as that quality which is perceptible to touch.

Other workers consider that texture applies to all foods and have endeavored to develop definitions that reflect a broad coverage. Some of these definitions are as follows:

> Texture means those perceptions that constitute the evaluation of a food's physical characteristics by the skin or muscle senses of the buccal cavity, excepting the sensations of temperature or pain (Matz, 1962).

> Texture is the composite of the structural elements of food and the manner in which it registers with the physiological senses (Szczesniak, 1963a).

> By texture we mean those qualities of food that we can feel either with the fingers, the tongue, the palate, or the teeth (Potter, 1968).

> Texture is the composite of those properties (attributes) which arise from the structural elements of food and the manner in which it registers with the physiological senses (Sherman, 1970).

> In its fullest sense the textural experience during chewing is a dynamic integration of mouthfeel, the prior tactile responses while handling the foodstuff, and a psychic anticipatory state arising from the visible perception of the food's overall geometry and surface features. . . . Texture should be regarded as a human construct. A foodstuff cannot have texture, only particular mechanical (and other) properties which are involved in producing sensory feelings or texture notes for the human being during the act of chewing the foodstuff (Corey, 1970).

> (Texture is) the attribute of a substance resulting from a combination of physical properties and perceived by the senses of touch (including kinesthesis and mouthfeel), sight, and hearing. Physical properties may include size, shape, number, nature and conformation of constituent structural elements (Jowitt, 1974).

> Texture is that one of the three primary sensory properties of foods that relates entirely to the sense of touch or feel and is, therefore, potentially capable of precise measurement objectively by mechanical means in fundamental units of mass or force (Kramer, 1973).

> Texture is the way in which the various constituents and structural elements of a food are arranged and combined in a micro- and macro structure and the external manifestations of this structure in terms of flow and deformation (deMan, 1975).

> (Texture comprises) those properties of a foodstuff, apprehended by the eyes and by the skin and muscle senses in the mouth, including roughness, smoothness, graininess, etc. (Anonymous, 1964).

> Texture (*noun*): All the rheological and structural (geometrical and surface) attributes of a food product perceptible by means of mechanical, tactile and, where appropriate, visual and auditory receptors (International Organization for Standardization, Standard 5492/3, 1979).

Although we do not have an entirely satisfactory definition of texture we can say with a high degree of certainty that texture of foods has the following characteristics:

1. It is a group of physical properties that derive from the structure of the food.
2. It belongs under the mechanical or rheological subheading of physical

properties. Optical properties, electrical and magnetic properties, and temperature and thermal properties are physical properties that are excluded from the texture definition.

3. it consists of a *group* of properties, not a single property.

4. texture is sensed by the feeling of touch, usually in the mouth, but other parts of the body may be involved (frequently the hands).

5. it is not related to the chemical senses of taste or odor.

6. objective measurement is by means of functions of mass, distance, and time only; for example, force has the dimensions MLT^{-2}, work has the dimensions ML^2T^{-2}, and flow has the dimensions L^3T^{-1}.

Since texture consists of a number of different physical sensations, it is preferable to talk about "textural properties," which infers a group of related properties, rather than "texture," which infers a single parameter. There are still many people handling foods who talk about the texture of a food as though it were a single property like pH. It is important to realize that texture is a multifaceted

TABLE 5

RELATIONS BETWEEN TEXTURAL PARAMETERS AND POPULAR NOMENCLATURE[a]

Mechanical characteristics Primary parameters	Secondary parameters	Popular terms
Hardness		Soft → firm → hard
Cohesiveness	Brittleness	Crumbly → crunchy → brittle
	Chewiness	Tender → chewy → tough
	Gumminess	Short → mealy → pasty → gummy
Viscosity		Thin → viscous
Elasticity		Plastic → elastic
Adhesiveness		Sticky → tacky → gooey
Geometrical characteristics **Class**		**Examples**
Particle size and shape		Gritty, grainy, coarse, etc.
Particle shape and orientation		Fibrous, cellular, crystalline, etc.
Other characteristics **Primary parameters**	**Secondary parameters**	**Popular terms**
Moisture content		Dry → moist → wet → watery
Fat content	Oiliness	Oily
	Greasiness	Greasy

[a]From Szczesniak (1963a); reprinted with permission of Institute of Food Technologists.

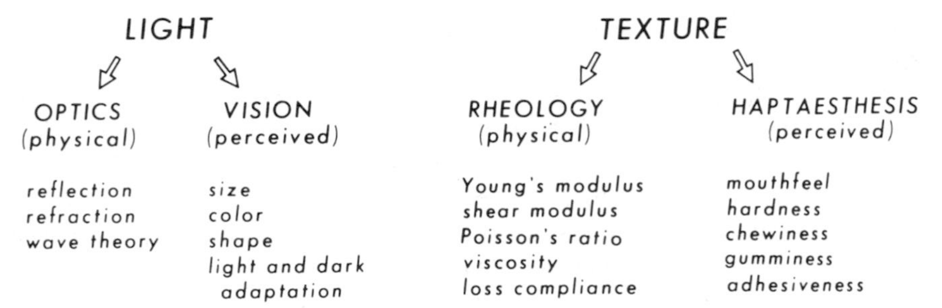

FIG. 1. Comparison of physical measurement and human perception of light and texture. (After Muller, 1969a.)

group of properties of foods. Table 5 lists some relations between textural parameters of foods and popular terms that are used to describe these properties.

These concepts lead to the following definition. The *textural properties* of a food are that group of physical characteristics that arise from the structural elements of the food, are sensed by the feeling of touch, are related to the deformation, disintegration, and flow of the food under a force, and are measured objectively by functions of mass, time, and distance.

Muller (1969a) claims that the term "texture" should be discarded because it is confusing. In present usage it means both an exact physical property and also a perceived property. He proposes two terms to take the place of the word texture: (a) *rheology,* a branch of physics that describes the physical properties of the food; and (b) *haptaesthesis* (from the Greek words meaning sensation and touch), a branch of psychology that deals with the perception of the mechanical behavior of materials.

Muller compares these two terms with the study of light, which has two distinct branches: (1) *optics,* the study of the physical properties of light, including reflection, refraction, wave theory, etc.; (2) *vision,* the study of the psychological and physiological human responses to light, such as the perception of objects, perception of color, light and dark adaptations, etc.

Figure 1 shows schematically the analogy that Muller uses to break the property of texture into two branches similar to that of light.

Other Definitions

Some other words that are used in a texture-related sense are described as follows:

Kinesthetics. "Those factors of quality that the consumer evaluates with his sense of feel, especially mouthfeel" (Kramer and Twigg, 1959). This word comes from the words ''kinein'' (the muscle sense to move) and ''aesthesis'' (perception).

Body. ''The quality of a food or beverage, relating variously to its consistency, compactness of texture, fullness, or richness'' (Anonymous, 1964). ''That textural property producing the mouthfeel sensation of substance'' (Jowitt, 1974). ''The quality of a food or beverage relating either to its consistency, compactness of texture, fullness, flavor, or to a combination thereof'' (American Society for Testing and Materials, Standard E253-78a).

Chewy. ''Tending to remain in the mouth without rapidly breaking up or dissolving. Requiring mastication'' (Anonymous, 1964). ''Possessing the textural property manifested by a low resistance to breakdown on mastication'' (Jowitt, 1974).

Haptic. ''Pertaining to the skin or to the sense of touch in its broadest sense'' (Anonymous, 1964).

Mealy. ''A quality of mouthfeel denoting a starchlike sensation. Friable'' (Anonymous, 1964). ''Possessing the textural property manifested by the presence of components of different degrees of firmness or toughness' (Jowitt, 1974).

Mouthfeel. ''The mingled experience deriving from the sensations of the skin in the mouth during and/or after ingestion of a food or beverage. It relates to density, viscosity, surface tension, and other physical properties of the material being sampled'' (Anonymous, 1964). ''Those textural characteristics of a food responsible for producing characteristic tactile sensation on the surfaces of the oral cavity; the sensation thus produced'' (Jowitt, 1974).

Getaway. ''That textural property perceived as shortness of duration of mouthfeel'' (Jowitt, 1974).

The following definitions were all developed by the International Organization for Standardization, Standard 5492/3, 1979:

Consistency. ''All the sensations resulting from stimulation of the mechanical receptors and tactile receptors, especially in the region of the mouth, and varying with the texture of the product.''

Hard (adjective). ''As a texture characteristic, describes a product which displays substantial resistance to deformation or breaking. The corresponding noun is hardness.''

Soft (adjective). ''As a texture characteristic, describes a product which displays slight resistance to deformation. The corresponding noun is softness.''

Tender (adjective). ''As a texture characteristic, describes a product which, during mastication, displays little resistance to breaking. The corresponding noun is tenderness.''

Firm (adjective). ''As a texture characteristic, describes a product which, during mastication, displays moderate resistance to breaking. The corresponding noun is firmness.''

Texture versus Viscosity

Viscosity is defined as the internal friction of a fluid or its tendency to resist flow. Both gases and liquids have viscosity, but since there are no gaseous foods (although some foods contain entrained gas) we will not discuss viscosity of gases.

At first sight the distinction between texture and viscosity seems simple—texture applies to solid foods and viscosity applies to fluid foods. Unfortunately, the distinction between solids and liquids is so blurred that it is impossible to clearly demarcate between texture and viscosity. While rock candy can definitely be considered as a solid and milk a liquid, there are many solid foods that exhibit some of the properties of liquids and many liquid foods that exhibit some of the properties of solids. Some apparently solid foods behave like liquids when sufficient stress is applied.

The indistinct separation between solids and liquids results in some confusion in the literature between food texture and viscosity and that confusion is reflected to some extent in this book. The author has followed the arbitrary distinction that foods that are usually considered to be solid or near-solid are discussed in Chapters 3 and 4 and foods that are usually considered to be liquid or near-liquid are discussed in Chapter 5. Some of the tests for solid foods described in Chapters 3 and 4 should really be discussed in Chapter 5 on viscosity, and some of the material in Chapter 5 could have been discussed in Chapters 3 and 4.

The nature of the overlap between solids and liquids should become more clear when the reader reaches the end of Chapter 5. At this point, the reader should be aware that the distinction between solids and liquids is not clearcut and that some inconsistencies in treatment are found because of this problem.

Texture and Food

Much food processing is directed to changing the textural properties of the food, generally in the direction of weakening the structure in order to make it easier to masticate. Wheat could be eaten as whole grains but most people find them too hard to be appealing. Instead, the structure of the wheat kernel is destroyed by grinding it into flour, which is then baked into bread with a completely different texture and structure than the grain of wheat. The texture of leavened bread is much softer and less dense than that of grains of wheat and is a more highly acceptable product, judging by the quantity of bread that is consumed.

The processing that is needed to develop desirable textural properties in foods can be expensive. In the United States the wholesale price of wheat is about 5–10 cents a pound while the retail price of bread is usually in the range of 50 cents to more than a dollar a pound. The wide disparity in price between bread and wheat indicates the high cost of the conversion of the wheat grain into bread and also the price that people are prepared to pay to obtain the type of textures they desire. Breakfast cereals made from wheat that has been rolled into flakes cost over $1 a pound, which is another indication of the price that people will pay to convert grains of wheat into a more texturally desirable form. One of the major reasons for cooking most vegetables before consumption is to soften them and make them easier to masticate.

Although much food processing is deliberately designed to modify textural properties, there are some instances where the textural changes are inadvertent, being a side result of processing for some other purpose. These textural changes are frequently undesirable. A good example of this is the extreme softening and severe textural degradation that results from canning, freezing, or irradiation preservation of fruits and vegetables. In some instances the damage to texture is so great that the resultant product is unsalable, in which case that processing method is not used on that commodity. For example, the dose of about two million rads required to sterilize horticultural crops causes such extreme softening of the tissue that it has eliminated the incentive to continue research to resolve questions on the safety of irradiation-sterilized fruit.

Foods might be classed into two groups, depending on the relative ease with which texture can be controlled:

1. *Native foods* are those foods in which the original structure of the agricultural commodity remains essentially intact. With these foods the food technologist has to take what nature provides in the form of fruit, fish, meat, vegetables, etc., and can only change the texture by processing methods such as heating, cooling, and size reduction. Usually there is almost no direct control over the composition of these foods, although with some of them it is possible to partially control the composition and texture by breeding, time of harvest, and cultural factors.

2. *Formulated foods* are those foods that are processed from a number of ingredients to make a food product that is not found in nature. Many native foods are transformed into ingredients for formulated foods, but in doing so the native plant or animal structure and organization is usually lost. Examples of this type of commodity are bread, ketchup, ice cream, jellies, mayonnaise, candy, and sausage. With this class of commodity it is possible to change the formulation by the number, amount, and quality of ingredients that are used in addition to processing variables, and hence there are more options available to control the texture of the finished product and to develop specified textures and structures not found in native foods.

Despite the wide range of options available, food technologists have experienced great difficulty in fabricating foods that closely simulate native foods because of their cellular structure and complex structural organization. The turgor that provides much of the crispness of many fresh fruits and vegetables arises from the physiological activity of the living tissue and is unlikely ever to be duplicated in a fabricated analog.

Textural properties are used as the basis of selection or rejection of certain parts of foods. Many children dislike the texture of bread crust and engage in various subterfuges to avoid eating it. Texture is the main reason why the skin of some fruits and vegetables is eaten while that of other fruits and vegetables is not eaten. The skin is usually eaten with the fleshy portion when it is tender or thin, as in the strawberry, cherry, green pea, and green bean. The skin is usually not eaten when it is texturally objectionable because it is thick, hard, tough, hairy, fibrous, or prickly, as in the grapefruit, pumpkin, mango, peach, banana, and pineapple. Of course, there are some borderline cases; some people peel their apples, figs, potatoes, and tomatoes before eating while others do not.

In recent years a great deal of attention has been given to "texturizing" vegetable proteins. Most people enjoy the chewy fibrous texture of muscle meat but this kind of texture is not found in vegetable proteins. However, vegetable proteins generally cost less than animal proteins because the biological conversion of vegetable protein into animal protein by the cow, pig, or chicken is inefficient, with, typically, 5–20% of the protein fed to the animal recovered as edible protein food. This inefficient conversion raises the cost of animal protein. In contrast, the direct conversion of vegetable protein into products with a meatlike chewy texture by modern processing technology is usually 70–90% efficient.

Considerable research attention is presently being given to imparting a meatlike texture to vegetable proteins in order to obtain the desirable chewy texture of meat coupled with the lower cost of the vegetable proteins and (for some people) avoidance of cholesterol and other undesirable features of meat. Substantial progress has been made in developing meatlike textures in vegetable proteins but more progress is needed before these products are equal to the meat in their overall textural properties.

The problem of imparting a desirable texture to a food is exemplified in the problems of fish protein concentrate (FPC). The production of FPC makes available for human consumption the protein from many species of fish that are normally not used. The general process is to remove the fat and moisture from the fish and grind the residue into a powder. The problems of developing a bland flavor and absence of fishy flavor, and obtaining stability and good nutritional value of the FPC have been solved, but the problem of utilizing FPC for food has not been satisfactorily solved. FPC is a dry powder and no more a food than is wheat flour a food. It is a food *ingredient* that must be fabricated into a food in much the same way as wheat flour is fabricated into bread, cookies, and similar products and this has proven to be an extremely difficult task. Dry FPC has such poor functional properties that it cannot be used to develop texture in formulated foods. At the present time the only satisfactory use for FPC is to add it to existing foods at levels that are so low that the textural properties of that food disguise the presence of FPC.

The problem of fabricating vegetable proteins into foods with acceptable texture is extremely difficult. Only those food technologists who have wrestled with this problem know how difficult it is. Several years ago a chemist, writing on future sources of food, wrote:

> The polymer chemist who has produced an almost endless variety of fibers, gels, gums, resins, and plastic products would encounter no major difficulty in incorporating synthetic food materials in products of nearly any desired consistency or texture, and could prepare highly acceptable counterparts of steak, Jell-o, cheese, or seafood.

This scientist should be sentenced to spend 10 years hard labor in the product development laboratory for making such a misleading statement! Acceptable texture has been a limiting factor in fabricated food development.

Rheology and Texture

Rheology is the study of the deformation and flow of matter. The science of rheology can be applied to any product and in fact was developed by scientists studying printing inks, plastics, rubber, and similar materials.

Food rheology "is the study of the deformation and flow of the raw materials, the intermediate products, and the final products of the food industry" (White, 1970). In this definition the term "food industry" should be broadly defined to include the behavior of foods in the home.

Psychophysics is "the study of the relationship between measurable stimuli and the corresponding responses" (International Organization for Standardization, Standard 5492/1, 1977).

Psychorheology. There are two types of definitions given to psychorheology. The first is a scientific definition: 1. Psychorheology is a branch of psychophysics dealing with the sensory perception of rheological properties of foods. Another definition, which might be called a people-centered definition, is the following: 2. Psychorheology is the relationship between the consumer preferences and rheological properties of foods.

Both of these definitions are meant to bridge the gap between the physical or rheological properties of foods and the sensing of those properties by the human senses (see Fig. 1).

The science of rheology has many applications in the field of food acceptability, food processing, and handling. A number of food processing operations depend heavily upon rheological properties of the product at an intermediate stage of manufacture because this has a profound effect upon the quality of the finished product. For example, the rheology of bread dough, milk curd, and meat emulsions are important aspects in the manufacture of high-quality bread, cheese, and sausage products. The agricultural engineer is interested in the ability of foods to be handled by machinery and in the creep and recovery of agricultural products that are subjected to stresses, particularly long-term stresses resulting from storage under confined conditions such as the bottom of a bulk container.

Viscometry, especially non-Newtonian viscometry, is an important component of the quality of most fluid and semifluid foods. The food engineer is interested in the ability to pump and mix liquid and semiliquid foods. Plasticity, pseudoplasticity, and the property of shear thinning are important quality factors in foods and the study of these properties is part of the science of rheology. A wide variety of foods, such as butter, margarine, applesauce, tomato catsup, mayonnaise, peanut butter, and many puddings are either plastic or pseudoplastic in nature. They are required to spread and flow easily under a small force but to hold their shape when not subjected to any external force other than gravity. All of these properties fall within the field of rheology.

When celebrating the golden anniversary of the founding of the field of rheology, the then president of the American Society of Rheology singled out for special comment the interesting rheological characteristics of foods in the following words:

> One of the world's greatest rheological laboratories is in the kitchen. Who can cease to wonder at the elasticity of egg white, or of the foam it forms when beaten with air? At the transformation of gelatin from a watery solution to an elastic gel? At the strange flow properties of mayonnaise, ketchup, peanut butter, or starch paste? Or at the way bread dough defies both gravity and centrifugal force as it climbs up the shaft of the beater? (Krieger, 1979).

Rheology is important to the food technologist because it has many applications in the three major categories of food acceptability:

1. *Appearance*. There is a small component of rheology in appearance because certain structural and mechanical properties of some foods can be determined by appearance; for example, we can see on the plate how runny the food is.

2. *Flavor*. Rheology has no direct part in this category, although the manner of food breakdown in the mouth can affect the rate of release of flavor compounds.

3. *Touch*. Rheological properties are a major factor in the evaluation of food quality by the sense of touch. We hold foods in the hand and from the sense of deformability and recovery after squeezing frequently obtain some idea of their textural quality. For example, fresh bread is highly deformable while stale bread is not; the flesh of fresh fish recovers quickly after squeezing while the stale fish does not. During the process of mastication a number of rheological properties

such as the deformation that occurs on the first bite and the flow properties of the bolus (the mass of chewed food with saliva) are sensed in the mouth.

The importance of rheology in foods has been well established in the preceding discussion. However, the science of rheology does not cover all of the aspects that should be included in the broad definition of food texture. Mastication is a process in which pieces of food are ground into a very fine state, but the process of size reduction (synonyms are comminution, disintegration, pulverization, and trituration) does not belong in the field of rheology. During mastication the size and shape of food particles and their surface roughness are sensed and these are important attributes of the overall textural sensation. Brandt *et al.* (1963) described the surface properties of food particles in the sensory terms of powdery, chalky, grainy, gritty, coarse, lumpy, beady, flaky, fibrous, pulpy, cellular, aerated, puffy, and crystalline. Bourne (1975a) suggested that the word "rugosity" or surface roughness is an important attribute of the food particles that are sensed in the mouth. But these attributes are not rheological properties of the food.

The ability of the food to wet with saliva and to absorb saliva or to release moisture or lipid are important textural sensations that do not belong in the field of rheology. Phase changes resulting from temperature changes occurring in the mouth are an important part of the texture sensation of some foods; for example, ice cream, chocolate, and jelly melt in the mouth while the oil in hot soup may solidify in the mouth during mastication. These changes are not rheological properties although they are frequently sensed by changes in rheological properties.

From this evidence we have to conclude that the field of food texture falls partly within the field of conventional rheology and partly outside this field. The food technologist certainly needs to define and measure certain rheological properties of foods, but he has interests where the classical science of rheology is of little help in his studies of the textural properties of foods and he is, therefore, forced to develop his own techniques.

Rheology defines and measures *properties* of foods. But the food technologist is also interested in the *process* of mastication and the changes in rheological and other textural properties that occur during mastication. The fact that fundamental rheological measurements usually do not correlate as well with sensory measurements of texture as do empirical tests may result from the incompleteness of the science of rheology to describe all of the changes, or perhaps even the most important changes that are actually sensed in the mouth and are of most interest to the food technologist.

One of the founders of the field of rheology stated,

The flow of matter is still not understood and since it is not mysterious like electricity, it does not attract the attention of the curious. The properties are ill defined and they are imperfectly measured if at all, and they are in no way organized into a systematic body of knowledge which can be called a science (Bingham, 1930).

Although this comment may not apply today to the field of rheology in general, it is fair to say that it still applies to the subfield of food rheology. Only a small number of research scientists devote their career to food rheology; there is a large volume of empirical information and a small volume of utilizable fundamental concepts. The author hopes that this book will help systematize the widely scattered body of knowledge in this field and hence promote the development of the field of food rheology into a rigorous scientific discipline.

Early History

It is not easy to decide where to begin citing the work of the early scientists who pioneered the development of the study of the texture and viscosity of foods. Robert Hooke (in England in 1660) enunciated the principle of elastic deformation of solids, giving rise to the descriptive term "Hookean solid" that is used today. A contemporary, Isaac Newton (in England in 1687), enunciated the law governing the flow of simple liquids, giving rise to the term "Newtonian fluid." However, the findings of these two eminent scientists did not apply specifically to foods.

Possibly, the first person to develop an instrument expressly for testing foods was Lipowitz (1861, Germany), who developed a simple puncture tester for measuring the firmness of jellies (see Fig. 2, p. 52). Carpi (1884, Italy) also developed a puncture tester for cooled olive oil and other fats. Schwedoff (1889, France) developed a deformation apparatus for jelly based on a torsion test and measured rigidity, viscosity, and relaxation.

Hogarth (1889, Scotland) obtained a patent for a device that measured the consistency of dough using the same principles as the modern Farinograph. Brabender, in Germany (1901–1970) developed a line of equipment for measuring the rheological properties of flour dough and founded companies in Germany and in the United States that still bear his name. Brabender (1965) recalled that an instrument for dough extensibility was developed in Hungary by Kosutány and Rejtö at the beginning of this century (Košutány, 1907). He also pointed out that, in 1905, another Hungarian, Professor Jenö von Hankóczy, designed an apparatus that measured the volume of air that could be blown into a disk of washed wheat gluten before it burst. This device was the forerunner of the Alveograph.

Wood and Parsons (1891, United States) describe a puncture test developed for measuring the hardness of butter. Brulle (1893, France) developed an *oléogrammétre* to measure the hardness of solid fats using the puncture principle. Sohn (1893, England), who was independently performing experiments similar to Brulle, felt he had been "scooped" when Brulle's publication appeared, and he hurried into print with a description of his apparatus accompanied by a list of seven rules that should be followed to avoid erroneous results. Perkins (1914, United States) continued the work of Brulle and Sohn in developing a puncture test to measure the hardness of fats. Kissling (1893, 1898, Germany) also studied penetration tests on greases and jellies by recording the time for rods of glass, zinc, or brass of various diameters to sink through the sample. Wender (1895, United States) studied the hardness of butter and margarine by measuring the viscosity of chloroform solutions of the fats in a U-shaped capillary viscometer that he called a "fluidometer." Lindsay (1901; Lindsay *et al.*, 1909; United States) measured the consistency of butter by measuring the depth that a mercury-weighted glass tube penetrated into butter when allowed to fall a standard height. Meyeringh (1911, Netherlands) also used a puncture test while Hunziker *et al.* (1912, United States) used a deformation test to measure butter hardness.

Cobb (1896, Australia) measured the hardness of wheat grains by measuring the force required to cut a grain of wheat in half by a pair of pinchers simulating biting between the front teeth. He defended his objective method against the skeptics by stating, "If the relative hardness here given differs from preconceived notions, so much the worse for the preconceived notions, unless it is shown that the methods adopted here are fallacious—an unlikely contingency." Roberts (1910, United States) used similar procedures to measure the hardness of wheat grains.

Waugh (1901, United States) clearly described a sensory deformation test as follows:

> Peaches and apricots are picked as soon as they show the first sign of ripening. The well-trained picker tests each fruit by taking it between his thumb and fingers and feeling it with the ball of his thumb. The fruit is not squeezed or bruised; but if it has the faintest feeling of mellowness its time has come, and the picker transfers it to his basket.

Leick (1904a, Germany) measured Young's modulus of elasticity of slabs of gelatin gels in tension and compression and showed that the modulus is approximately proportional to the square of the gelatin concentration. How to measure the firmness of jellies was a matter of interest to a number of early researchers, including Alexander (1906), who was awarded a United States patent (Alexander, 1908) for his apparatus; E. S. Smith (1909), who was also awarded a United States patent; Valenta (1909); Hulbert (1913); Sindall and Bacon (1914); Low (1920); C. R. Smith (1920); Sheppard *et al.* (1920); Oakes and Davis (1922); Freundlich and Seifriz (1923); Sheppard and Sweet (1923); Poole (1925); and Tracy (1928). Bloom (1925) was awarded a United States patent for a "machine for testing jelly strength of glues, gelatins and the like." This became the Bloom Gelometer, which is still used by the gelatin industry to measure the jelly grade of gelatins. Tarr (1926, United States) developed the

Tarr–Baker Jelly Tester, a puncture test that measured the firmness of pectin jellies. Sucharipa (1923, United States) attempted to measure the firmness of pectin jellies by means of compressed air.

Goldthwaite (1909, 1911, United States) described the texture of a fruit jelly as follows:

> The ideal fruit jelly . . . will quiver, not flow, when removed from its mold; a product with texture so tender that it cuts easily with a spoon, and yet so firm that the angles thus produced retain their shape; a clear product that is neither syrupy, gummy, sticky, nor tough; neither is it brittle and yet it will break, and does this with a distinct beautiful cleavage which leaves sparkling characteristic faces.

It is clear from this description that Goldthwaite understood the multifaceted nature of texture.

Washburn (1910, United States) also struggled to define differences in textural properties, going to some effort to distinguish between "body" and "texture" of ice cream.

Lehmann (1907a, Germany) devised an apparatus called the "Dexometer" to measure the toughness of meat and used the same instrument to measure the softening of vegetables during cooking (Lehmann, 1907b). This was probably the first objective test to measure meat toughness.

Willard and Shaw (1909, United States) give results from a puncture test that was used to measure the strength of egg shells but did not describe the equipment.

Professor Morris of Washington State University developed the first puncture tester for measuring the firmness of fruit in 1917 but did not publish his results for several years (Morris, 1925). In the meantime, other workers became aware of his work and developed their own designs of fruit pressure testers, sometimes publishing before Morris (e.g., Lewis *et al.*, 1919; Murneek, 1921; Magness and Taylor, 1925).

A graduate student at Kansas State College by the name of Lyman Bratzler was assigned by his advisor, Professor Warner, a research problem involving toughness of meat. He developed a mechanical shearing device whose principle of operation is well known today as the Warner–Bratzler Shear (Warner, 1928; Bratzler, 1932, 1949). Tressler (1894–1981), who has made numerous contributions to the field of food technology, developed a tenderness test for meat based on the puncture principle, which he considered to be superior to the Warner–Bratzler Shear (Tressler *et al.*, 1932; Tressler and Murray, 1932). He called the Warner–Bratzler Shear "the mousetrap," possibly because of the manner in which it snaps back into place when a tough piece of meat finally shears. Pitman (1930) developed a shear test somewhat similar to the Warner–Bratzler Shear for measuring the firmness of almonds. Tauti *et al.* (1931, Japan) developed a physical test for measuring the firmness of raw fish.

Bingham (1914) developed a U-tube viscometer with applied air pressure that he called a "plastometer." This apparatus was used by Herschel and Bergquist (1921) to measure the consistency of starch pastes, and by Porst and Moskowitz (1922) for processed corn products.

Davis (1921, United States) devised the three parallel bar test for measuring the breaking strength or shortness of cookies, calling it a "shortometer." This was later improved by Fisher (1933). Hill (1923, 1933, United States) developed the Hill Curd Tester for measuring the firmness of cheese curd; Babcock (1922, United States) developed the falling plummet test for measuring the firmness of whipped cream; Vas (1928, Netherlands) developed a penetrometer for measuring the firmness of cheese curd; and Knaysi (1927) developed a falling-ball viscometer to measure the viscosity of buttermilk.

Stewart (1923) found that the volume of popped popcorn correlates well with popcorn quality. Sayre and Morris (1931, 1932) measured the volume of juice that could be expressed from sweet corn and concluded that it was a satisfactory test for physical quality of sweet corn. This procedure eventually developed into the Succulometer test (Kramer and Smith, 1946).

The number of scientists in the field began to multiply in the 1930s and 1940s and continues to multiply. Most of these scientists are still living. Their names are referenced throughout the pages of this book. Shortage of space demands that the recounting of history stop at this point. However, one contemporary must be singled out for special mention—Dr. George W. Scott Blair (1902–). Dr. Scott Blair, an Englishman, and one of the founders of the science of rheology, is world renowned for his pioneering contributions to food rheology and also the rheology of soils, plastics, and biological fluids. He authored over 250 publications on rheology and is author or editor of seven books. Because of his early work on flour (Scott Blair *et al.*, 1927) and later on dairy products and psychorheology in the 1930s to 1950s, he is considered to be the "father" of food rheology. In 1929, while on a sabbatic leave at Cornell University he attended a meeting in Washington, D.C., that resulted in the official adoption of the term "rheology" and the formation of the (American) Society of Rheology. He was also a founding member and president of the British Society of Rheology. A special issue of *Journal of Texture Studies* (Vol. 4, No. 1, 1973) took the form of a festschrift honoring Dr. Scott Blair on his seventieth birthday.

We hope the time will come when a historian will piece together the contributions of persons from every continent who have collectively brought the field of food texture measurement and the science of food rheology up to its present level of accomplishment.

Suggestions for Further Reading

The *Journal of Texture Studies,* published quarterly by Food and Nutrition Press, 1 Trinity Square, Westport, Connecticut 06880, publishes original research, reviews, and abstracts on rheology, psychorheology, physical and sensory testing of foods, and pharmaceuticals. It is the best single source of information on developments in the field of food rheology, texture, and viscosity.

The following books and articles contain much useful information:

Behrens, D., and K. Fischbeck, eds. 1974. "Lebensmittel-Einfluss der Rheologie," DECHEMA Monograph, Vol. 77, No. 1505–1536. Verlag Chemie, Weinheim.

Brennan, J. G. 1980. Food texture measurement. *In* "Development in Food Analysis Techniques, Vol. 2" (R. D. King, ed.), pp. 1–78. Appl. Sci., London.

Corey, H. 1970. Texture in foodstuffs. *CRC Crit. Rev. Food Technol.* **1,** 161–198.

deMan, J. M., P. W. Voisey, V. F. Rasper, and D. W. Stanley. 1976. "Rheology and Texture in Food Quality." Avi, Westport, Connecticut.

Kramer, A., and A. S. Szczesniak, eds. 1973. "Texture Measurements of Foods." Reidel Publ., Dordrecht, Netherlands.

Matz, S. A. 1962. "Food Texture." Avi, Westport, Connecticut.

Mohsenin, N. N. 1970. "Physical Properties of Plant and Animal Materials." Gordon & Breach, New York.

Muller, H. 1973. "An Introduction to Food Rheology." Crane Russak, New York.

Rha, C. H., ed. 1974. "Theory, Determination, and Control of Physical Properties of Food Materials." Reidel Publ., Dordrecht, Netherlands.

Scott Blair, G. W., ed. 1953. "Foodstuffs: Their Plasticity, Fluidity, and Consistency." Wiley (Interscience), New York.

Scott Blair, G. W. 1958. Rheology in food research. *Adv. Food Res.* **8,** 1–61.

Sherman, P. 1970. "Industrial Rheology with Particular Reference to Foods, Pharmaceuticals, and Cosmetics." Academic Press, New York.

Sherman, P., ed. 1979. "Food Texture and Rheology." Academic Press, New York.

Society of Chemical Industry. 1960. "Texture in Foods," Monograph No. 7. Soc. Chem. Ind., London.

Society of Chemical Industry. 1968. "Rheology and Texture of Foodstuffs," Monograph No. 27. Soc. Chem. Ind., London.

Sone, T. 1973. "Consistency of Foodstuffs." Reidel Publ., Dordrecht, Netherlands.

Van Wazer, J. R., J. W. Lyons, K. Y. Kim, and R. E. Colwell. 1963. "Viscosity and Flow Measurement. A Laboratory Handbook of Rheology." Wiley (Interscience), New York.

White, G. W. 1970. Rheology in food research. *J. Food Technol.* **5,** 1–32.

CHAPTER 2

Body–Texture Interactions

The properties of texture and viscosity are perceived by the human senses. Hence, in order to understand texture and viscosity it is necessary to know something about how the human body interacts with food. Most people are well aware of the structure and function of the teeth, and everybody is familiar with the process of mastication. Nevertheless, a brief review of these topics is needed to introduce the discussion of the sensing of texture and viscosity.

Mastication is a process in which pieces of food are ground into a fine state, mixed with saliva, and brought to approximately body temperature in readiness for transfer to the stomach where most of the digestion occurs. After some residence time in the stomach the food passes to the small intestine where digestion continues and from whence the nutrients are absorbed into the bloodstream and distributed throughout the body. Pulverization of food is the main function of mastication, but it also imparts pleasurable sensations that fill a basic human need. Table 1 summarizes the degree of size reduction that must occur before food can be absorbed and utilized by the body. The process of mastication is an early step in the process of size reduction to small molecules. Mastication usually reduces particle size by two to three orders of magnitude before passing to the stomach where another approximately 20 orders of magnitude of size reduction are accomplished by chemical and biochemical action. If food cannot be reduced to particles of the order of a few multiples of 10^{-22} g, it is not absorbed and utilized but is excreted.

Some Definitions

Masticate. To chew, grind, or crush with the teeth and prepare for swallowing and digestion. Note: Mastication is a *process.*

TABLE 1

STEPS IN THE COMMINUTION OF FOOD BEFORE ABSORPTION BY THE BODY

State	Approx. particle mass (grams)	Process	Location	Implement
Large cookie				
Whole cookie	20	Biting off	Mouth	Incisors
Mouthsize portion	5	Grinding, crushing	Mouth	Molars
Swallowable paste (bolus)	1×10^{-2}	Biochemical attack	Stomach, intestine	Acid, enzymes
Hexose sugar molecules	3×10^{-22}	Absorption	Intestines	—
Whole dressed steer				
Whole carcass	3×10^{5}	Sawing and cutting	Butcher shop	Saw, knives
Cooked steak	3×10^{2}	Cutting	Plate	Knife and fork
Mouthsize portion	5	Shearing, grinding	Mouth	Teeth
Swallowable paste	1×10^{-2}	Biochemical attack	Stomach, intestines	Acid, enzymes
Amino acid molecules	2×10^{-22}	Absorption	Intestines	—

Bolus. A mass of chewed food in the mouth.

Deglutition. The act or process of swallowing food. Deglutition tips the food into the esophagus (gullet), the tube which leads down to the stomach. Deglutition ends the voluntary portion of the digestive process. The rest of the digestive process is involuntary and automatic.

Since textural properties of foods are perceived primarily in the mouth there is a need to know something about the structure of the organs and tissues of the mouth and the actions that occur during mastication.

1. *Teeth (dentes)* are the main agent for masticating foods and breaking them into small pieces. They also play an important role in clear speech and facial structure and appearance. Crooked, decayed, or missing teeth cause disfigurement and sometimes self-consciousness. From the external viewpoint teeth consist of two parts: (1) the *crown* is that part that protrudes above the gums and is visible in the mouth and (2) the *root* is that portion that is not visible in the mouth but is buried in the gums and serves to anchor the teeth in the jawbone.

A cross-sectional cut through a tooth shows that it is composed of several layers of tissues (Fig. 1). The enamel is the very hard external layer that covers the crown of the tooth and contacts the food during mastication. Underneath the enamel is the dentin, which is hard tissue forming the body of the tooth and which constitutes the principal mass of the tooth. The cementum is a bonelike tissue that covers the root. The pulp is a soft tissue that occupies the central portion of the tooth called the pulp chamber. It contains nerves, arteries, veins, and lymph vessels. These vessels enter the tooth through small openings at the tip of the root. The periodontal ligament (membrane) is the layer of connective tissue that lies between the cementum and the jawbone and helps to hold or support the tooth in its place. Small elastic fibers are connected to the tooth via

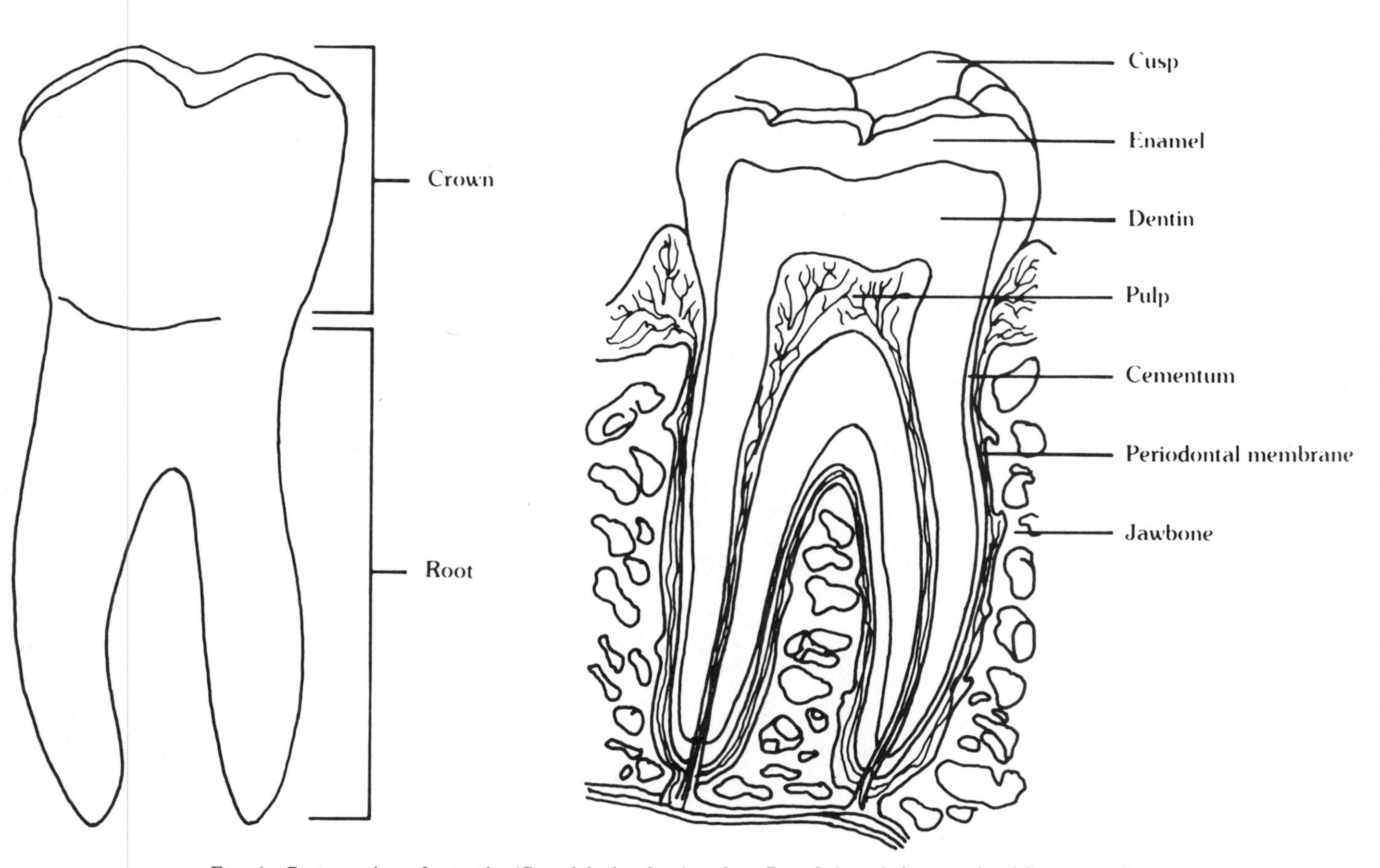

FIG. 1. Cross section of a tooth. (Copyright by the American Dental Association, reprinted by permission.)

the cementum along the entire surface of the root. A cusp is a pointed or rounded surface on the crown of the tooth and is the main contact surface for breaking up the food. The teeth are composed principally of calcium phosphate. Teeth are not bones; they are much harder and more dense than bones.

Teeth may be classified according to their shape and the function they perform (Fig. 2). The incisors, located in the center front of the mouth, are wedge-shaped and have a sharp flat edge which is used to cut or incise foods. The cuspids, which are located at the corners of the mouth, have a long heavy root and a crown with a single pointed cusp. They are often called the "eye" teeth or canines. These are used to tear foods. The premolars (bicuspids) are located behind or in back of the cuspids and have two cusps and one or two roots. They are used to both tear and crush foods. The molars are located at the back of the mouth. Each molar has two or three roots and several cusps that occlude with the

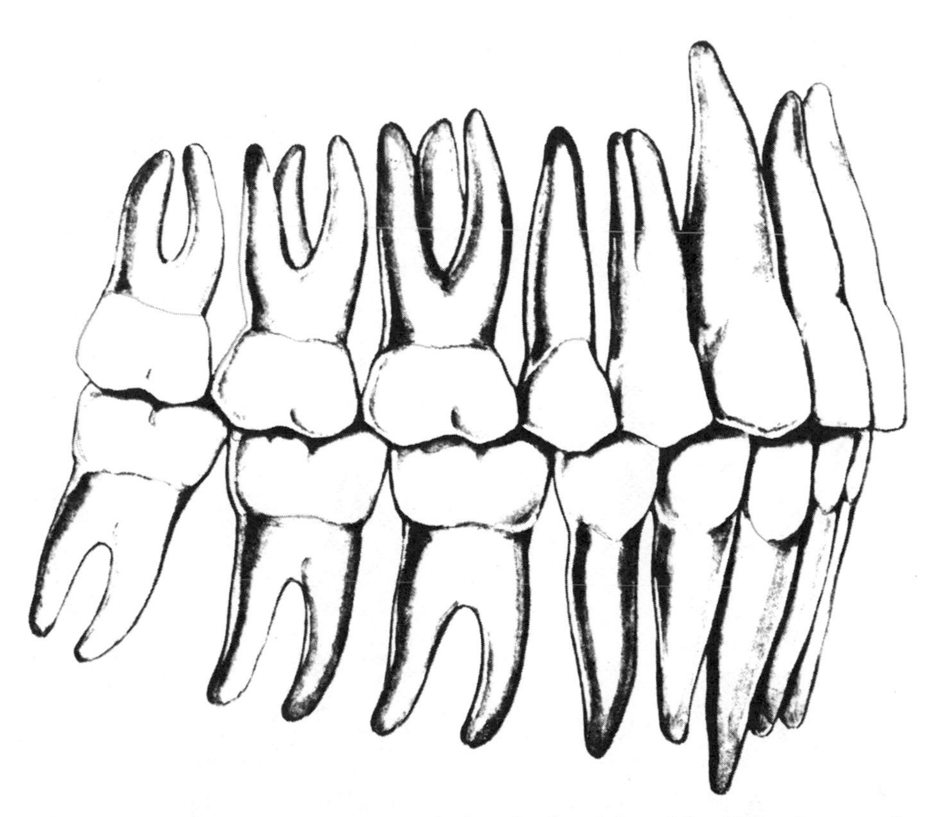

FIG. 2. Normal occlusion of permanent teeth. In order from left to right: third molars, second molars, first molars, second bicuspids, first bicuspids, cuspids, lateral incisors, central incisors. (Copyright by the American Dental Association, reprinted by permission.)

opposing molars. Their broad crowns are used to grind and crush the food with a grinding millstone-type of action.

A child in full dentition has 20 teeth (Fig. 3). The 10 teeth in each jaw comprise 2 central incisors, 2 lateral incisors, 2 cuspids, 2 first molars, and 2 second molars. These primary (deciduous, milk, or "baby") teeth appear between 6 and 24 months in the average child and are shed between 6 and 12 yr to be replaced by the adult or permanent teeth as the jaw increases in size sufficient to accommodate the larger size and increased number of adult teeth.

Full dentition in the adult consists of 32 teeth, 16 in each jaw (Fig. 3). Each jaw contains 2 central incisors, 2 lateral incisors, 2 cuspids, 2 first bicuspids, 2 second bicuspids, 2 first molars, 2 second molars, and 2 third molars. These teeth erupt from the age of approximately 6 to 21 yr. The third molars (wisdom teeth) generally appear in the late teens or early twenties and complete full dentition in the adult. The normal times for eruption and shedding of teeth are shown in Table 2.

Partial or full dentures (artificial teeth) may be fitted to offset the loss of

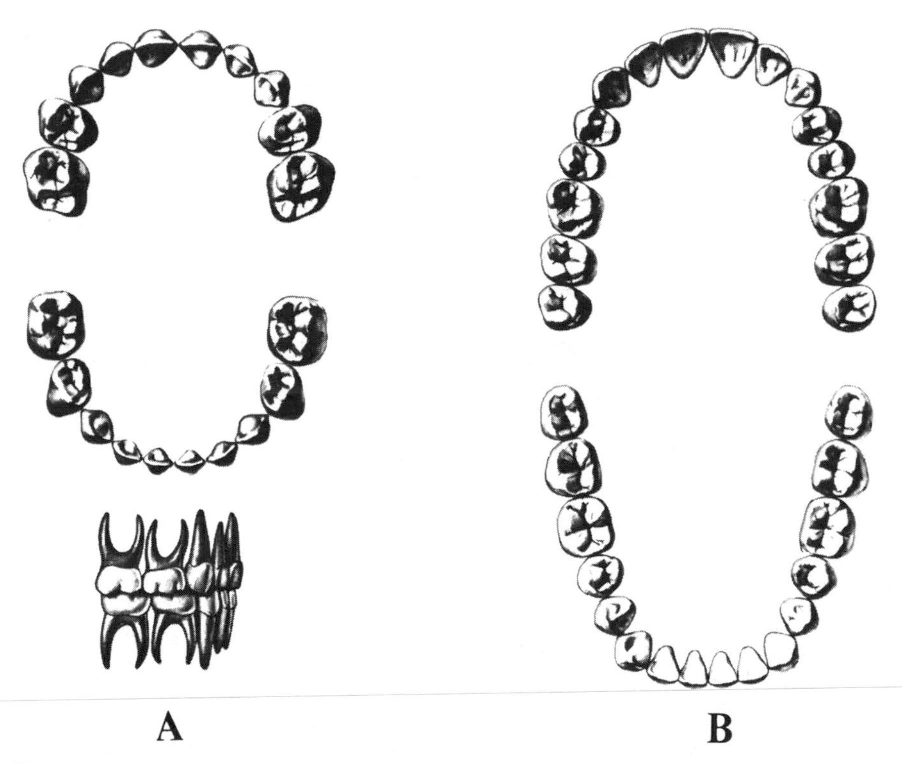

FIG. 3. A, eruption and shedding of the primary teeth. B, eruption of the permanent teeth. (Copyright by the American Dental Association, reprinted by permission.)

TABLE 2

ERUPTION AND SHEDDING OF HUMAN TEETH

Primary teeth			Permanent teeth	
Upper	Eruption (months)	Shedding (yr)	Upper	Eruption (yr)
Central incisor	7½	7½	Central incisor	7–8
Lateral incisor	9	8	Lateral incisors	8–9
Cuspid	18	11½	Cuspid	11–12
First molar	14	10½	First bicuspid	10–11
Second molar	24	10½	Second bicuspid	10–12
			First molar	6–7
			Second molar	12–13
			Third molar	17–21
Lower			Lower	
Central incisor	6	6	Central incisor	6–7
Lateral incisor	7	7	Lateral incisor	7–8
Cuspid	16	9½	Cuspid	9–10
First molar	12	10	First bicuspid	10–12
Second molar	20	11	Second bicuspid	11–12
			First molar	6–7
			Second molar	11–13
			Third molar	17–21

natural teeth, but they do not perform as well as healthy natural teeth (see Tables 3 and 5, pp. 36, 39). People with reduced masticatory efficiency caused by incomplete dentition or dentures often compensate for the deficiency by selecting foods that are easier to chew and present less challenge to masticatory function.

2. The *lips (labia oris)* are the two highly mobile fleshy folds that surround the orifice of the mouth and admit food and liquid into the oral cavity. The lips also prevent the loss of food from the mouth between masticatory strokes. They have a variety of sensory receptors that can judge the temperature and some of the textural properties of foods. The lips have a high acuity to touch; they are even more sensitive than the tips of the fingers.

3. *Cheeks (buccae)* form the sides (lateral walls) of the mouth and face and are continuous with the lips. They consist of outer layers of skin, pads of subcutaneous fat, muscles associated with chewing and facial expression, and inner linings of stratified squamous epithelium. The cheeks keep the food within the oral cavity and return the food between the teeth between bites.

4. The *tongue (lingua)* is a strong, mobile, muscular organ with its base and central part attached to the floor of the mouth. It nearly fills the oral cavity when

the mouth is closed. It is a very active organ during the act of mastication, working in close proximity to the teeth but seldom caught between the teeth. Skillful coordinated neuromuscular functions between the tongue and teeth are required for painless mastication. It returns food between the teeth between chews and is actively involved in mixing the bolus with saliva and in moving the bolus toward the pharynx during swallowing. It is used to break up soft foods against the hard palate without the help of the teeth and is the organ most responsible for sensing the surface or geometrical properties of foods because of its ability to perceive minute differences in particle size, shape, firmness, and roughness. The tongue demonstrates a more acute tactile sensibility than any other part of the body. Two-point sensibility is the shortest distance between two points that can be perceived as two separate stimuli. For the tongue this is 1.4 mm, for the fingertip 2 mm, and for the nape of the neck 36.2 mm. The tongue is also the principal organ of taste and an important organ of speech.

5. The *palate* (roof of the mouth) consists of two sections. To the front of the mouth (anterior) lies the hard palate (*palatum durum),* which consists of a bony skeleton covered with a thin layer of soft tissue. It separates the oral cavity from the nasal cavity and presents a hard surface against which foods can be pressed by the tongue to break them up, spread them out, or mix with saliva.

The soft palate (*palatum molle*), which lies at the back of the mouth (posterior), consists of a thick fold of muscular membrane containing muscles, vessels, nerves, lymphoid tissues, and mucous glands. During swallowing or sucking it is elevated to close the opening to the nasal cavity, thus preventing food from entering the nasal cavity from the oral cavity.

6. The *gums (gingivae)* are composed of dense fibrous tissue that surround the teeth and help anchor them.

7. *Salivary glands* provide the saliva that hydrates foods, lubricates the bolus, and begins the digestion of carbohydrates. There are three pairs of salivary glands: the *sublingual* (beneath the tongue), *submandibular* (beneath the jaw), and *parotid* (beneath the ear). Secretions from the salivary glands enter the oral cavity through narrow tubes called salivary ducts.

8. The *upper jaw (maxilla)* serves to anchor the upper teeth and is fairly immobile during mastication. The teeth in the maxilla can be likened to the anvil against which the food is pressed to break and crush it by the lower teeth.

9. The *lower jaw (mandible)* is a horseshoe-shaped bone that anchors the lower teeth and articulates (moves) primarily in a reciprocating vertical motion with approximate sinusoidal speed. A variable amount of lateral (sideways) motion is also present, depending on the nature of the food. Foods that are easily crushed (e.g., snack foods such as potato chips) require little lateral motion whereas foods that are tough (such as meat) require a rather large amount of lateral motion to masticate the food.

A number of muscles are responsible for articulating the mandible. The most powerful of these is the masseter muscle, which is capable of generating high compressive forces aided by two other powerful muscles, the temporal and medial pterygoid. Lateral and protrusive movements of the mandible are largely controlled by the lateral pterygoid and suprahyoid muscles. The jaws are operated by the most complex muscular system in the body; five different movements are available, most of them generating high forces.

Articulation of the mandible occurs about a highly specialized and complex composite joint called the temperomandibular joint, which allows five different movements—far more than any other joint in the body. The joint contains two compartments. During normal chewing action it acts as a hinge joint in which the mandible rotates in a vertical direction in the first compartment (Fig. 4). When the jaw is opened very wide or protruded forward, the mandible glides out of the first compartment into the second compartment, which is a movable sliding

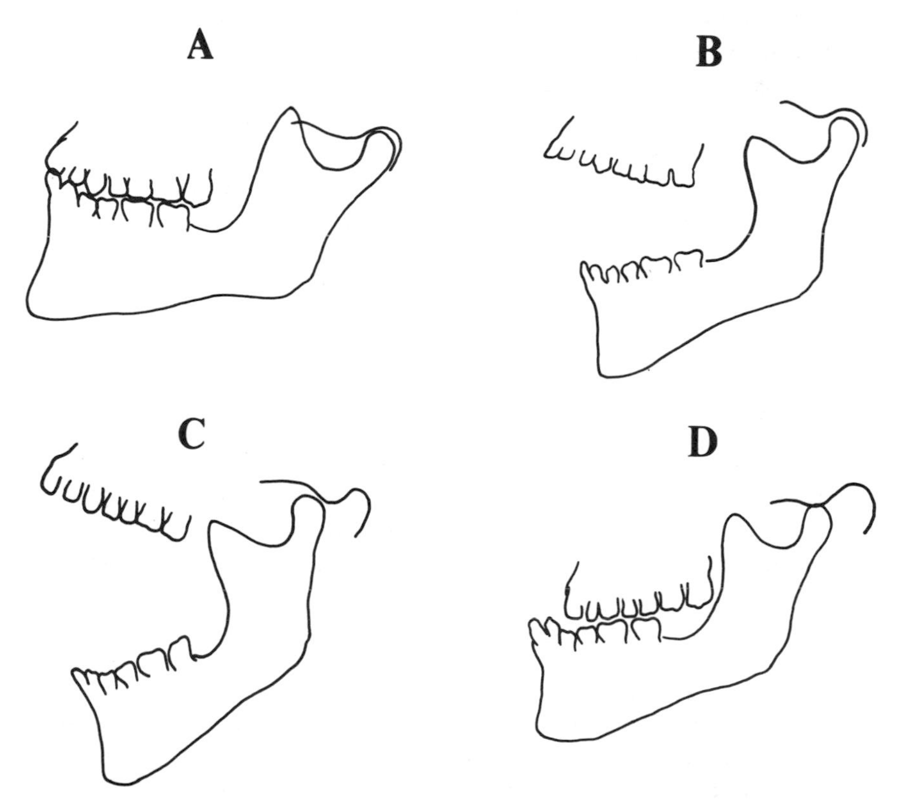

FIG. 4. The temperomandibular joint. A, in occluded position; B, opened in pure hinge movement; C, opened as wide as possible; D, in protruded position. Notice how the mandible slips out of its cup in positions C and D.

socket. The mandible can be moved laterally when in either compartment, although the extent of the sideways movement is limited by the temperomandibular ligaments. From a position in which the incisors are in contact, the mandible may be moved downward to open the jaw, laterally for a sideways swing, forward in protrusion, and backward in retrusion. The temperomandibular joint is remarkable in its flexibility and in the variety of movements it can accomplish with great force. It is the key to the various chewing modes that are available to masticate foods with widely differing combinations of physical properties.

Many people can feel the temperomandibular joint move by lightly placing the tips of the fingers on the jaw just in front of the ears. The hinge action can be felt by opening and closing the jaw to a moderate degree and the lateral motion can be felt by swinging the jaw sideways. When the jaw is protruded or opened very wide, one can feel the joint glide forward from the pure hinge compartment into the socket compartment.

10. The *oral cavity (cavium oris proprium)* is the space bounded by the lips and cheeks, by the palate above, and the muscular floor below. It contains the teeth and tongue.

11. The *pharynx* is the cavity at the back of the mouth that connects the nasal and oral cavities with the larynx (voice box) and the esophagus (tubular passageway to the stomach). When the bolus is pushed into the pharynx by the tongue, the swallowing reflex is initiated and the following responses occur in rapid succession:

a. The soft palate is raised, preventing the bolus from entering the nasal cavity.
b. The larynx is elevated to prevent the bolus from entering the trachea (windpipe).
c. The tongue presses up against the soft palate, sealing off the oral cavity from the pharynx while the pharynx moves upward toward the bolus.
d. The muscles at the lower end of the pharynx relax and open the esophagus.
e. The muscles of the upper end of the pharynx contract, forcing the bolus into the esophagus. Peristalsis (alternate contractions and relaxation of the muscles along the esophagus that cause a contraction ring to move along the esophagus) moves the bolus down the esophagus to the stomach. When the peristaltic waves reach the stomach, the muscles that guard its entrance relax and allow the bolus to enter. These muscles contract after the bolus has entered the stomach, closing off the entrance and preventing regurgitation of the acid stomach contents into the esophagus.
f. The muscles and organs return to their normal position.

12. *Other.* The arm, neck, and shoulder muscles may be brought into use at times, especially when biting off a piece of tough food.

Occlusion refers to the manner in which the upper and lower teeth meet and fit together as the jaw is closed. In good occlusion the cusp surfaces of the upper teeth fit closely to the lower teeth. The medial occlusal position (or intercuspal position) is that position in which the mandible returns when the jaws are snapped shut automatically from a wide opening when no food is in the mouth. In this position the upper and lower molars and cuspids are in direct contact and the cusps fit together to give an uneven line while the upper incisors lie in front of, and partially cover the lower incisors (see Fig. 2).

From the medial occlusal position the jaw can be protruded (moved forward after slight opening) so that the upper and lower incisors meet in readiness for biting off; in this position the molars do not contact each other. The mandible can be protruded even farther forward by pressing the temperomandibular joint forward into its second compartment. It can also be retruded until the lower incisors are well behind the upper incisors. From the medial occlusal position the lower jaw can also be pulled sideways to the right or left (lateral movement). The ability of the mandible to be moved in all directions from the medial occlusal position allows a wide range of chewing techniques to be employed.

Malocclusion (bad closing) occurs when the cusps of the upper and lower teeth do not fit well when the mandible is in the closed position. This is a problem to which dentists devote much attention.

Mastication refers to the entire complex of processes that occurs as the food is chewed and brought into a condition ready to be swallowed. It may be a voluntary or involuntary act. This is an extremely complex set of processes that is generally not well understood or appreciated. Mastication is a biting–chewing–swallowing action that is a complex stimulation–motor feedback process in which a constant stream of stimuli travels from mouth to brain and a corresponding stream of instructions travels from the brain to the mouth instructing it how to proceed (Fig. 5). This complexity has been well described by Yurkstas (1965) as follows:

> We sometimes fail to appreciate the complexity of the chewing apparatus. It is truly remarkable that most people perform this function daily, with little or no forethought. Mastication involves the use of forces that sometimes exceed 100 lb and pressures that are probably 10,000 lbs in^{-2}. One hundred blows per minute are often delivered for periods of one-half to one hour at a time. These blows are automatically controlled and are precise to within a few hundredths of an inch, since a mistimed blow or misguided stroke can cause intense pain or result in considerable damage.

The Sequence of Mastication

The time devoted to masticating a food, number of chews, and type of chewing motion varies considerably from person to person, and from one food to

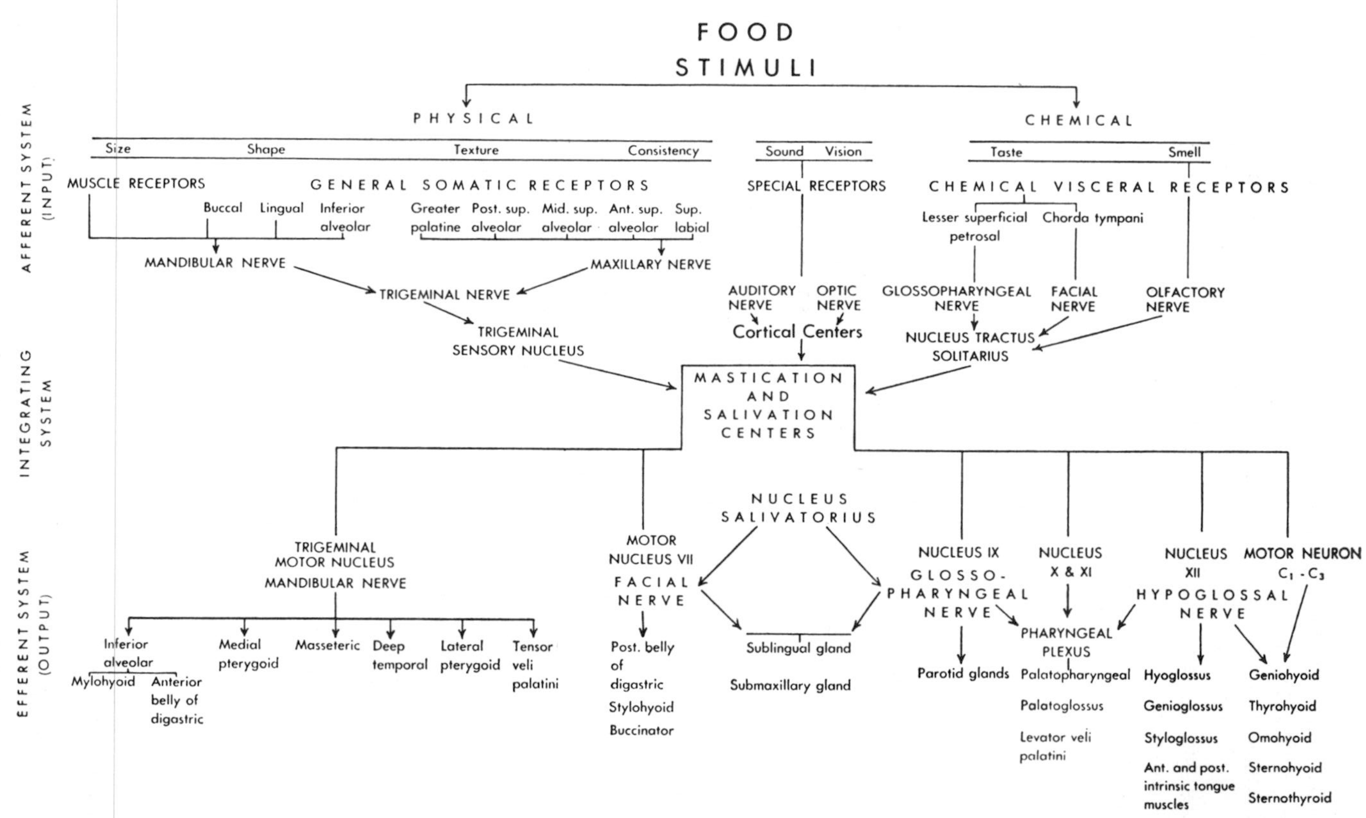

FIG. 5. The complex neuromuscular mechanism involved in the act of chewing food. The senses of sound, vision, taste, and smell participate indirectly in chewing through their influence on the salivary secretion. (From Kapur *et al.*, 1966; reprinted by permission of W. B. Saunders Co.)

another. The sequence below is the most common sequence found with the majority of foods.

1. Bite off a piece of food with the incisors. Soft foods are usually wiped off the spoon with the lips instead of using the incisors.

2. Cut into small pieces with the incisors when necessary.

3. Puncture or tear apart with cuspids and bicuspids as necessary.

4. Grind into small particles with the molars, simultaneously mixing the food into a paste with the saliva using both tongue and teeth. Soft, smooth foods are manipulated by the tongue more than by the teeth. This process is mainly one of mixing the food with saliva when there are no hard pieces to be broken down by the teeth.

5. Swallow the liquid portion and fine particles, retaining the insufficiently chewed portion in the mouth.

6. Continue the grinding, mixing, and swallowing sequence until the bolus has disappeared and the mouth is empty and ready to bite off the next piece. Pierson and LeMagnen (1970) showed that there is only one deglutition for very soft and liquid foods and that the number of deglutitions increased as the hardness, dryness, or compactness of the food increased.

The first few chews on a piece of food are generally slow as one manipulates the piece within the mouth to soften it with saliva or cut it into smaller pieces with the incisors. When the bolus reaches a consistency that can be readily managed, the chewing rate is stepped up to the normal chewing rate, which then remains fairly constant for the remainder of that chewing cycle.

The size of the pieces of food that are swallowed is known as the "swallowing threshold." Yurkstas (1965) studied this and concluded:

> The results show that the swallowing threshold was directly related to masticatory performance, the correlation coefficient being significant to the 1% level. Thus, people with superior masticatory ability attained a finer degree of food pulverization at the swallowing threshold than did people who possessed dentitions that were less efficient. The person with the diminished ability to chew compensated for his dental handicap by swallowing larger particles of food. . . . People who had poor dentitions did not compensate for their dental handicap by chewing for a longer period of time or by increasing the number of masticatory strokes."

Table 3 shows the effect of missing teeth on masticatory performance.

Kapur *et al.* (1964) showed that the chewing process by natural teeth is preferential; that is, the coarse particles are ground more rapidly than fine particles as chewing proceeds, while mastication in subjects with complete dentures is nonpreferential—all particles are pulverized at random.

It should be noted that the forces exerted by the teeth provide the *stress* on the food while the movement of the jaw provides the *strain* on the food during mastication. (These two terms are defined in the next chapter.)

TABLE 3
EFFECT OF MISSING TEETH ON MASTICATORY PERFORMANCE

	Chewing efficiency[a]	
	Mean	Range
Complete dentition	88	75–97
Third molar missing	78	45–92
Third and one other molar missing	55	17–83
Dentures	35	9–57

[a]Chewing efficiency is defined as the percent of food passing through a 20-mesh screen after 20 chews. Data from Yurkstas (1965).

The rate at which people chew depends partly on the food and partly on the person. Each time the author teaches his class in food rheology he gives the students sticks of a well-known brand of chewing gum and asks them to measure their chewing rate once the gum has been brought to a "steady-state" condition. In this classroom situation the mean chewing rate is approximately 60 chews per minute with a range of 24–105 chews per minute (Table 4). Using informal tests on a number of people on sticks of the same brand of chewing gum the author has found a chewing rate as low as 26 chews per minute to a high of 132 chews per minute (Bourne, 1977).

The effect of the food on the chewing rate follows a complex pattern. What seems to happen is demonstrated schematically in Fig. 6. The chewing rate remains approximately constant as one moves from foods of low toughness to foods of moderate toughness. This constancy is achieved by increasing the power output of the jaw (power is the rate of doing work). As the food continues to increase in toughness the limit of comfortable power output is reached. Beyond

TABLE 4
CHEWING RATES ON STICKS OF CHEWING GUM

	Chews per minute			
Year	Mean	Maximum	Minimum	Number of respondents
1973	64.5	98	45	34
1975	54.8	105	27	23
1977	60.4	84	38	20
1979	70.3	105	48	28
1981	65.5	100	24	30

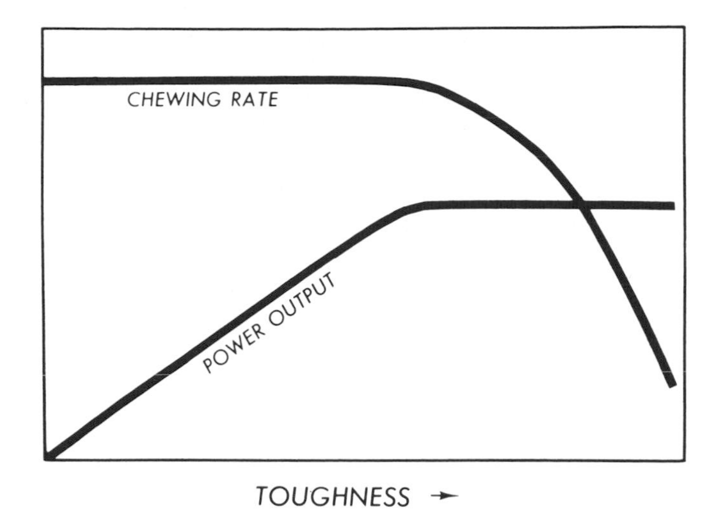

Fig. 6. Schematic representation of the relationship between chewing rate and power output on foods of increasing toughness.

this point the power output remains approximately constant, and this is achieved by slowing the rate of mastication. One chews tough meat and chewy caramels more slowly than foods that require less energy for mastication.

The chewing pattern is completely changed with extremely hard foods such as rock candy and nuts in the shell. These foods are usually placed carefully between the molars where the maximum leverage is available and the force is steadily increased until the food cracks or shatters. In these cases the compression rate before breaking is almost zero. The chewing mode is that of constant rate of increase in force application. This is a stress-dependent type of mastication in contrast to the usual strain-dependent type.

Rate of Compression between the Teeth

The rate of compression between the teeth varies over a wide range and is affected by several factors. Table 4 indicates the wide range of chewing speeds from person to person on a standard product. It has been noted above that the first few chews on a piece of food are frequently slower than the regular chewing rate and that tough foods are masticated more slowly than tender foods.

How widely the jaw is opened affects the compression speed. Some people make short strokes of the jaws while others make longer strokes. People who make long strokes will have a higher compression rate if they use the same number of chews per minute because the average compression speed is the product of the number of chews per minute by twice the distance between the teeth at the point of maximum opening.

The mandible articulates in approximately the arc of a circle around the temperomandibular joint. The teeth that are closer to this joint move a smaller distance than the teeth that are farther from the joint. The incisors are the farthest from the joint and move at about twice the speed of the molars. Even among the molars the first molar moves at a faster rate than the third molar because of its greater distance from the temperomandibular joint.

The rate of movement of the jaw follows approximately a sine curve. The actual rate of compression will vary continuously throughout each masticatory stroke, reaching a maximum speed at approximately midstroke and falling to zero at the end of the stoke.

If we assume 60 chews per minute as the average chewing rate and an average stroke length of 10 mm, then the average compression rate is 1200 mm min^{-1}, or 20 mm sec^{-1}. As noted above, there will be substantial variations from this "average" figure.

Soothing Effect of Mastication

Mastication has been found to have a pronounced soothing effect. Chewing "soothes the nerves." Fidgeting activities such as finger tapping, leg swinging, pipe smoking, adjusting the hair or mustache, etc., greatly decline in frequency when mastication is taking place. The sucking and chewing that a fretful baby gives to its thumb or a pacifier is another example of the soothing effect of mastication. Chewing gum is a harmless way to use the soothing effect of mastication to relieve tension. For this reason it would be desirable to allow students to chew gum during examinations!

Saliva

The flow of saliva that is generated by the salivary glands (see item 7, p. 30) lubricates the bolus, softens dry foods, flushes away food particles, initiates the first phase of digestion through its ptyalin content, and aids deglutition. The act of chewing stimulates the flow of saliva. One study found that the mean saliva flow among a group of people at times of nonstimulation was 26 ml h^{-1} with a range of 2.5–110 ml h^{-1} (Jenkins, 1978). When the saliva flow was stimulated in the same people by giving them flavored wax to chew, the saliva flow increased to 46–249 ml h^{-1}. It should be noted that factors other than mastication can stimulate the flow of saliva; for example, the smell or sight of food or talking about food. There is the classical example of Pavlov's dogs that were conditioned to salivate at the sounding of a bell (Pavlov, 1927).

Saliva generally consists of approximately 99.5% water and 0.5% solids, but these figures can vary widely from person to person and from day to day within the same person. The main constituent of saliva is a glycoprotein called *mucin* which imparts a slimy mucus character to the saliva, thus assisting in the lubrication of the bolus. Saliva contains the enzyme ptyalin (an amylase) which assists in the biochemical breakdown of the food. The parotid gland secretion is rich in ptyalin but watery because it is low in mucin. The submandibular gland secretion is viscous because it is rich in mucin but is low in ptyalin. The secretion from the sublingual gland is mixed, containing both ptyalin and mucin. Total ash is approximately 0.25%. In a study of 3400 people it was found that the pH of saliva ranged from 5.6 to 7.6 with a mean of 6.75 (Jenkins, 1978).

Forces Generated between the Teeth

In general the maximum force exerted between the teeth is 15 kg between the incisors, 30 kg between the cuspids, and 50–80 kg between the molars. This range in readings is undoubtedly due to the leverage effect—the molars are much closer to the fulcrum of the mandible than are the incisors.

Oldfield (1960) noted that Borelli measured the total force exerted by the jaw in 1681 by hanging weights on the lower jaw, and found that the maximum weight that could be supported was about 100 lb. Table 5 shows the maximum forces that can be exerted between the teeth among a primitive tribe (Eskimos) and a civilized tribe (Americans). The wide differences between the two tribes shown in this table are undoubtedly due to the fact that the Eskimos eat a great amount of tough, hard foods and they chew on animal skins to improve the quality of the pelt, which develops the masseter and other muscles of the jaw,

TABLE 5

FORCES EXERTED BETWEEN TEETH (IN POUNDS)[a]

	Male		Female	
Subject	Mean	Maximum	Mean	Maximum
Eskimo	270	348	200	326
American (natural teeth)	120	200	85	165
American (dentures)		~60		

[a]Data taken from Waugh (1937) and Klatsky (1942).

while Americans, eating mostly soft foods, are never required to develop the muscular strength of the Eskimos. It is interesting to notice that the mean value of 200 lb for female Eskimos is equal to the maximum value of 200 lb for the American male. The author leaves his readers to draw whatever conclusions they want from these figures!

Another national difference in chewing ability is shown in Table 6, where the number of chews required to bring food to the point of deglutition is shown for a trained American texture panel and a trained Filipino texture panel. In each case the Filipinos required more chews than the Americans for the same type of food. This difference is probably due to the fact that the Filipino diet is basically cooked rice, which is soft and requires little mastication. Other foods that are used in the Filipino diet are generally cut into small pieces before being brought to the table so that the diet on the whole is not challenging or demanding from the textural standpoint.

Table 7 shows the average force *per tooth* required to masticate some common foods by two wearers of full upper and lower dentures. These data show a sixfold range from 0.3 kg for boiled beets to 1.8 kg for a French roll. It would be interesting to see whether the same force levels were exerted by persons possessing their natural teeth.

The distribution of forces that were applied to three teeth by two denture wearers fitted with full upper and lower dentures is shown in Table 8. The authors of this report (Yurkstas and Curby, 1953) state that

> Mastication of hard rolls resulted in almost equal force distribution among the three teeth studied. This was due to the fact that initially the rolls were masticated in the first and second bicuspid area, and, as they were softened, the molar area was utilized to a greater degree. The ingestion of liquids with rolls resulted in a slight posterior distribution of force. The raw vegetables studied were divided into soft and hard categories. Softer raw vegetables showed relatively equal force distribution on all three teeth, while tougher ones such as carrots were masticated in the first bicuspid area. The cooked vegetables showed a definite trend toward the posterior area. When liquids were ingested simultaneously with bread, there was a definite shift towards the first molar area in preference for mastication. Softer meats such as hamburger

TABLE 6
AVERAGE CHEW COUNTS ON SELECTED FOODS

	US-trained panel[a]	Filipino-trained panel[b]
Frankfurter	17.1	22.1
Jelly beans	25.0	34.0
Steak	31.8	56.6
Caramels	37.3	61.6

[a]Unpublished data from A. S. Szczesniak.
[b]Unpublished data from M. C. Bourne.

TABLE 7

AVERAGE FORCE PER TOOTH DURING MASTICATION[a]

Food	Force (kg)	Food	Force (kg)
French roll	1.8	Ham on white bread	1.0
Tender steak	1.4	Lobster	0.8
Pear (hard)	1.4	Apple (Macintosh)	0.7
Celery (raw)	1.3	Cucumber (raw)	0.7
Carrot (raw)	1.3	Raised doughnut	0.7
Bologna on roll	1.2	Broccoli	0.6
White bread (with crusts)	1.1	Potato (boiled)	0.6
Tomato	1.1	Crabmeat	0.5
Hard rye bread	1.0	Tuna fish	0.5
Hamburger (broiled)	1.0	Shrimp	0.5
Coleslaw	1.0	Cake	0.5
Lettuce	1.0	Cabbage (boiled)	0.4
Orange section	1.0	Carrot (boiled)	0.4
French roll (with liquid)	1.0	Beets (boiled)	0.3

[a]Data from Yurkstas and Curby (1953).

were definitely masticated in the molar area, whereas steaks were generally chewed in the bicuspid region.

Reasons for Masticating Food

It is worth noting why food is masticated. The major reasons are set out in the following:

1. *Gratification.* Chewing is an enjoyable sensory experience that gives great satisfaction. It is one of the few sensory pleasures that lasts from the cradle to the grave. This point is especially significant for the older person for whom many other sources of pleasure are diminishing. Foods should be selected for the elderly that will give them the maximum masticatory pleasure while satisfying their nutritional needs, and yet not go beyond the limits set by their reduced chewing ability.

2. *Comminution.* Breaking the food into smaller pieces makes swallowing possible.

3. *Mix with saliva.* This lubricates the bolus and softens many hard, dry foods, making them easier to swallow. The enzymes in the saliva start digestion of starches.

4. *Temperature adjustment.* The human race likes to consume much of its food and drink in a cold or hot condition. The mouth seems to be able to

TABLE 8

PERCENT FORCE DISTRIBUTION ON THREE TEETH DURING MASTICATION[a]

Food	First bicuspid (tooth no. 4)	Second bicuspid (tooth no. 5)	First molar (tooth no. 6)
Hard rolls	30	40	30
Rolls plus liquid	24	45	32
Raw vegetables (tough)	41	33	26
Raw vegetables (soft)	37	31	32
Cooked vegetables	15	39	46
Breads	32	26	43
Bread plus liquid	20	27	58
Tough meat	19	48	33
Tender meat	19	29	52
Fish	28	39	33

[a]Data from Yurkstas and Curby (1953).

withstand a wider temperature range than most other parts of the body and the residence time during mastication brings the food close to normal body temperature before sending it on to the stomach.

5. *Release flavor.* Many substances responsible for odor and taste sensations are released as the food is pulverized, causing a stronger stimulus to the chemical receptors in the oral and nasal cavities.

6. *Increase surface area.* The chemical and biochemical attack on the food in the stomach occurs at the surface of each food particle. Mastication greatly increases the surface area available to digestion and also decreases the thickness of each food particle, thus promoting rapid digestion.

Nonoral Methods for Sensing Texture

Although most of the sensing of texture occurs in the mouth and with the lips, it is possible to measure textural properties outside the mouth, most commonly with the fingers and the hand. It is a common practice to hold and squeeze foods in the hand, and this frequently gives a good method for assessing the textural quality of the food. The food may be squeezed between the forefinger and the opposed thumb or between two, three, or four fingers and the opposed thumb. It may be squeezed by pressing with the whole palm on top of the food which is resting on a firm surface such as a table, or the two palms may be placed at opposite ends of the food and squeezed. The size of the object frequently determines the method that is used. The forefinger and opposed thumb are generally used for small objects while the entire hand or two hands are used on large objects such as a loaf of bread. While the hand is usually used to touch foods, it is possible to use other parts of the anatomy such as cheeks, elbows, and feet to obtain some index of the textural qualities of foods.

The viscosity of fluid and semifluid foods is usually assessed by manipulating the food with the tongue (Shama and Sherman, 1973b) or by slurping into the mouth from a spoon (Szczesniak *et al.*, 1963). Stirring with a spoon or finger or tilting the container and watching the rate of flow are also used to measure viscosity (Shama *et al.*, 1973).

A visual manifestation of texture can be found according to the rate and degree that foods spread or slump. One observes the fluidity of a food by the ease with which it pours from a container or flows across the plate. With more solid foods one observes how far the food slumps; for example, a firm jelly holds its shape well while a soft jelly sags to a greater degree.

The sensory measurement of texture is discussed in more detail in Chapter 6.

CHAPTER 3

Principles of Objective Texture Measurement

Introduction

There is such a wide range in types of foods and the types of textural and rheological properties that they exhibit, and such a wide variety of methods used to measure these properties, that it becomes necessary to attempt to classify them into groups in order to understand the system. Several classification systems have been propounded.

It is possible to classify texture measurements according to the commodity that is being tested; for example, tests that are used for cereals, meat, fish, poultry, vegetables, fruit, dairy products, fats, confectionery, beverages, legumes and oilseeds, and miscellaneous foods.

Matz (1962) classified foods on the basis of their textural properties into liquids, gels, fibrous foods, agglomerates of turgid cells, unctuous foods, friable structures, glassy foods, agglomerates of gas-filled vesicles, and combinations of these. Amerine *et al.* (1965) classified foods into four groups: (a) liquid, (b) fruits and vegetables, (c) meats, and (d) other foods. Sone (1972) classified foods on the basis of their textural properties as liquid foods, gel-like foods, fibriform foods, cellular-form foods, edible oils and fats, and powdered foods.

The classification of texture measurements on the basis of commodity or the type of textural properties is useful but what is probably a better type of classification is based on the type of test that is used, because many tests are applicable to more than one type of food. When food is placed in the mouth, the structure is destroyed by the act of mastication until it is ready to be swallowed. The basic process of mastication occurs regardless of what kind of food is in the mouth.

TABLE 1
TYPES OF TESTS FOR MEASURING FOOD TEXTURE

Objective		Subjective	
Direct	Indirect	Oral	Nonoral
Fundamental	Optical	Mechanical	Fingers
Empirical	Chemical	Geometrical	Hand
Imitative	Acoustical	Chemical	Other
	Other		

Therefore, it seems logical to concentrate on the type of test rather than the nature of the food.

Drake (1961) developed a classification system based on the geometry of the apparatus as follows: (1) rectilinear motion (parallel, divergent, convergent); (2) circular motion (rotation, torsion); (3) axially symmetric motion (unlimited, limited); (4) defined other motions (bending, transversal); and (5) undefined motions (mechanical treatment, muscular treatment).

Table 1 lists the type of tests that are used for measuring food texture. These may be divided into objective tests that are performed by instruments and subjective tests that are performed by people. Objective tests can be divided into direct tests that measure real textural properties of materials, and indirect tests that measure physical properties that correlate well with one or more textural properties. Subjective tests can be classified into oral (those tests that are performed in the mouth) and nonoral (in which some part of the body other than the mouth is used to measure the textural properties). Subjective tests will be discussed in Chapter 6.

Fundamental Tests

These tests measure well-defined rheological properties. Before attempting to use this class of test on foods, it should be borne in mind that they were developed by scientists and engineers interested in the theory and practice of materials of construction, and they may not be very useful in measuring what is sensed in the mouth when food is masticated. The outlook of the materials scientist and the food technologist are opposite. One wants to measure the strength of materials in order to design a structure that will withstand the forces applied to it under normal use without breaking. The other wants to measure the strength of food, and frequently weakens its structure deliberately so that it will break down into a fine state suitable for swallowing when subjected to the limited crushing forces of the teeth, imparting pleasurable sensations during the process of comminution.

When a test piece is broken into two pieces the materials scientist normally stops his test because he has all the information he needs about the material. In contrast, the food technologist considers that a test has barely begun when a food is broken into two pieces, and he continues the test in order to break it down into progressively smaller pieces. Hence, food texture measurement might be considered more as a study of the *weakness* of materials rather than strength of materials.

The most commonly used fundamental tests are listed below. The first four apply to solids while the fifth applies to fluids.

1. Young's modulus of elasticity (E) $= \dfrac{\text{stress}}{\text{strain}} = \dfrac{F/A}{\Delta L/L}$. (in compression or tension)

2. Shear modulus (G) $= \dfrac{\text{shearing stress}}{\text{shearing strain}} = \dfrac{F/A}{\gamma/L}$.

3. Bulk modulus (K) $= \dfrac{\text{hydrostatic pressure}}{\text{volume strain}} = \dfrac{P}{\Delta V/V}$.

4. Poisson's ratio $\mu = \dfrac{\text{change in width per unit width}}{\text{change in length per unit length}} = \dfrac{\Delta D/D}{\Delta L/L}$.

When the volume is unchanged during test, $\mu = \frac{1}{2}$. If volume decreases, $\mu < \frac{1}{2}$.

5. Viscosity $= \sigma/\dot{\gamma}$,

where F is applied force, A is cross-sectional area, L is unstressed length, ΔL is change in length caused by the application of force F, γ is displacement (shear modulus), P is pressure, V is volume, D is diameter, σ is shear stress (viscosity), and $\dot{\gamma}$ is shear rate (viscosity).

Note that the stress is always a *force* measurement while strain is always a *distance* measurement. Strain is the change in dimensions of a test specimen caused by the application of a stress. *Stress* and *strain* are not synonyms. In simple uniaxial compression (the type of test most frequently used by food technologists), stress is the force per unit area (measured in newtons) and strain is the change in length per unit length (measured in millimeters).

Rheological theory shows that the first four moduli are interrelated, as follows:

$$G = 3EK/(9K - E),$$
$$K = E/3(1 - 2\mu) = EG/(9G - 3E) = G[2(1 + \mu)]/3(1 - 2\mu),$$
$$E = 9GK/(3K + G) = 2G(1 + \mu) = 3K(1 - 2\mu),$$
$$\mu = (E - 2G)/2G = (1 - E/3K)/2.$$

Fundamental tests generally assume (1) small strains (1–3% maximum); (2) the material is continuous, isotropic (exhibiting the same physical properties in every direction), and homogeneous; and (3) the test piece is of uniform and regular shape. Most textural tests made on foods fail to comply with the three assumptions listed above. The conversion equations given above generally work well for engineering materials such as steel and glass, but they are not effective for interconversion of the moduli for rubber, and they probably do not apply to moduli that are measured on most foods.

Fundamental tests are generally slow to perform, do not correlate as well with sensory evaluation as do empirical tests, and use expensive equipment. They are not used to any great extent in the food industry but they do have a place in some research laboratories. Szczesniak (1963b) aptly described the usefulness of fundamental tests as follows:

> Since most foodstuffs do not have simple rheological properties that are independent of stress and strain conditions, and since rheological properties once measured and defined are not meaningful in a practical sense unless related to functional properties, fundamental tests serve the greatest value to the food technologist by providing bases for the development of more meaningful empirical tests.

Table 2 shows the results obtained with apples and peaches by a fundamental test and an empirical test as compared with sensory evaluation of firmness. The stiffness coefficient, which is essentially on index of Young's modulus of elasticity and is a fundamental test (see p. 92), gives consistently lower correlations with sensory measurements than does the Magness–Taylor test, which is an empirical type of measurement. In view of the fact that the Magness–Taylor pressure tester costs less than $100 and one test can be performed in about 30 sec while the acoustic spectrometer that is used to measure the stiffness coefficient costs over $10,000 and requires about 15 min to make a test, it can be seen why the food industry prefers empirical tests.

TABLE 2

Correlation Coefficient between Sensory and Instrumental Firmness Measurements[a]

	Stiffness coefficient[b] (f^2m)	Magness–Taylor[c]
Red delicious apples		
October 1968	0.84	0.92
March 1969	Not significant	0.71
October 1969	0.68	0.89
March 1970	0.44	0.86
Elberta peaches	0.87[d]	0.957

[a]Data from Finney (1971a) and Finney and Abbott (1972).
[b]Sonic resonance test, a fundamental test.
[c]Puncture test, an empirical test.
[d]Highest value from 22 experiments.

Muller (1969b) surveyed the types of food texture measurements that are used in the United Kingdom and concluded that of the rheological tests used, "it is striking that with a few exceptions the methods employed are empirical. This might support the jest that theoretically sound instruments do not work in practice and those that are theoretically unsound do."

Empirical Tests

These tests measure parameters that are poorly defined, but from practical experience are found to be related to textural quality. This is the most widely used class of instruments in the food industry. The tests are usually easy to perform, rapid, and frequently use inexpensive equipment. Problems with this type of test are the poor definition of what is being measured, the arbitrariness of the test, frequently no absolute standard is available, and the tests are usually only effective with a limited number of commodities. Since empirical tests are frequently successful in measuring textural properties of foods and are the most widely used in the food industry, this book will deal with them extensively. It is the author's opinion that these tests should be studied in order to understand the reasons for their successes and the principles on which they operate in order to find how to make them more effective and to make them scientifically more rigorous.

Imitative Tests

These tests imitate the conditions to which the food material is subjected in practice. This class may be considered as a type of empirical test because the tests are not fundamental tests. Examples of this kind of test are the General Foods Texurometer that imitates the chewing action of the teeth, the Farinograph and other dough-testing apparatus that imitate the handling and working of bread dough, the Bostwick Consistometer and Adams Consistometer that measure the flow of semifluid foods across the plate, and butter spreaders.

Figure 1 shows schematically the relationships among empirical, fundamental, and imitative tests, and Table 3 lists the advantages and disadvantages of each type. The ideal texture measuring apparatus should combine the best features of the fundamental, empirical, and imitative methods and eliminate the undesirable features of each of these. At the present time there is no ideal texture measuring equipment or system. Empirical methods are used almost entirely. The future direction of the research should be to move from the empirical into the ideal by including more of the fundamental and imitative aspects in empirical tests. The ideal texture measuring technique will probably be some combination of the present empirical, fundamental, and imitative methods.

Another method of classification of food texture instruments is on the basis of the variable or variables that are measured in the test. Table 4 gives such a

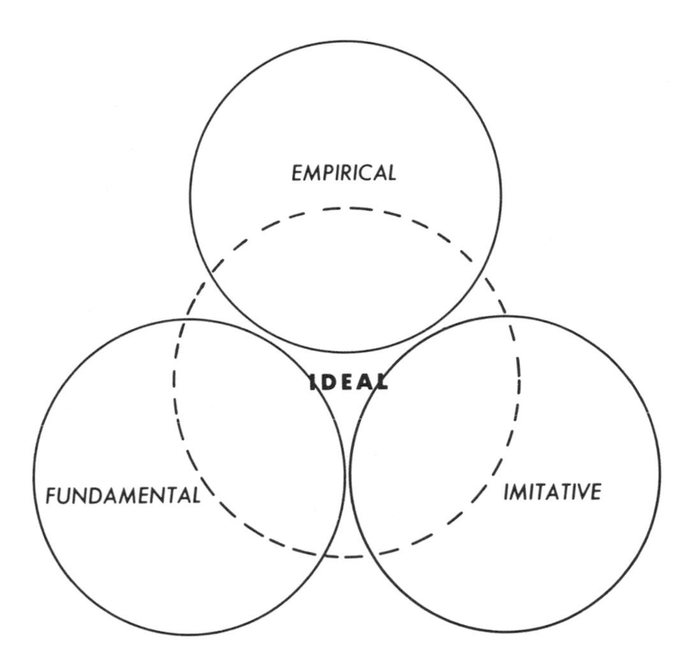

FIG. 1. Schematic representation of the ideal texture measuring apparatus and its derivation from empirical, fundamental, and imitative instruments. (From Bourne, 1975b; reprinted with permission from D. Reidel Publ. Co.)

classification and is the system that will be used throughout this book to discuss and classify the principles of objective measurements of food texture. The system classifies according to the principle of the test, not according to the kind of food. It rests on the assumption that since all foods are ground into a fine state during mastication, there must be many common elements in their textural properties that are not restricted to any one commodity group. The principles of the tests will be discussed in this chapter. Chapter 4 discusses commercially available instruments and their operation.

Force Measuring Instruments

Force measuring instruments are the most common of the texture measuring instruments. Force has the dimensions mass $\times$ length $\times$ time^{-2}. The standard unit of force is the newton (N). Because of their multiplicity this heading is broken into the subclassifications (a) puncture, (b) compression–extrusion (c), shear, (d) crushing, (e) tensile, (f) torque, and (g) bending and snapping.

TABLE 3

Comparison of Different Systems of Objective Texture Measurement of Food[a]

System	Advantages	Disadvantages
Empirical	Simple to perform	No fundamental understanding of the test
	Rapid	Incomplete specification of texture
	Suitable for routine quality control	Arbitrary procedure
	Good correlation with sensory methods	Cannot convert data to another system
	Large samples give averaging effect	Usually "one point" measurement
Imitative	Closely duplicates mastication or other sensory methods	Unknown physical equivalent measurement
	Good correlation with sensory methods	Evaluation of graphs slow
	Complete texture measurement	Not suitable for routine work
		Restricted to "bite-size" units
Fundamental	Know exactly what is measured	Poor correlation with sensory methods
		Incomplete specification of texture
		Slow
Ideal	Simple to perform	None
	Rapid	
	Suitable for routine work	
	Good correlation	
	Closely duplicates mastication	
	Complete texture measurement	
	Know exactly what is measured	
	Can use large or small size samples	

[a]From Bourne, 1975c; reprinted with permission of D. Reidel Publ. Co.

Puncture Testing

The puncture test measures the force required to push a punch or probe into a food. The test is characterized by (a) a force measuring instrument, (b) penetration of the probe into the food causing irreversible crushing or flowing of the food, and (c) the depth of penetration is usually held constant.

Puncture testers embody one of the simplest types of texture measuring instruments and one of the most widely used. The first food puncture tester was probably the one developed by Lipowitz (1861), who placed a flat disk 1 or 2 in. in diam on the surface of a gelatin jelly in a beaker (Fig. 2). The flat disk was connected to a funnel by means of a vertical iron rod, and lead shot was slowly poured into the funnel until there was just sufficient weight to make the disk

TABLE 4

OBJECTIVE METHODS FOR MEASURING FOOD TEXTURE[a]

Method	Measured variable	Dimensional units	Examples
1. Force	Force (F)	mlt^{-2}	
a. Puncture	F	mlt^{-2}	Magness–Taylor
b. Extrusion	F	mlt^{-2}	Shear press, Tenderometer
c. Shear	F	mlt^{-2}	Warner–Bratzler Shear
d. Crushing	F	mlt^{-2}	—
e. Tensile	F	mlt^{-2}	—
f. Torque	F	mlt^{-2}	Rotary Viscometers
g. Snapping	F	mlt^{-2}	Brabender Struct-o-Graph
h. Deformation	F	mlt^{-2}	
2. Distance			
	a. Length	l	Penetrometers
	b. Area	l^2	Grawemeyer Consistometer
	c. Volume	l^3	Bread volume
3. Time	Time (T)	t	Ostwald Viscometer
4. Energy	Work ($F \times D$)	ml^2t^{-2}	—
5. Ratio	F or D or T measured twice	Dimensionless	Specific gravity
6. Multiple	F and D and T	mlt^{-2}, l, t	Instron, Ottawa Texture Measuring System, GF Texturometer
7. Multiple variable	Anything	Unclear	Durometer
8. Chemical analysis	Concentration	Dimensionless (%)	Alcohol insoluble solids
9. Miscellaneous	Anything	Anything	Optical density, crushing sounds, etc.

[a]Adapted from Bourne (1966a); reprinted from *J. Food Sci.* **31,** 1114, 1966. Copyright by Institute of Food Technologists.

penetrate into the jelly. The total weight of the shot, funnel, rod, and disk was used as a measure of jelly consistency. This early test, although primitive, contains the essential elements of the puncture test: namely, a punch that penetrates into the food, application of an increasing force (lead shot), and measurement of the yield point force (by scales not shown in figure). This apparatus evolved into the well-known Bloom Gelometer.

The second food puncture tester was probably the one developed by Carpi (1884), who measured the weight required to force a 2-mm-diam iron rod 1-cm deep into hardened oils. Brulle (1893) used a similar principle for measuring the hardness of butter, and Sohn (1893) spelled out the procedure necessary to obtain

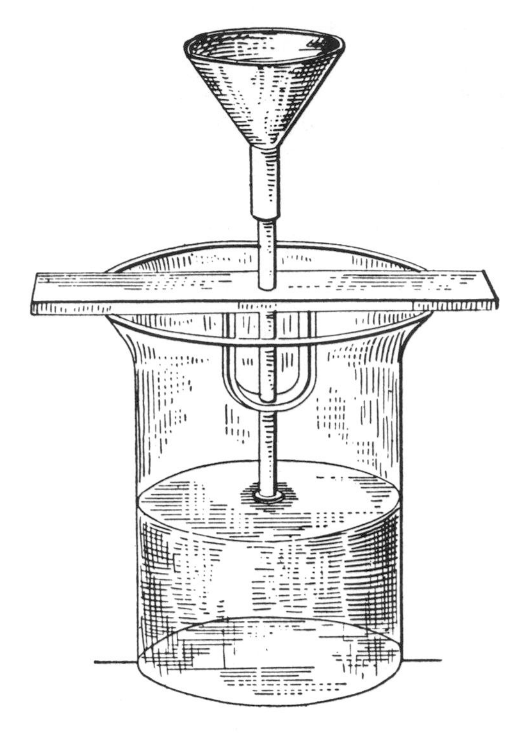

FIG. 2. The Lipowitz Jelly Tester. (From Lipowitz, 1861.)

reproducible results with the Brulle instrument. This developed into the Van Doorn butter tester. The first puncture tester for horticultural products was developed by Professor Morris in the state of Washington (Morris, 1925). This evolved into the well-known Magness–Taylor, Chatillon, and EFFI-GI fruit pressure testers. Tressler *et al.* (1932) performed a puncture test on meat, which evolved into the Armour Tenderometer (Hansen, 1971, 1972).

Puncture testing instruments are all *maximum-force* instruments. They may be classed into *single-probe* instruments, such as the Magness–Taylor, EFFI-GI, Chatillon, the University of California Fruit Firmness tester, the Bloom Gelometer, and the Marine Colloids Gel tester, and the *multiple-probe* instruments such as the Armour Tenderometer, the Christel Texture Meter, and the Maturometer.

Puncture testing instruments might also be classified by the manner in which the force is applied. A constant rate of application of force is used for the majority of these instruments (e.g., Magness–Taylor and other fruit pressure testers, the Armour Tenderometer, and the Bloom Gelometer). Other puncture testing instruments use a constant rate of travel of the probe, including the Marine Colloids Gel tester, the Christel Texture Meter, and the Maturometer.

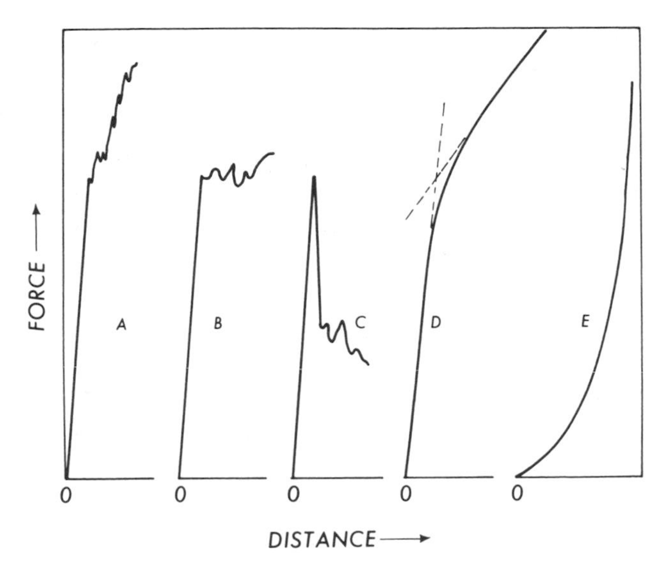

FIG. 3. Schematic representation of the five different types of force–distance curves that are obtained in puncture tests. [From Bourne, 1979b; copyright Academic Press Inc. (London) Ltd. with permission.]

When a punch is mounted in an instrument that automatically draws out a force–distance or force–time curve (such as the Instron), five basic types of curves are obtained, as shown schematically in Fig. 3. In types A, B, and C there is an initial rapid rise in force over a short distance of movement as the pressure tip moves onto the commodity. During this stage the commodity is deforming under the load; there is no puncturing of the tissues. This stage ends abruptly when the punch begins to penetrate into the food, which event is represented by the sudden change in slope called the *yield point,* or sometimes "bio-yield point." The initial deformation stage is not of great concern in puncture testing.

The yield point marks the instant when the punch begins to penetrate into the food, causing irreversible crushing or flow of the underlying tissues and is the point of greatest interest in puncture testing. Mohsenin *et al.* (1963) showed that this is the point where crushing and bruising begins on fruits such as apples. Considerable work has been done on the implications of the yield point and this will be discussed below.

The third phase of the puncture test, namely, the direction of the force change after the yield point and during penetration of the punch into the food, separates the puncture curves into three basic types: A, the force continues to increase after the yield point; B, the force is approximately constant after the yield point; C, the force decreases after the yield point. There is a continuous change in slope, from positive slope in type A curves to approximately zero slope in type B curves to negative slope in type C curves. Type A curves merge into type B curves, depending on the steepness of the slope of the force–distance curve after the yield point, and, likewise, type B curves merge into type C curves. There are occasions when one needs to use subjective criteria to decide whether a curve is type A or B, or type B or C. The sensory and physical meaning of the difference between type A, B, and C curves is presently not well understood. Friction of the food along the sides of the punch accounts for a slightly increasing positive slope in a limited number of cases, but there are cases (e.g., freshly harvested apples) where friction cannot account for the increase in force after the yield point has been passed.

A fourth type of curve, shown in curve D, is obtained on some starch pastes and whipped toppings and foams. It is essentially a type A curve except that the yield point is not sharply delineated by an abrupt change in slope; rather there is a gradual change in slope. The intersection formed by extrapolating the two straight-line portions of type D curves is usually a precise and reproducible point that can be used as a yield point figure; hence, a type D curve may be considered as a special case of a type A curve.

The type E curve is found with some starch pastes. This type of commodity shows no yield point, behaves essentially as a viscous liquid, and is unsuited to the puncture test because no meaningful results can be extracted from a type E puncture test curve at the present time.

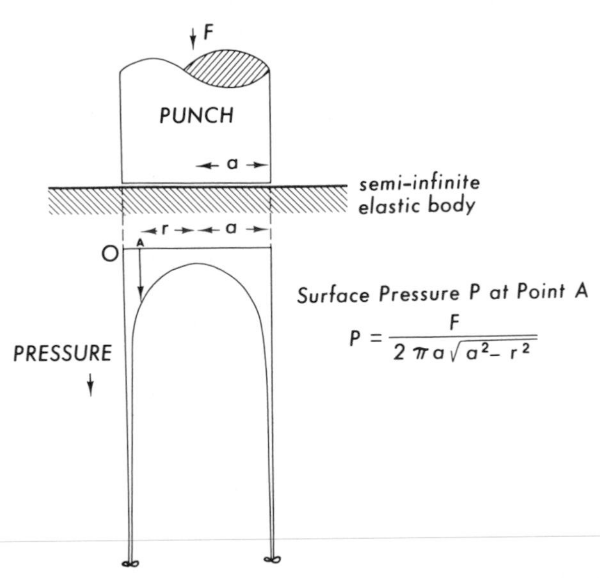

FIG. 4. Theoretical stress distribution in a semiinfinite elastic body compressed under a rigid plunger: P, surface pressure at A; F, total force; a, radius of punch; r, distance to A. [Redrawn from Morrow and Mohsenin, 1966; copyright Academic Press Inc (London) Ltd. with permission.]

Morrow and Mohsenin (1966) showed that the theoretical stress distribution under a rigid die acting against a semiinfinite elastic body follows the Boussinesq equation:

$$P = F/2\pi a\,(a^2 - r^2)^{1/2}, \tag{1}$$

where P is the pressure at any point under the punch, F is the total force applied to punch, a is the radius of punch, and r is the distance from center of punch to stressed area.

According to this equation the stress in the food is highest at the perimeter of the punch and lowest at the center of the punch. This is demonstrated graphically in Fig. 4. This is a theoretical stress distribution and there are probably substantial deviations from this equation in practical situations. This equation only applies *before* the yield point is reached; that is, during the deformation stage. The Boussinesq equation does not apply during or after the yield point. The point of major interest in this equation is that the distribution of the stress under the punch is uneven with the highest stresses at the perimeter.

Using the theory of contact stresses between two bodies pressing against each other Yang and Mohsenin (1974) developed an equation for the initial slope in the puncture of Rome variety apples as follows:

$$F = -(2\pi/3)DRh_1(a\xi)[(a^2/R^2)(1 - a^2/R^2)^{1/2}$$
$$-\frac{5(1 - a^2/R^2)^2 - 14(1 - a^2/R^2)^3 + 9(1 - a^2/R^2)^4}{(1 - a^2/R^2)^{1/2}(2 - 3(a^2/R^2)^2}$$
$$+ \tfrac{1}{2}ln(1 - a^2/R^2) - 2(1 - a^2/R^2) + \tfrac{3}{4}(1 - a^2/R^2)^2 + \tfrac{1}{2}a^4/R^4 + \tfrac{5}{4}], \tag{2}$$

where F is the force; D, the deformation at axis of symmetry; R, the radius of curvature of punch; h_1, a complex function; a, the radius of surface of contact; and ξ, a value between 0 and 1 (put at 0.7 by authors).

Bourne (1966b) has shown that the yield-point force is proportional to both the area and perimeter of the punch, and to two different textural properties of the food being tested. Figure 5 shows schematically what happens at the point of penetration of the punch into the food. There is compression of the food under the punch which is proportional to the area of the punch and shearing around the edge of the punch which is proportional to the perimeter. This relationship can be expressed in the form of the equation

$$F = K_cA + K_sP + C, \tag{3}$$

where F is the force on the punch (in newtons but sometimes it is measured in kg or lb); K_c, the compression coefficient of commodity (N mm^{-2}); K_s, the shear

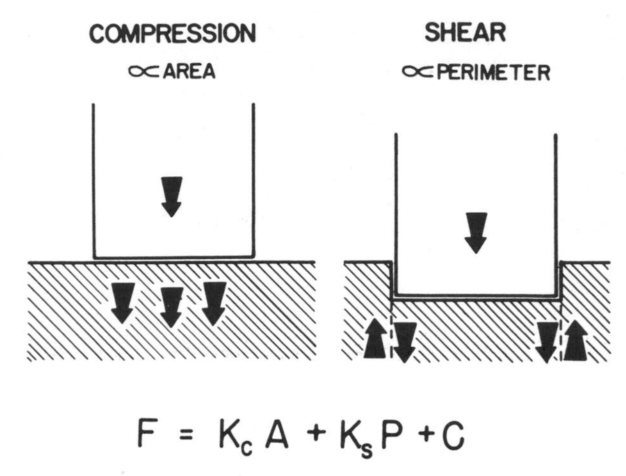

$$F = K_c A + K_s P + C$$

FIG. 5. Schematic representation of a puncture test. (Reprinted from *J. Food Sci.* **31,** 284, 1966; copyright by Institute of Food Technologists.)

coefficient of commodity (N mm^{-1}); A, the area of the punch (mm^2); P, the perimeter of the punch (mm); and C, a constant (N).

The validity of the above equation was proved by means of two sets of flat-faced rectangular-shaped punches: one set had constant perimeter with area varying from 0.25 to 1.00 cm^2 and the second set had constant area with perimeter varying from 4.0 to 8.5 cm. Two circular punches were included: one with a cross-sectional area of 1.00 cm^2 and the other with a perimeter of 4.0 cm. These are shown in Fig. 6.

Each of these punches was pressed into foamed polystyrene board and the yield points were measured by means of an Instron machine. Figure 7, which plots the mean puncture force against punch area for the constant perimeter punches, shows a rectilinear relationship between puncture force and punch area. From the equation it follows that the slope of this line gives the numerical value of the compression coefficient K_c and the intercept on the Y axis gives the value $(K_sP + C)$.

Figure 8 shows that a plot of puncture force against punch perimeter is rectilinear provided the area is kept constant. From the equation it follows that the slope of this line gives the numerical value of the shear coefficient (K_s) and the intercept on the Y axis gives the value $(K_cA + C)$ Since the values of K_s and K_c can be obtained from the slopes of these plots it follows that the constant C can also be obtained by taking the intercept value and substituting known values for either K_sP or K_cA and calculating the value for C. Thus, it is possible to evaluate all the parameters in this equation from the force measurements made with this series of punches.

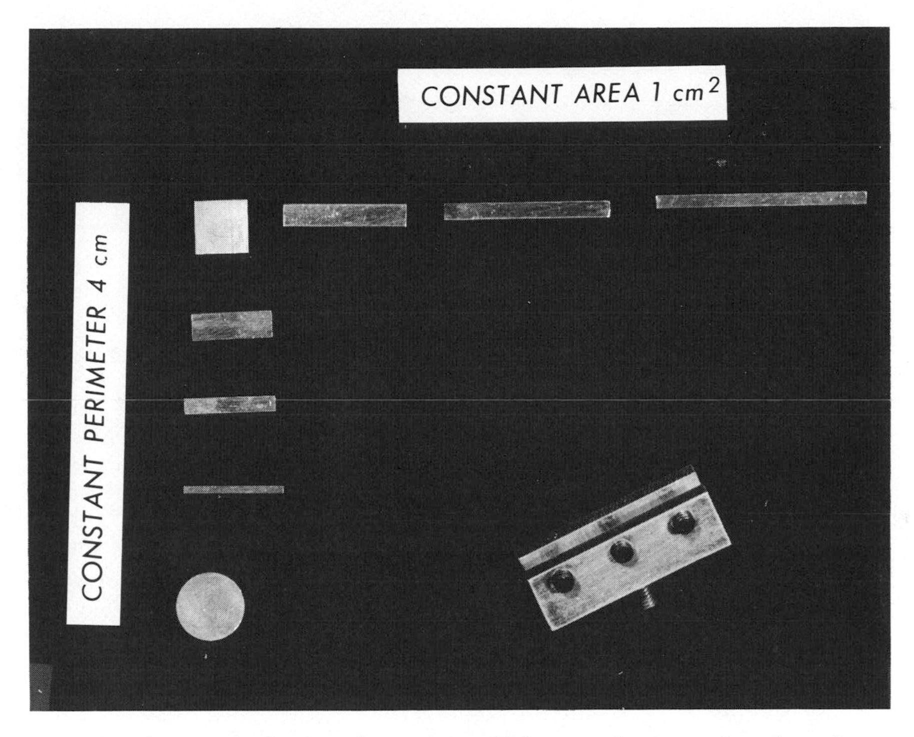

FIG. 6. Set of rectangular-faced punches used to establish area- and perimeter-dependence of puncture force. The device in the lower right-hand corner holds the various punches. (Reprinted from *J. Food Sci.* **31,** 285, 1966; copyright by Institute of Food Technologists.)

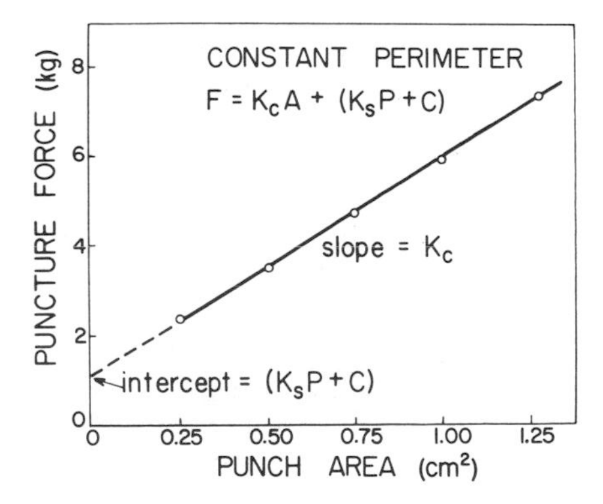

FIG. 7. Puncture force versus punch area with constant perimeter on polystyrene board using the rectangular-faced punches shown in Fig. 6. (Reprinted from *J. Food Sci.* **31,** 286, 1966; copyright by Institute of Food Technologists.)

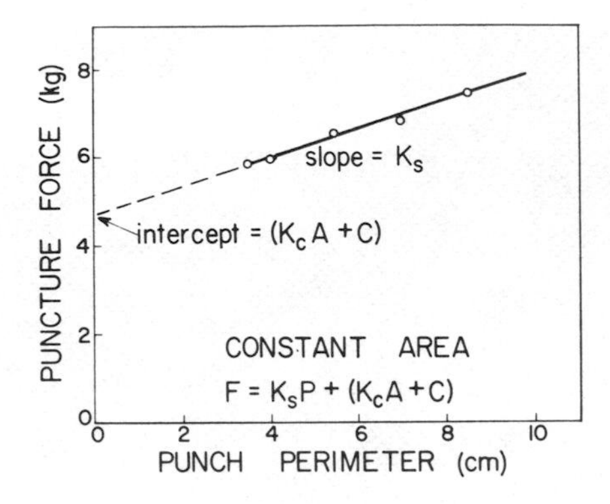

FIG. 8. Puncture force versus punch perimeter with constant area on polystyrene board using the rectangular-faced punches shown in Fig. 6. (Reprinted from *J. Food Sci.* **31,** 286, 1966; copyright by Institute of Food Technologists.)

This relationship has been found to apply to a wide variety of foods. Figure 9 shows the puncture-force/punch-area relationships for a number of foods. In each case a rectilinear relationship was found. A similar rectilinear relationship was found between puncture force and punch perimeter for these same foods (Bourne, 1966b).

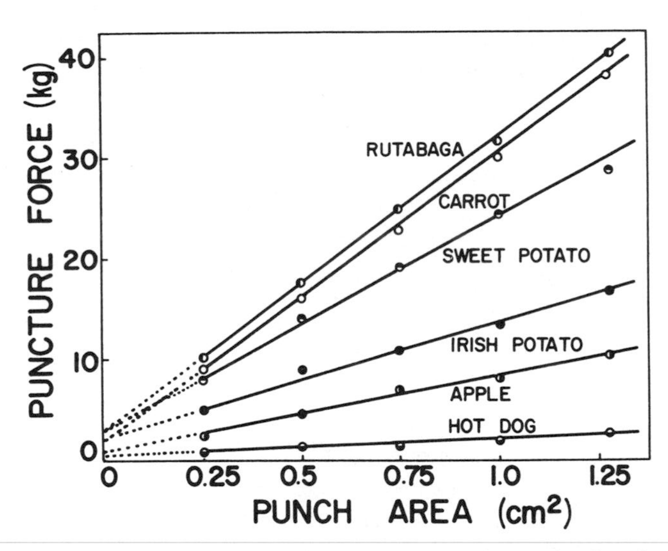

FIG. 9. Puncture force versus punch area with constant perimenter on various foods using the rectangular-faced punches shown in Fig. 6. (Reprinted from *J. Food Sci.* **31,** 286, 1966; copyright by Institute of Food Technologists.)

TABLE 5

NUMERICAL VALUES OF PUNCTURE TEST COEFFICIENTS FOR VARIOUS COMMODITIES[a]

Commodity	Compression coefficient K_c ($N\ mm^{-2}$)	Shear coefficient K_s ($N\ mm^{-1}$)	Constant C (N)
Expanded polystyrene	0.477	0.333	−2.26
High-density polystyrene	1.29	2.16	−26.9
Polyurethane	0.350	0.284	−4.61
Apples (raw, Limbertwig variety)	0.737	0.157	0.294
Apples (raw, Fr. vonBerl variety)	0.631	0.0686	3.92
Banana (ripe, yellow)	0.0422	0.0588	−0.588
Creme-filled wafers	0.104	0.137	6.28
Carrot (uncooked core tissue)	2.75	−0.0294	21.4
Wiener (cold)	0.166	0.00392	1.47
Potato (Irish, uncooked)	1.06	0.509	5.88
Rutabaga (uncooked)	2.90	0.843	−1.47
Sweet potato (uncooked)	1.94	0.883	3.43
1% agar gel	0.0147	0.0049	−0.0981
2% agar gel	0.0618	0.0284	−0.196
3% agar gel	0.119	0.157	−3.24

[a]Adapted from Bourne (1966b).

Table 5 lists the numerical values of coefficients for a number of food commodities. The physical meaning of the constant C would be interpreted from the punch force equation as being the force required to puncture a commodity with a punch of zero area and zero perimeter. Constant C has a value close to zero for most of the commodities tested, and in such cases could be neglected without introducing any great inaccuracies. Some commodities have a value for C that is numerically too high to be attributed to experimental error, and these C values are usually negative. In these cases it seems probable that there is a zone of influence around the punch such that the actual compression area on the commodity is larger than the area of the punch. However, the real meaning of the value of C in these instances has not yet been elucidated with certainty.

DeMan (1969) confirmed the fact that the puncture force is dependent upon both the area and perimeter of the punch with processed cheese. With butter and margarine, however, he found that the shear coefficient is zero, which causes the term K_sP to fall out of the puncture equation and makes the puncture test on these commodities dependent on area only. DeMan considered that with fats there is flow rather than compression under the punch, and for these commodities he postulated the equation

$$F = K_f A, \tag{4}$$

where F is the puncture force; A, the area of the punch; and K_f, the flow coefficient (replacing K_c, the compression coefficient).

For circular punches the area and perimeter can be substituted by functions of diameter that follow from the geometry of circles to give the following equation:

$$F = (\pi/4)K_cD^2 + \pi K_sD + C, \tag{5}$$

where D is the diameter of the punch.

The puncture equation explains why a simple doubling of the area of a circular punch usually fails to double the puncture force. When the area of a circular punch is increased by a factor of 2.0, the perimeter is increased by a factor of $\sqrt{2} = 1.41$. The puncture force will then be doubled only if the shear coefficient K_s is zero or if the shape of the punch changed so that both perimeter and area are doubled.

In designing punches for a test device it is possible to give added or less weight to the shear component by increasing or decreasing the perimeter/area ratio of the punch. Figure 10 shows two methods of manipulating the perimeter/area ratio of a punch. The first single punch at the bottom of Fig. 10 has an area of 1.00 cm^2 and a perimeter of 3.55 cm. The nest of four punches immediately above it has a combined area of 1.00 cm^2 and a combined perimeter of 7.10 cm. The nest of four punches will normally give a higher puncture force reading than a single punch, even though the areas are equal because the amount of shearing with the nest of four punches is double that of the single punch. The second circular punch above the nest of four punches has an area of 0.469 cm^2 and a perimeter of 2.42 cm, while its star-shaped partner immediately above has the same area but a perimeter of 3.78 cm.

It is possible to obtain numerical values for the shear and compression coefficients of a food by using a set of circular-shaped punches (Bourne, 1975b). Dividing the basic puncture equation [Eq. (3) above] through by the area and converting the perimeter and area into functions of the diameter gives the following equation:

$$F/A = 4K_p/D + K_a + 4C/\pi D^2. \tag{6}$$

According to this equation the plot of F/A against $1/D$ should be rectilinear with a slope equal to $4K_p$ and an intercept on the y axis of $K_a + 4C/\pi D^2$. Dividing the Eq (3) through by perimeter and converting the area into a function of diameter gives the equation

$$F/P = K_aD/4 + K_p + C/\pi D. \tag{7}$$

According to this equation a plot of F/P versus diameter of the punch should be rectilinear with a slope equal to $K_a/4$ and an intercept equal to $K_p + C/\pi D$. The validity of these equations has been shown to hold quite well for foods, provided punches larger than approximately 2 mm diam are used (Bourne, 1975b).

The section of a puncture curve beyond the yield point (see Fig. 3) represents

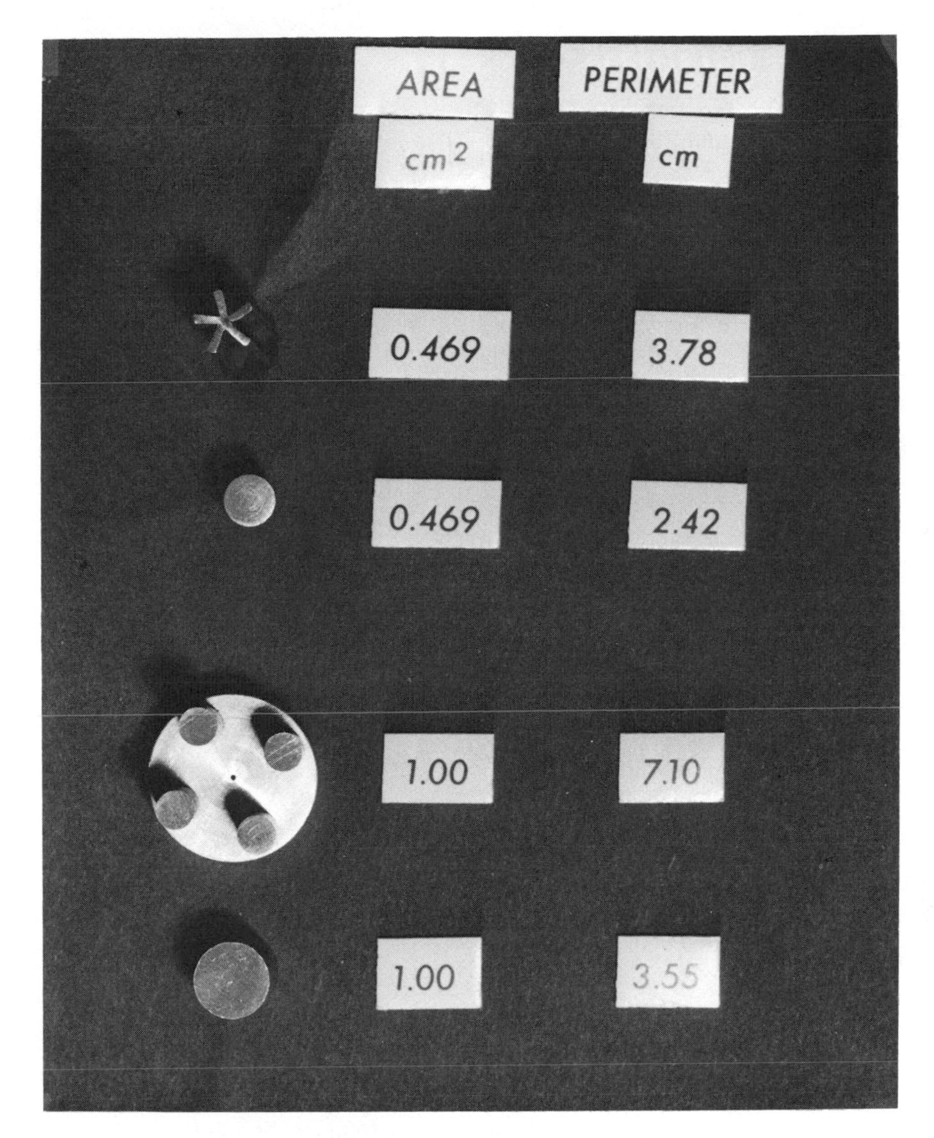

FIG. 10. Two pairs of punches. Each pair has equal face area but differing perimeters. (Reprinted from *J. Food Sci.* **31,** 288, 1966; copyright by Institute of Food Technologists.)

the force required to penetrate into the food. In a type A curve the penetration force beyond the yield point increases with penetration depth; this type of curve is characteristic of freshly picked apples. In a type B curve the force of penetration is approximately constant; this is typical of many apples that have been held in cold storage for periods of several months and soft ripe fruits such as peaches and pears. In type C curves the force to penetrate is lower than the yield-point force. This type of curve is almost always found with raw vegetables.

Yang and Mohsenin (1974) used contact stress theory to develop an equation for the penetration of the punch into Rome apples:

$$F = -\pi a^2 k[\sqrt{3} + nC_a \ln(D_0/2a)], \tag{8}$$

where F is the force; a, the radius of the surface of contact; k, the shearing strength; n, a correction factor; C_a, a correction factor; and D_0, the diameter of the assumed cylinder.

Yang and Mohsenin found a good match between the experimental data on Rome apples and the above theoretical equation. It is worth noting however that Yang and Mohsenin's experimental curve is essentially a B type curve where the penetration force is approximately constant. Since this was obtained with Rome apples it seems almost certain that the experiments were performed in the spring on apples that had been held in cold storage for several months. It is unlikely that this equation would apply to freshly harvested Rome apples which give an A type curve. There is no experimental data to show whether or not Eq. (8) applies to foods other than the Rome variety of apples.

The significance of type A, type B, and type C curves, as far as measuring sensory textural characteristics of foods, remains to be worked out. With apples, the type A curve is typical of a freshly harvested juicy crisp apple while the type B curve is typical of dry, softer, and mealy-textured apples that have been in storage for several months. The type C curve is frequently found in raw vegetables and some apples and seems to be associated with a "woody" type of texture. Much work, however, remains to be done in this area.

Compression–Extrusion Testing

The compression–extrusion test consists of applying force to a food until it flows through an outlet that may be in the form of one or more slots or holes that are in the test cell. The food is compressed until the structure of the food is disrupted and it extrudes through these outlets. Usually the maximum force required to accomplish extrusion is measured and used as an index of textural quality. This type of test is used on viscous liquids, gels, fats, and fresh and processed fruits and vegetables. Since extrusion requires that the food flow under pressure, it seems reasonable to use it on food that will flow fairly readily under an applied force and not to use it on those foods that do not flow easily, such as bread, cake, cookies, breakfast cereals, and candy.

A simple type of compression–extrusion test is shown in Fig. 11, in which the food is placed in a strong metal box with an open top. A loose fitting plunger is then forced down into the box until the food flows up through the space between the plunger and the walls of the box. This space is called the annulus.

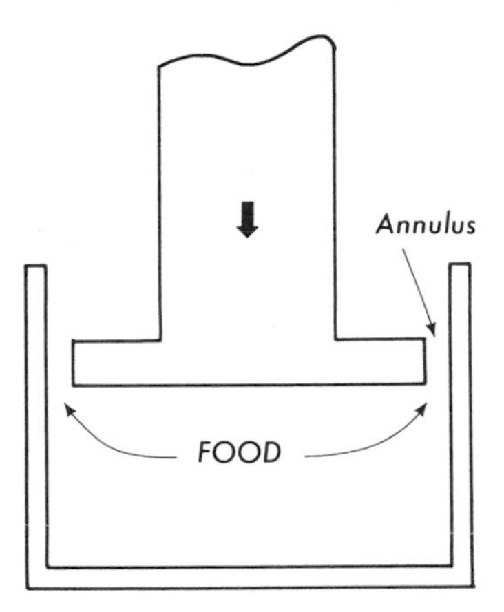

FIG. 11. Schematic diagram of a simple cell for pure compression–extrusion. (Reprinted with permission from D. Reidel Publ. Co.)

Figure 12 demonstrates the compression–extrusion test on peas. In section A the peas have been placed in a beaker and the compressing plunger has just contacted the surface of the peas. In section B the peas have been packed down solid so that the air between the peas has been removed and replaced with liquid expressed from the peas. Section C shows the actual process of extrusion where the peas are forced to flow around the space between the edge of the compressing plunger and the inside wall of the beaker.

FIG. 12. Compression–extrusion test on fresh green peas: A, packing down; B, compression of solid pack; C, extrusion through the annulus. [Reprinted from "Rheology and Texture in Food Quality" (deMan, Voisey, Rasper, and Stanley, eds.), p. 266; with permission of AVI Publ. Co.]

A typical force–distance curve obtained from such an apparatus is shown in Fig. 13. From A to B the peas are deformed and compressed to pack more and more tightly into the diminishing space available under the descending plunger; there is almost no rupture or breaking of the peas. At approximately the point B the peas are packed solid and liquid begins to be pressed from the peas filling the interstices. At point B or soon afterwards the pack is solid except for small amounts of entrapped air, and the force increases steeply from B to C pressing out more juice in the process. At point C the peas begin to rupture and flow up through the annulus, and this process continues to point D when the compressing plunger is reversed in direction and the force falls to zero. Point C gives the force necessary to begin the process of extrusion, and the plateau CD shows the force needed to continue extrusion. From B to C represents the increasing force being applied to an almost incompressible mixture of solids and liquid.

The shape and magnitude of the compression–extrusion curve is influenced by the elasticity, viscoelasticity, viscosity, and rupture behavior of the material; sample size, deformation rate, sample temperature, type of test cell; sample test size; and homogeneity of the sample (Voisey *et al.*, 1972). With most processed fruits and vegetables and many fresh products the plateau CD is horizontal or nearly so. The unevenness of the plateau is caused by variations in the firmness or toughness of the particles that are passing through the annulus zone at any particular time.

In general, the slope of the curve during the process of extrusion is approximately horizontal, but there are times when it will show a steadily increasing or

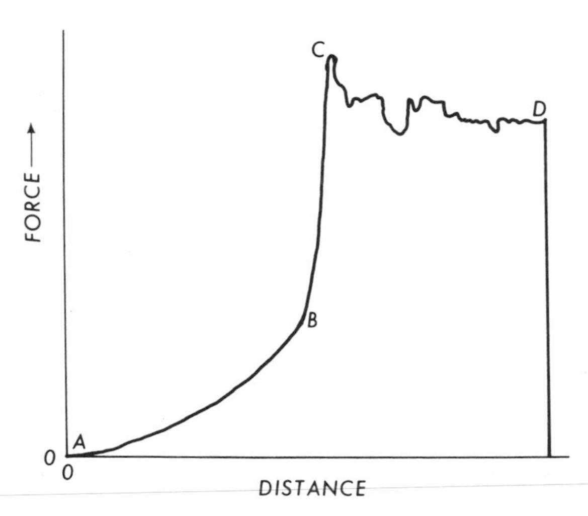

FIG. 13. Typical force–distance curve obtained with a simple compression–extrusion test (compare with Fig. 12). [Reprinted from "Rheology and Texture in Food Quality (deMan, Voisey, Rasper, and Stanley, eds.), p. 265; with permission of AVI Publ. Co.]

decreasing slope. According to Voisey *et al.* (1972), the slope of the extrusion part of the curve can indicate four different behavior patterns.

1. The force reduces rapidly with further compression. This indicates that the sample was compressed until a catastrophic failure occurred, indicating that resistance to shearing is the dominant mechanism of this test.
2. The force decreases slowly, indicating some shearing resistance combined with some extrusion and possibly adhesion of the sample to test cell.
3. An approximately horizontal plateau indicates either shearing of successive layers of the sample or a combination of shearing, extrusion, and adhesion occurring simultaneously.
4. The force steadily increases as extrusion proceeds. This indicates further compression of the sample in addition to various amounts of adhesion, extrusion, and shearing.

For the type of test cell shown in Fig. 11 the extrusion force is inversely proportional to the width of the annulus, as shown in Fig. 14 where the maximum extrusion force for a uniform sample of sized, graded fresh green peas is plotted against the annulus width. With a wide annulus, a small change in annulus width has a small affect upon the extrusion force, but as the annulus width narrows, the extrusion force increases steeply.

The annulus width also affects the evenness of the force in the plateau region. As the annulus width becomes narrower, the force plateau becomes less even, until at very narrow annulus widths the force fluctuates rapidly and violently along the plateau. Figure 15 shows force–distance plots for extrusion of fresh

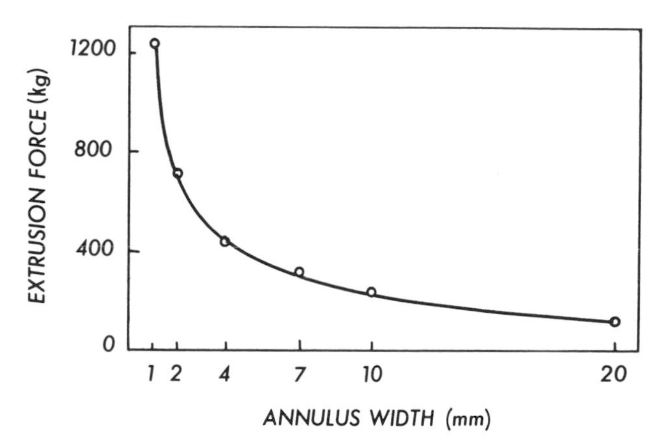

Fig. 14. Effect of annulus width on maximum extrusion force on fresh green peas. (From Bourne and Moyer, 1968; reprinted from *Food Technol.* **22,** 1015, 1968. Copyright by Institute of Food Technologists.)

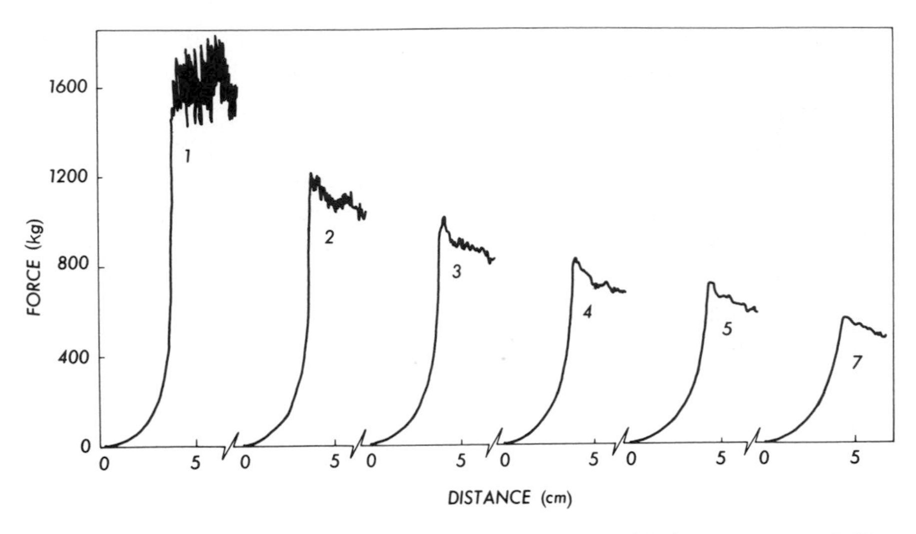

FIG. 15. Typical force–distance curves obtained from a uniform lot of fresh green peas extruded in a simple extrusion cell with annulus widths of 1, 2, 3, 4, 5, 7 mm. (From Bourne and Moyer, 1968; reprinted from *Food Technol.* **22,** 1016, 1968. Copyright by Institute of Food Technologists.)

green peas in a simple extrusion cell. A layer of food material moves up between the plunger and the wall of the extrusion cell as extrusion is taking place. When the annulus is wide a large number of peas can be accommodated within the extrusion zone at any instant of time and the extrusion force is the mean force required to crush and extrude a large number of peas. This averaging effect results in a uniform force along the plateau even though the plateau itself is decreasing slowly. As the annulus width is narrowed fewer peas can be accommodated within the extrusion zone at any instant, and the averaging effect is reduced. A single hard pea will therefore make the force rise to a high level while it is passing through the extrusion zone, and likewise a single soft pea will make the force fall a considerable amount when a narrow annulus width is used.

An important question concerning the extrusion testing technique arises: What width of annulus gives the greatest resolving power in discriminating between two samples of food that are nearly equal in their textural properties? Figure 16 shows a plot of the maximum extrusion force of green peas with annulus widths ranging from 1 to 10 mm against the alcohol insoluble solids (AIS) of the peas. (Alcohol insoluble solids is a well-established index of maturity of green peas and allows evaluation of the textural properties of the peas independently of the extrusion force.) Figure 16 shows that the slopes of the extrusion force versus AIS curves increase as the annulus width decreases. However, we need to take into account the fact that as the annulus width decreases the test is moving into higher force ranges. This problem is overcome by plotting the logarithm of the

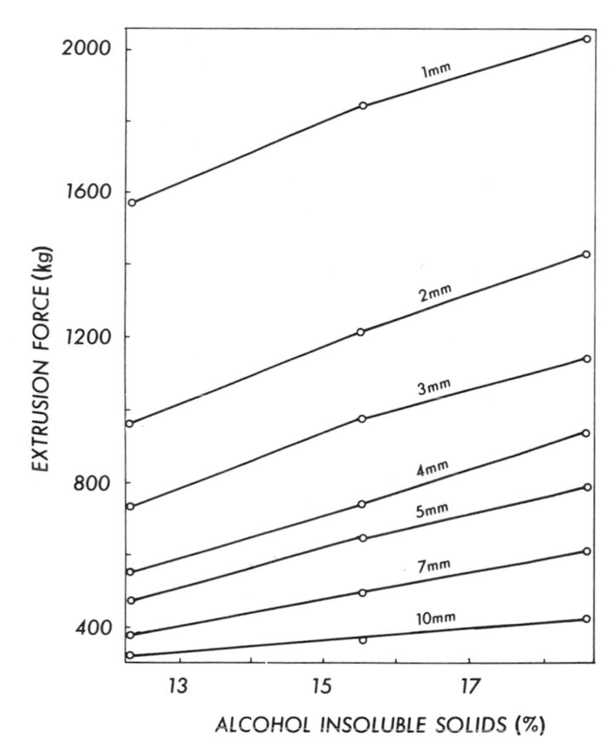

FIG. 16. Extrusion force as a function of alcohol insoluble solids in compression–extrusion test on fresh green peas using annulus widths of 1, 2, 3, 4, 5, 7, 10 mm. (From Bourne and Moyer, 1968; reprinted from *Food Technol.* **22,** 1016, 1968. Copyright by Institute of Food Technologists.)

extrusion force (which normalizes the forces against AIS). In the normalized log-force–AIS relationship (Fig. 17), the slope increases steadily from 1 to 4 mm annulus and then decreases again. There is a fairly broad plateau from about 4 to 7 mm annulus. The greatest resolving power for green peas occurs with a 4-mm-wide annulus. It is interesting to note that the FMC Pea Tenderometer and the Texture Press both use slits ⅛ in. wide (3.2 mm), which comes close to the optimal width for the cylindrical extrusion cell on green peas.

The FMC Pea Tenderometer and the standard multibladed test cell of the Food Technology Texture Test System (Texture Press) (see Fig. 7, pp. 135–141) were originally considered to be based on the principle of shearing under pressure. Bourne and Moyer (1968) pointed out that this class of instrument was basically an extrusion test on materials such as green peas. Szczesniak *et al.* (1970) concluded from extensive study that different foods undergo different types of disintegration in the Texture Press. They made the following postulates:

In compression, force is proportional to (sample weight)^2.
In shear, force is proportional to sample weight.
In extrusion, force is independent of sample weight.

By measuring the forces generated on 24 different foods at various sample test weights and calculating standard errors for the various models they were able to identify the type of disintegration most likely to occur for that food in the Texture Press.

The results of this analysis are summarized in Table 6. More than one model was appropriate for many foods. It is noteworthy that pure compression was not an appropriate model for any of the foods tested while pure shear was an appropriate model for only two foods. Extrusion in combination with compression or shear, or compression plus shear was an appropriate model for 21 foods. Only white bread and peanuts did not fit a model that included extrusion, and these two commodities do not flow under pressure. This study concluded that extrusion is an important component in the testing of most foods in the Texture Press.

Voisey *et al.* (1979) developed an extrusion test accessory for the Ottawa Texture Measuring System that measures the softness of cake frostings (icing) packed in collapsible tubes. Two rollers are pulled over opposite sides of the tube forcing the frosting through the nozzle, which behaves as an extrusion rheometer because the mass-produced tubes are of uniform dimensions and shape. A high

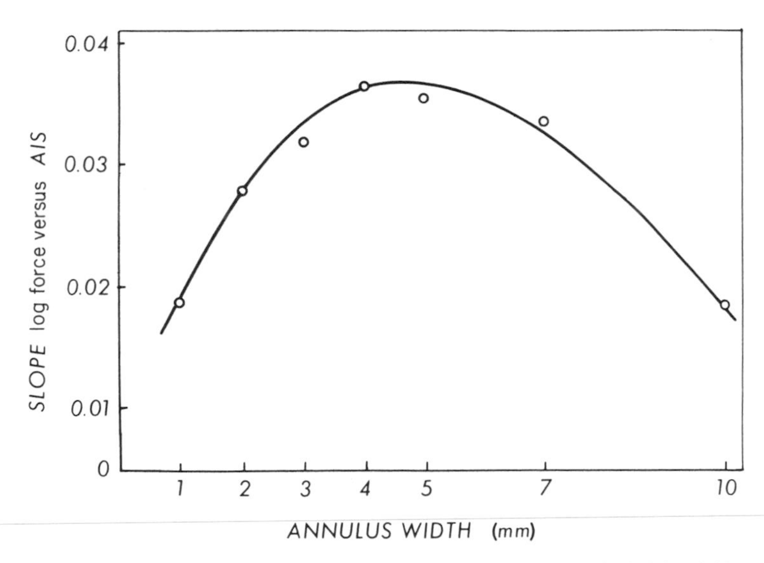

FIG. 17. Effect of annulus width on the slope of the log force versus alcohol insoluble solids in compression–extrusion test on fresh green peas. (Plotted from data of Bourne and Moyer, 1968.)

TABLE 6

POSSIBLE CLASSIFICATION OF PRODUCTS' BEHAVIOR IN THE TEXTURE PRESS[a]

Product	Compression	Shear	Compression and shear	Compression and extrusion	Shear and extrusion	Compression shear and extrusion
Apples, (Rome variety)	[b]	[b]	[c]	[d]	[d]	[c], poor fit
Apples, (Lyons variety)	[b]	[b]	[c]	[d]	[d]	[c]
Bananas, ripe	[b]	[b]	[c]	[d]	[d]	[c]
Bananas, overripe	[b]	[b]	[c]	[d]	[d]	[c]
String beans, raw	[b]	[b]	[c]	[b]	[d]	[c]
String beans, cooked	[b]	[b]	[c]	[b]	[d]	[d]
Common white beans, cooked	[b]	[b]	[c]	[b]	[b]	[d]
Lima beans, frozen, cooked	[b]	[b]	[c]	[b]	[d]	[d]
Beets, canned	[b]	[b]	[c]	[d]	[d]	[c]
Bologna	[b]	[b]	[c]	[b]	[d]	[c]
White bread, sliced	[b]	[b]	[d]	[b]	[c]	[c]
White bread, cubed	[b]	[b]	[d]	[b]	[c]	[c]
Sponge cake	[b]	[d]	[d]	[d]	[c]	[c]
Carrots, raw	[b]	[b]	[c]	[d]	[d]	[c]
Carrots, canned	[b]	[b]	[c]	[d]	[d]	[d]
American cheese	[b]	[b]	[c]	[d]	[b]	[c]
Cucumbers	[b]	[b]	[c]	[d]	[d]	[d]
Meat	[b]	[b]	[c]	[b]	[d]	[c]
Peas, canned	[b]	[b]	[c]	[b]	[b]	[d]
Peas, frozen, cooked	[b]	[b]	[c]	[b]	[d]	[d]
Peanuts	[b]	[b]	[d]	[b]	[c]	[c], poor fit
Raisins	[b]	[b]	[c]	[d]	[b]	[c]
Rice, converted	[b]	[b]	[c]	[b]	[b]	[d]
Rice, precooked	[b]	[d]	[c]	[b]	[d]	[c]

[a]From Szczesniak *et al.* (1970); reprinted from *J. Texture Stud.* with permission of Food and Nutrition Press.

[b]Almost certainly not an appropriate model.

[c]Has one or more negative parameters, certainly not an appropriate model.

[d]Possibly an appropriate model.

correlation ($r = -0.96$) was found between maximum extrusion force and sensory evaluation.

Benbow (1971) found that the extrusion of ceramic pastes through a short circular hole could be related to the die geometry by the expression

$$P = Y \ln(A_0/A),$$

where P is the pressure for flow through a short circular hole; A_0, the area of the barrel; A, the die area; and Y, a material parameter. Benbow (1981) suggested that this equation would probably hold for food pastes such as cake frosting.

Shear Testing

To an engineer "shear" means the sliding of the contiguous parts of a body relative to each other in a direction parallel to the plane of contact under the influence of a force tangential to the section on which it acts. The food technologist sometimes uses shear in this sense but more often uses the word "shear" to describe any cutting action that causes the product to be divided into two pieces. This cutting action is not the same as true shear, but since it is widely described as shear among food technologists, we shall use the word shear to denote this commonly accepted meaning of "cutting across." The difference between true shear and cutting is shown schematically in Fig. 18.

A new term needs to be coined to describe cutting action; this will preserve the

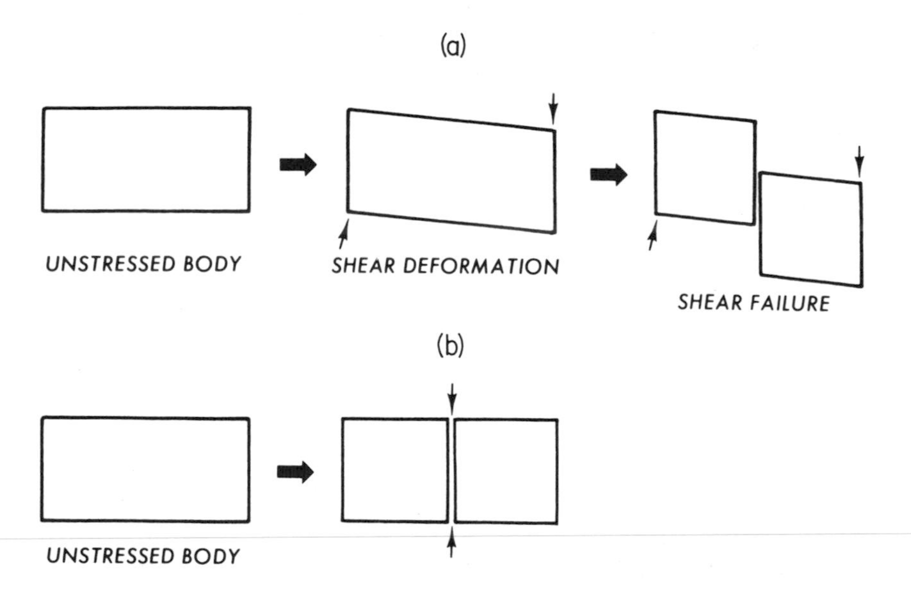

FIG. 18. Comparison of (a) true shear failure with (b) cutting shear failure.

purity of the meaning of the word "shear" and will prevent the confusion that occurs when attempts are made to apply the theory of shear tests to cutting tests.

The best known shearing apparatus is the Warner–Bratzler Shear (Warner, 1928). The working part of this apparatus consists of a stainless steel blade 0.040 in. thick in which a hole, consisting of an equilateral triangle circumscribed around a 1-in.-diam circle is cut and the edges rounded off to a radius of 0.02 in. (In some publications the Warner–Bratzler Shear is misrepresented as having a rectangular-shaped hole in the blade.) A sample of meat, usually a cylinder 0.5 or 1 in. in diam, is placed through the hole and two metal anvils, one on each side of the blade, move down forcing the meat into the V of the triangle until it is cut through (see Fig. 10, p. 145). A force gauge measures the maximum force encountered during this cutting action. This is a cutting action rather than a true shear.

Voisey and Larmond (1974) mounted the Warner–Bratzler shear blade and various adaptations of the blade in the Instron and studied the effects of changing the dimensions of the blade using wieners as the test material. Figure 19 shows how the cylindrical piece of wiener is compressed by the descending anvil and changes cross-sectional shape to conform to the shape of the hole in the blade. Eventually the sample fills all the available area. The type of failure appears to be principally tension as the sample is stretched around the blade. However, a complex stress pattern is established which is a combination of tension, compression, and shear. It seems probable that most of the so-called shear tests used on foods are similar in pattern and result in what is predominately tensile failure of the specimens.

Voisey and Larmond (1974) studied the effect of changing the angle of the cutting edges of the blade. The shearing force increases as the angle of the blade widens from 30 to about 70° after which further widening of the angle causes no increase in force (Fig. 20). These authors also studied the effect of changing the

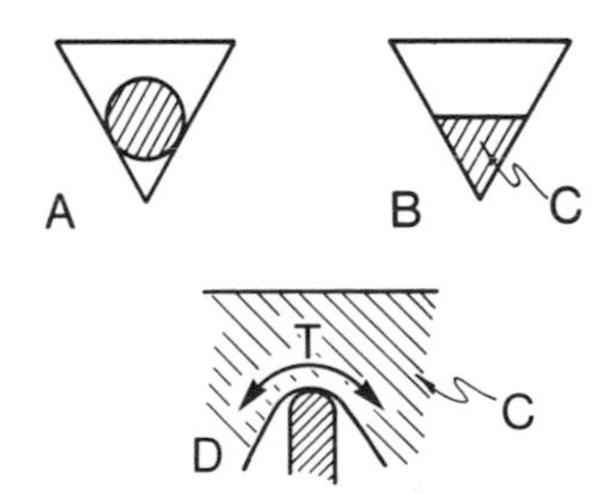

FIG. 19. Schematic of shearing in the triangular blade of the Warner–Bratzler Shear: A, unstressed circular test piece; B, sample deforms under compression to fill available space; C, sample; D, tension failure occurs around the edge of the shear blade; T, tensile stress. (Courtesy of P. W. Voisey; reprinted with permission from Canadian Institute of Food Science and Technology.)

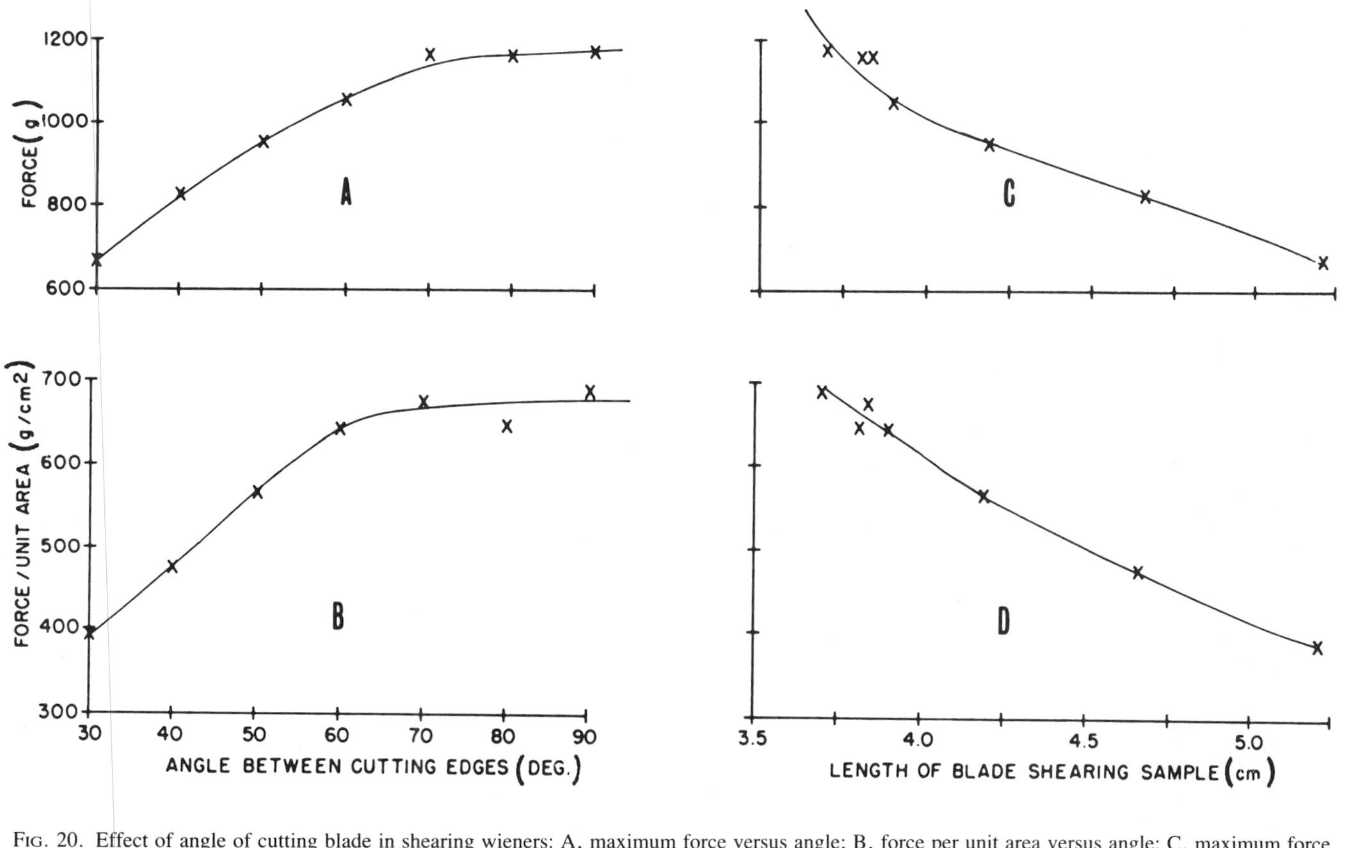

FIG. 20. Effect of angle of cutting blade in shearing wieners: A, maximum force versus angle; B, force per unit area versus angle; C, maximum force versus length of blade cutting sample; D, force per unit area versus length of blade cutting sample. (Courtesy of P. W. Voisey; reprinted with permission from Canadian Institute of Food Science and Technology.)

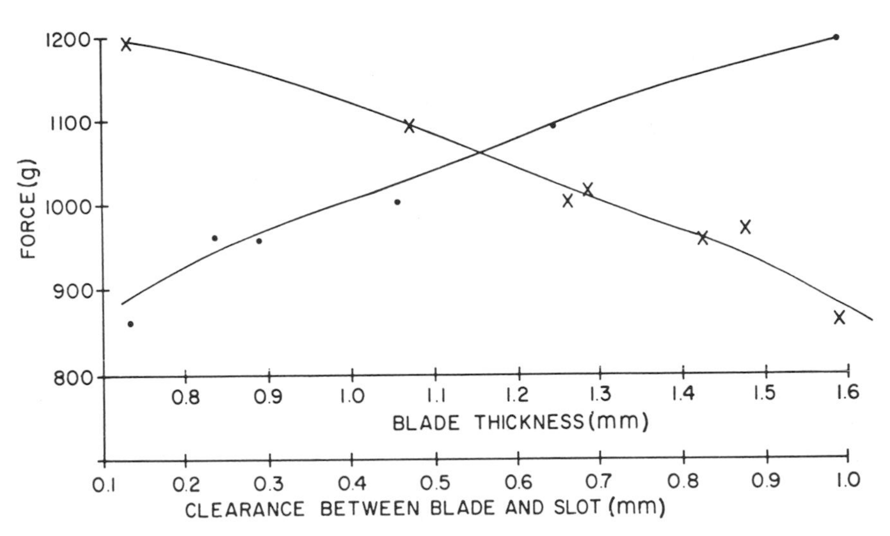

FIG. 21. Effect of blade thickness (•) and clearance (×) between blade and anvil in shearing wieners. (Courtesy of P. W. Voisey; reprinted with permission from Canadian Institute of Food Science and Technology.)

thickness of the blade and the width of the clearance between the blade and the moving anvil. Figure 21 shows that the shear force increases as the thickness of the blade increases and it decreases with increasing clearance. These authors also found that the rate at which the test is performed introduces significant differences in the rupture force and other parameters measured in these tests. In subsequent work Voisey and Larmond (1977) showed that changing the rate of travel of the anvil did not significantly increase the correlation between sensory tenderness rating and the Warner–Bratzler shear rating.

Although no similar work has been reported on muscle meat with its intact fibers, the work of Voisey and Larmond cited above indicates the importance of standardizing test conditions.

The relationship between diameter of the test piece and Warner–Bratzler shear force is not clear at the present time. Kastner and Henrickson (1969) found with cooked pork chops a nonlinear relationship between diameter and Warner–Bratzler shear force. When their data is recalculated as shear force versus $(\text{diam})^2$ the plot appears to be linear; that is, the shear force is directly proportional to the cross-sectional area. Pool and Klose (1969) found with cooked turkey meat that the shear force was proportional to $(\text{diam})^{1.2}$ and they pointed out that the fibers failed in tension. Davey and Gilbert (1969), using a wedge-type shear on beef, found the shear force to be proportional to the square root of the area; that is, proportional to the diameter. Culioli and Sale (1981) sheared spun fababean protein fibers in a rectangular blade double-shear apparatus and found that the maximum force increased linearly with the initial thickness of the sample over the range 2–13 mm. In view of the uneven results between different researchers the only conclusion that can be made at this time is that the diameter of the test piece should be standardized for any one study.

Volodkevich (1938), sometimes spelled Wolodkevich, described a shear test for meat consisting of two wedges, each with a 2.5-mm radius, that compress and shear the meat. MacFarlane and Marer (1966) developed a similar double-wedge apparatus for measuring the tenderness of lamb, calling their apparatus the MIRINZ Tenderometer. Smith and Carpenter (1973) developed a similar type of apparatus that was hand operated, calling it the NIP Tenderometer.

Rhodes *et al.* (1972) mounted the Volodkevich wedge in the Instron and studied 10 parameters that were extracted from the force–distance curves using computer analysis. They concluded that

> none of the correlations between sensory data and the single instrumental measurement derived from force–deformation curves has improved significantly on those already reported in the literature; nor has the use of multivariate statistical techniques . . . produced any more than marginal advantages.

A number of single-blade shear apparatus are described in the literature; for example, Wiley *et al.* (1956), whose single-blade shear is now an optional accessory to the Food Technology Texture Test System. There are several reports in the literature of shear testers that are based on the principal of a wire cutting the product; for example, Wilder (1947), Gould (1949), and Vanderheiden (1970).

The standard cell of the Food Technology Texture Test System (frequently called Kramer Shear Press) has some elements of shear in the test. This feature of this test cell is discussed more fully on p. 68, 69.

Compression and Crushing

There are two main types of compression tests:

1. Uniaxial compression. The sample is compressed in one direction and is unrestrained in the other two dimensions.

2. Bulk compression. The sample is compressed in three dimensions, usually by means of hydraulic pressure. Figure 22 illustrates the principle of uniaxial and bulk compression.

Uniaxial compression applied to a Hookean solid of uniform cross-sectional areas as a small strain before rupture gives rise to the property known as Young's modulus of elasticity.

$$\text{Young's modulus of elasticity } E = \frac{\text{compressive stress}}{\text{compressive strain}} = \frac{F/A}{\Delta L/L},$$

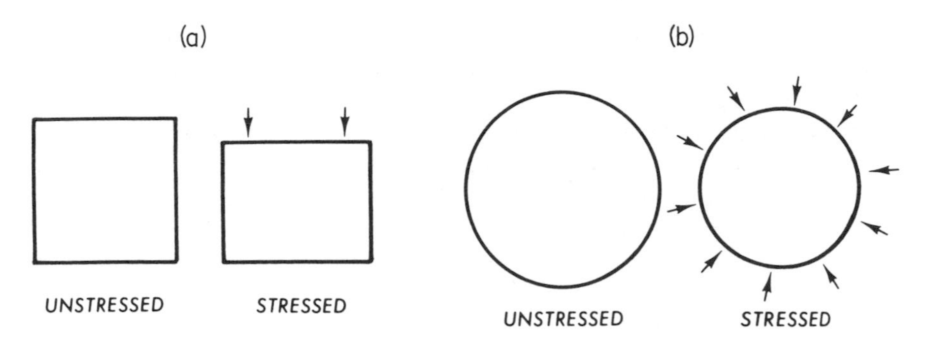

FIG. 22. Uniaxial compression (a) and bulk compression (b) compared.

where F is the applied force; A, the cross-sectional area; L, the unstressed height; and ΔL, the change in height due to F. Young's modulus of elasticity, which is the slope of the stress–strain curve, is a measure of stiffness and is a property that is widely used in engineering. But since most foods are viscoelastic rather than elastic and are usually subjected to large compressions in testing, the strict definition of Young's modulus seldom applies to food materials. However the concept of Young's modulus of elasticity is frequently used to express the stress–strain ratio of the food, at least under moderately light compressions and in the area of the force–compression curve that is reasonably linear. Mohsenin and Mittal (1977) proposed the term "modulus of deformability" for foods in place of modulus of elasticity in order to preserve the purity of the meaning of the latter term. This is an excellent suggestion that should be put into practice.

A high degree of uniaxial compression causes the product to rupture, spread, fracture, or break into pieces and even the concept of modulus of deformability no longer applies. Texture profile analysis, which is discussed on p. 114, utilizes the principle of uniaxial compression to crush foods.

When a cylinder of material is compressed uniaxially, its diameter usually increases. Poisson's ratio is defined as the ratio of the fractional increase in diameter (transverse strain) and the fractional decrease in height (axial strain):

$$\text{Poisson's ratio } \mu = \frac{\text{change in width per unit width}}{\text{change in length per unit length}} = \frac{\Delta D/D}{\Delta L/L},$$

where D is the diameter of unstressed material; ΔD, the change in diameter caused by stress; L, the height of unstressed material; and ΔL, the change in height caused by stress.

Poisson's ratio is 0.5 for materials in which no volume change occurs under compression. The potato has a Poisson's ratio of 0.45 to 0.49, indicating there is a small change in volume as it changes shape. Apple flesh has a Poisson's ratio of 0.21 to 0.34, indicating that its volume decreases substantially during uniaxial compression. Fresh bread probably has a Poisson's ratio close to zero because of its high compressibility.

The calculation of Poisson's ratio assumes uniform distribution of stress, and stresses that are below the proportional limit of the material. These conditions usually do not prevail during the testing of foods. Several researchers have measured Poisson's ratio of various foods (Hammerle and McClure, 1971; Segars *et al.*, 1977), but thus far it has not been seriously considered as an index of quality by the food industry.

The definition of the bulk modulus is

$$\text{Bulk modulus } K = \frac{\text{hydrostatic pressure}}{\text{volume strain}} = \frac{P}{\Delta V/V},$$

where P is the pressure; V, the unstressed volume; and ΔV, the change in V caused by P. This equation applies for Hookean solids but has also been applied to foods. Bulk compression causes a change in volume without a change in shape. Bulk compression is seldom used in testing foods, probably because of the slowness and difficulty of performing a test under conditions where the force is applied by means of hydraulic pressure. Many foods contain variable amounts of entrapped air or other gas within their structure which can profoundly affect the bulk compressibility of the food. White and Mohsenin (1967) describe a low-pressure bulk compression apparatus and Sharma and Mohsenin (1970) give results from bulk compression of apples. Finney and Hall (1967) performed bulk compression tests on potato tubers.

Tensile Tests

Tensile tests are not widely used with foods, which is understandable because the process of mastication involves compression, not tension, of the food between the molars. It has already been pointed out that food fails in tension in many cutting-shear tests. Nevertheless, a few tensile tests are performed. One of earliest tensile tests was that of Howe and Bull (1927), who endeavored to measure the tensile strength of meat but encountered difficulty in making clamps that would hold the meat so that the meat would not tear at the clamps. Platt and Kratz (1933) cut pieces of bread and cake to a standard shape, held them between large spring paper clips, and ran water into a small bucket attached to the lower clip until the piece of bread broke and then measured the volume of water. Personius and Sharp (1938) used a similar simple apparatus for measuring the tensile strength of potato. Halton and Scott Blair (1937) experienced difficulty in performing tensile tests on bread dough because of the sagging of the dough and overcame the problem by supporting the dough on a bath of mercury, performing tensile tests in a horizontal plane instead of the customary vertical plane.

Tschoegl *et al.* (1970) overcame the problem of dough sag during tensile testing by suspending doughnut-shaped pieces of dough in a fluid of equal density. This technique was used by Rasper *et al.* (1974) and Rasper (1975).

Nowadays, instruments such as the Instron, Ottawa Texture Measuring System, or the Food Technology Texture Test System are generally used to perform tensile tests.

A conventional tensile test assumes that the sample fractures almost instantaneously in a plane that is approximately perpendicular to the plane of the applied tension. The maximum force is the tensile strength of the material. Many foods subjected to tension do not fail suddenly; fracture begins with a small crack that slowly spreads across the sample over a comparatively long period of time and the crack may or may not be perpendicular to the plane of the applied tension. Several cracks may appear and spread simultaneously. This type of break makes it difficult to obtain a meaningful interpretation of the tensile force measurement.

Another problem with many foods is that of holding the sample so that the break occurs within the sample and not at the jaws that hold the sample. This problem is often solved by cutting out dumbbell-shaped test pieces and holding the sample at the wide ends. The sample is then more likely to break in the narrow center portion of the test piece. Pool (1967) devised a unique solution to the problem of holding the sample in tensile tests by using a fast-acting strong adhesive (Eastman 910, methyl 2-cyanoacrylate) to cement the ends of cylinders of chicken meat to metal plates. The cement forms a bond stronger than the tensile strength of the chicken in a minute or two. Pool then mounted the two metal plates containing the piece of chicken in the Instron to perform tensile tests.

Gillett *et al.* (1978) developed an attachment that allows the tensile strength of sliced processed meats to be measured in a horizontal plane. Two horizontal plates, each with four rows of protruding vertical metal spikes, are used to hold and extend the sample. A cable runs around a pulley to the load cell of a recording texturometer and pulls the plates apart when the instrument is operating. This attachment is available as an accessory for the Food Technology Texture Test System.

Tensile tests are used to measure the adhesion of a food to a surface. In this type of test the sample of food has a disk pressed onto it after which the force required to pull it off is measured. Jansen (1961) and Claassens (1958, 1959a,b) used this technique to measure the stickiness or hesion of butter. The Texture Profile Analysis parameter of adhesiveness measures the maximum force required to pull the compression surface from the test piece after the first compression and therefore contains one element of tensile testing (Friedman *et al.*, 1963). Henry and Katz (1969; Henry *et al.*, 1971) used this technique in an Instron to measure the adhesiveness of puddings and toppings and developed and identified several tensile parameters from the force–distance curve that was so obtained.

Torsion

In a torsion test a force is applied that tends to rotate or twist one part of the object around an axis with respect to the other parts. The tendency of a force to produce rotation about an axis is called the torque T with dimensions mass $\times$ length^2 $\times$ time $^{-2}$. If a force of F newton is applied to a body at R meter from the axis of rotation (Fig. 23),

$$\text{Torque } T = FR \text{ newton meter.}$$

Torque is often expressed in non-SI units. These can be converted to the SI unit of newton-meter (N m) by using the following multiples:

1 N m torque equals 1.000×10^{-7} dyn-cm,
980,600 g force-cm,
9.806 kg force-m,
0.00706 oz force-in.,
0.1129 lb force-in.,
1.355 lb force-ft.

The major application of this test principle is in the rotary viscometers that are widely used to measure the viscous properties of foods. This application will be discussed in Chapter 5, which deals with viscosity and consistency measurements.

The Farinograph and the Mixograph are torque measuring instruments (see pp. 147–148).

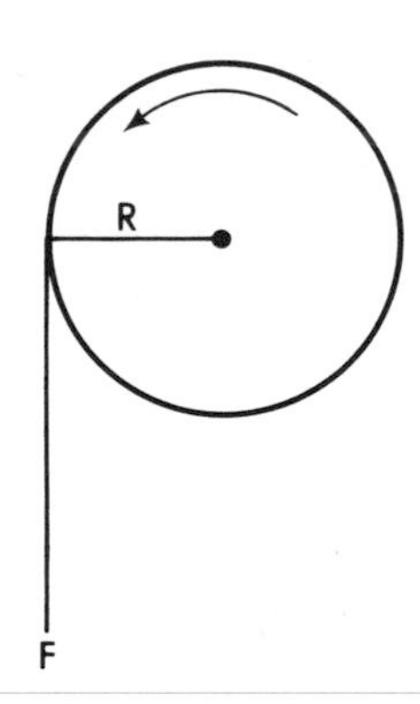

FIG. 23. The torque principle: force F applied at radius R from center of body causes a tendency for rotation.

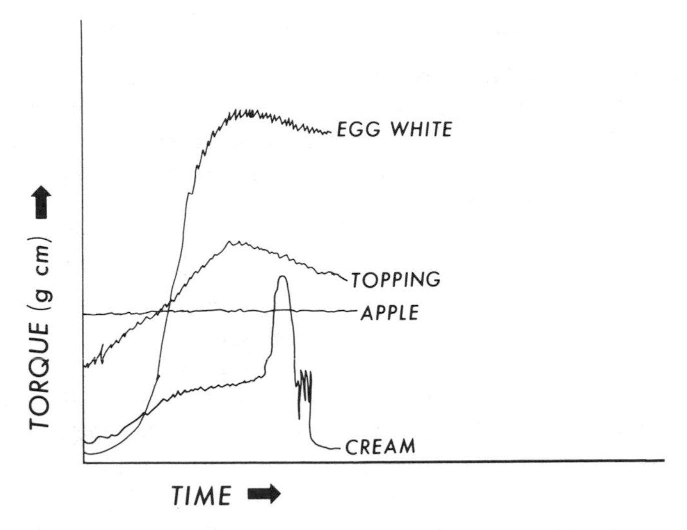

FIG. 24. Changes in torque as a function of whipping time for egg white, dessert topping, applesauce, and cream. (Drawn from data of Voisey and deMan, 1970.)

Nemitz *et al.* (1960) and Nemitz (1963) developed a laboratory apparatus that twisted a whole fish and used it to measure the progress of rigor mortis in fish. Karacsonyi and Borsos (1961) used a torsion device to measure the strength of dry spaghetti and found it useful for the detection of hidden failures. Scott Blair and Burnett (1963) developed a laboratory torsiometer for measuring the coagulation of renneted milk in a cheese vat called a "cheese curd torsiometer." Hashimoto *et al.* (1959) used the principle of torsonial strain to measure the firmness of sausage.

Templeton and Sommer (1933), Mueller (1935), and Voisey and deMan (1970) used the torque principle to measure the change in consistency of products that are beaten. Figure 24 shows how this apparatus can be used to study the behavior of products that increase in viscosity as they are whipped. Applesauce maintains a constant torque throughout the whipping time, while cream reaches a peak viscosity and then lowers in viscosity as it breaks down to form butter. The figure also shows the development of viscosity in egg whites and a topping mix as they are beaten.

Diehl *et al.* (1979) designed a torsion test attachment for use in the Instron and used it to measure structural failure in apple, potato, and honeydew melon. They found tension failure in the torsion tests but shear failure in simple compression tests and concluded that torsion was sometimes preferable to uniaxial compression in measuring breakdown of these foods.

Bending and Snapping

Bending and snapping tests are usually applied to food that is in the shape of bar or sheet. The two most common types of apparatus are shown schematically in Figure 25. On the left side is the triple beam apparatus in which the piece of food rests on two supports and a third compressing bar moves down between the two supports bending the food until it snaps.

The Bailey Shortometer is used in the baking industry to measure the shortness and snapping properties of crackers and cookies (Bailey, 1934). The Struc-O-Graph manufactured by the Brabender Company uses the triple-beam principle for measuring snapping and bending in food (see p. 149).

The right-hand side of Fig. 25 shows a cantilever beam where the product is held at one end and is allowed to bend freely throughout its length. The amount of bending is measured either by the distance that the unsupported end of the food moves or by the angle it subtends to the horizontal plane. Sterling and Simone (1954) used the cantilever beam principle to measure bending and crispness in almonds. Several laboratory-made instruments that use this principle have been described in the literature; for example, the R.P.C. Droopmeter that measures the bending of french fries under gravity (Anonymous, 1966).

Snappy foods are typified by a rigid unbending texture that breaks suddenly once the fracture force has been reached. Bruns and Bourne (1975) studied snapping in foods and found that the force required to snap a test specimen of uniform cross section complies with mathematical models derived from engineering theory. For uniform bars with a rectangular cross section the snapping equation is as follows:

$$F = \tfrac{2}{3}\sigma_c bh^2/L, \qquad (9)$$

where F is the snapping force; σ_c, the failure stress; b, the width of beam; h, the height of beam; and L, the length of beam between supports. For uniform bars with a cylindrical cross section the snapping equation is

$$F = \sigma_c \pi R^3/L, \qquad (10)$$

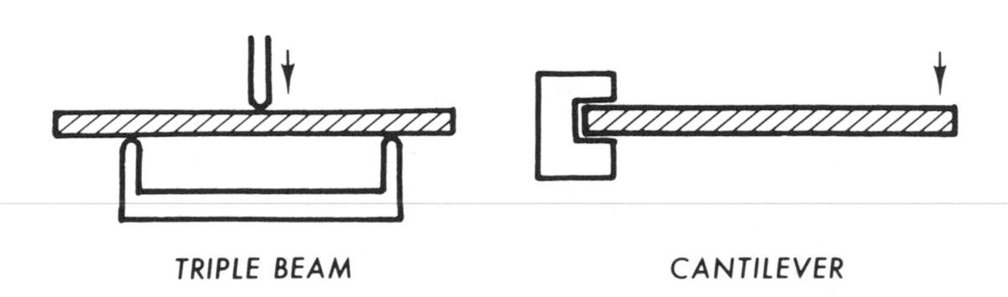

FIG. 25. Two ways to perform bending and snapping tests.

where F is the snapping force; σ_c, the failure stress; R, radius of beam; and L, the length of beam between supports.

The above equations establish the importance of using samples of uniform size and shape in this type of test. Particular attention needs to be paid to the thickness of the sample because the snapping force is proportional to the square of the thickness or cube of the diameter. If it is impossible to obtain test pieces of uniform size and shape, it is advisable to correct the force data according to the equations above.

Leighton *et al.* (1934) used the sagging of ice cream bars resting on two supports as a means of measuring the apparent viscosity. They used the equation

$$\eta = 5gmL^3/1152RI, \qquad (11)$$

where η is the viscosity; g, the gravity constant; m, the mass of sample; L, the length between supports; R, the rate of sag in centimeters per second; and I, the moment of inertia of a cross section ($0.049D^4$ for a beam of circular cross section with diameter D).

These workers found that the apparent viscosity of ice cream was in the range of 2×10^8 P. This seems to be a useful method for measuring extremely high viscosities. Coulter and Combs (1936) used Leighton's method to determine the sagging properties of butter.

Distance Measuring Instruments

Distance measurements may be divided into three classes: (1) linear measurement with dimension of length, (2) area measurement with dimension lenth^2, and (3) volume measurement with dimension length^3.

Linear Measuring Instruments

A number of simple testing apparatus that are based upon distance measurements are known, including the Bostwick Consistometer, which measures the distance catsup and fruit puree flow along a horizontal trough; the Hilker–Guthrie Plumit, which measures the depth a cylindrical metal rod falls into sour cream and yogurt; the Ridgelimeter, which measures the sag of fruit jellies as a means of obtaining the grade of pectin; and the Haugh egg quality meter, which measures the height the white of an egg stands up after breaking out from the shell.

The principle of measurement of these instruments is so simple that it needs no analysis. The instruments themselves will be described in the following chapter.

Robertson and Emani (1974) injected a stream of liquid dye at high velocity (10^3–10^4 cm/sec) through a small orifice (0.254 mm diam) for a short time (0.1

sec) into bread. The energy of the stream is absorbed by the bread; hence, the distance the stream of dye penetrated is an index of bread texture. Good correlations are reported between sensory tests of bread aging and the liquid jet penetrometry data.

The Penetrometer consists of a cone and vertical shaft assembly that is allowed to sink into a solid fat under the force of gravity for a standard time after which the depth of penetration is measured. The left-hand side of Fig. 26 demonstrates the principle. According to Haighton (1959) the following formula applies to cone penetration on solid fats of a wide range of hardness:

$$C = KW/p^{1.6}, \tag{12}$$

where C is the yield value of the product; K, a constant whose value depends on the cone angle; W, the weight of cone assembly in grams; and P, the penetration depth after 5 sec.

Mottram (1961) derived a similar formula from studies of penetrometer tests on printing inks:

$$S_0 = KMg/h^n, \tag{13}$$

where S_0 is the yield value; M, the mass of cone assembly in grams; g, a gravity constant; h, the depth of cone penetration; n, $\simeq 2$ but varies according to the type of material; and K, a constant depending on cone angle.

Although Mottram's terminology is not the same as Haighton's, Eqs. (12) and (13) are similar in their essential features.

Mottram published an equation for the constant K as follows:

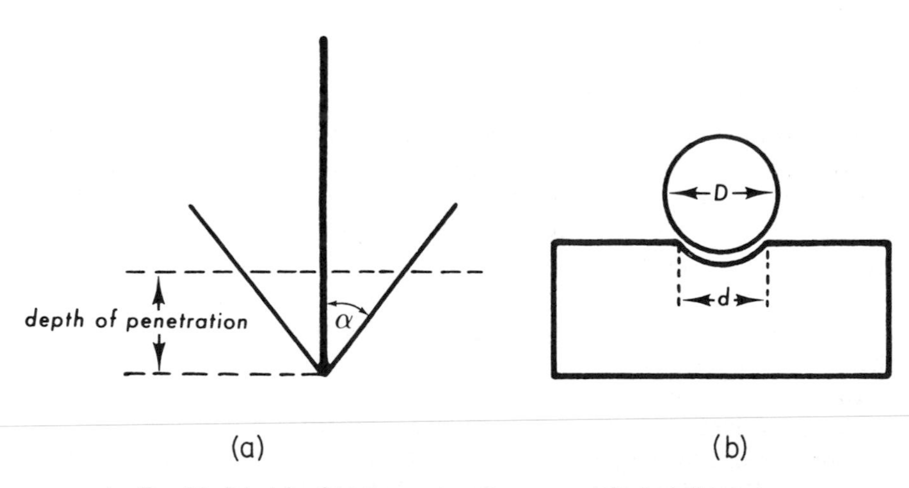

FIG. 26. Principle of (a) the penetrometer cone and (b) the ball indenter.

$$K = (\cos^2 \alpha \cot \alpha)/\pi, \tag{14}$$

where α = ½ cone angle (see Fig. 26).

Dixon and Parekh (1979) used a cone penetrometer to measure the firmness of butter and developed the equation

$$\text{butter firmness} = \frac{\text{cone mass} \times (\text{cone angle})^{-1.65}}{(\text{penetration depth})^2}.$$

These researchers found that this equation accounted for 96.9% of the variation in observations and 91% of the variation in sensorially perceived spreadability of butter.

Another type of penetration test that is used on fats and similar products is the ball indenter that is the principle of the SURDD tester (see Fig. 26b). A steel ball is pressed onto the material with a constant force for a given time, then it is removed and the diameter of the impression measured (Feuge and Guice, 1959). This test is similar in principle to the Brinell and Vickers hardness testers that are used to measure hardness of metals. The following equation applies to this test:

$$H = 100P/(\pi D/2)[D(D^2 - d^2)^{1/2}], \tag{15}$$

where H is the hardness index in kilograms per square centimeter; P, the kilogram force on ball; D, the diameter of ball in millimeters (usually in the range 3–12 mm for fats); and d, the diameter of impression in material.

The same principle is used in the Smetar hardness tester, which measures the hardness of wheat grains by the size of the impression made by a diamond indenter (Smeets and Cleve, 1956). It is also used in the adaptation of the Barcol Impressor to measure the hardness of wheat grains (Katz *et al.*, 1959, 1961).

Rebound Distance

When dry, mature peas that have been soaked and cooked are bounced off an inclined surface, the horizontal distance they rebound is a function of their elasticity coefficient and is related to textural quality (Crean and Haisman, 1965). Firm peas bounce farther than soft peas. Asymmetry of rheologically identical peas causes some scatter, but the method can fractionate peas into several fractions on the basis of their firmness. The principle of the test is shown in Fig. 27. The relationship between rebound range and firmness as measured by the Maturometer is shown in Fig. 28. These authors developed the following equation for perfect spheres having known coefficients of elasticity:

$$\text{rebound range} = tv \cos(\phi + \alpha - 90), \tag{16}$$

where $t = [-v \sin(\phi + \alpha - 90) \pm (v^2 \sin^2(\phi + \alpha - 90) + 2gh)^{1/2}]g^{-1}$; $v = u(\sin^2 \alpha + e^2 \cos^2\alpha)^{1/2}$; ϕ is the angle at which pea leaves the plate and equals $\cot^{-1}(e \cot \alpha)$; α, the angle of the plate to the horizontal; u, the velocity of pea

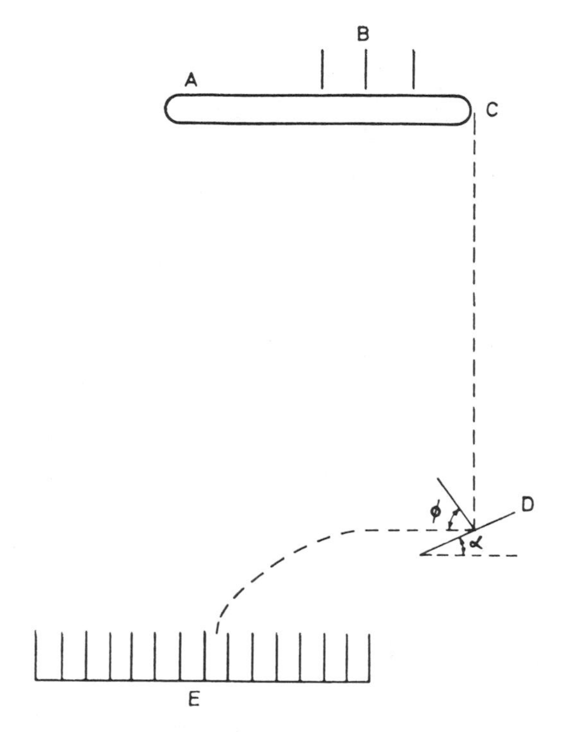

FIG. 27. Principle of the rebound test: individual peas sorted into single file by gates B fall off conveyor belt AC onto inclined surface D and bounce into compartments at E. (Reprinted with permission from Blackwell Scientific Publications, Ltd.)

just before hitting the plate; e, the coefficient of elasticity of the pea; h, the height of the point of impact above the collecting box; and g, a gravitational constant. The principle of rebound distance has been used for more than a century by the cranberry industry of northeastern United States to separate soft and substandard cranberries from sound fruit. Bryan *et al.* (1978) used the principle of rebound distance to separate cull oranges that have a low coefficient of restitution from sound fruit that have a higher coefficient of restitution.

Hamann and Carroll (1971) used a vibratory technique that depends upon elasticity to separate muscadine grapes into maturity grades based on firmness and to separate bruised blueberries from sound blueberries (Hamann *et al.*, 1973).

Deformation

Deformation is the change in height or diameter of a food under the application of a force. Implicit in this definition is the assumption that this is a nondestruc-

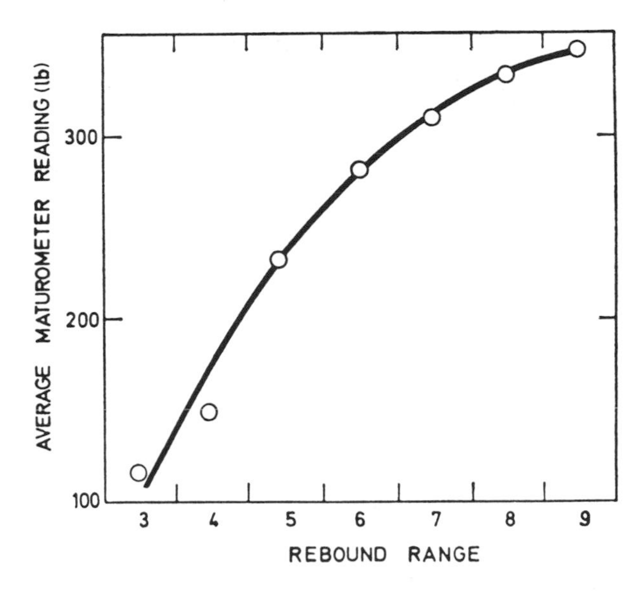

FIG. 28. Maturity of peas versus rebound range in the bounce test (see Fig. 27). (Reprinted with permission from Blackwell Scientific Publications, Ltd.)

tive test; that is, the amount of force applied is less than that required to break or rupture the commodity. This physical property is usually measured sensorially by squeezing the food in the hand. Deformation is generally considered to be one method of measuring the "firmness" of a commodity. In fact it is preferable to consider it as a measurement of "softness" because the firm product gives a lower reading than the soft product.

Figure 29 shows schematically a straight-line force–compression relationship for a firm, medium, and soft commodity. The deformation test usually measures the deformation under a standard force. The application of force F to three ideal commodities gives deformations f, m, and s, respectively, for the firm, medium, and soft product (Fig. 29a). Occasionally the force required to achieve a standard deformation is used (Fig. 29b) where standard deformation D is achieved by forces s, m, and f. In this case the firm product will give a higher numerical value than the soft product, but there is a risk that the firm product will break under the high force required to achieve the standard deformation.

Although deformation is widely used sensorially for measuring firmness of foods, there is little in the way of instrumentation available for this test. The Baker Compressimeter is a commercially available instrument that is used for measuring the firmness of bread crumb, the Ridgelimeter is used to measure the quality of pectin in fruit jellies (Cox and Higby, 1944), and the Marius egg deformation tester is available for measuring the deformability of whole eggs.

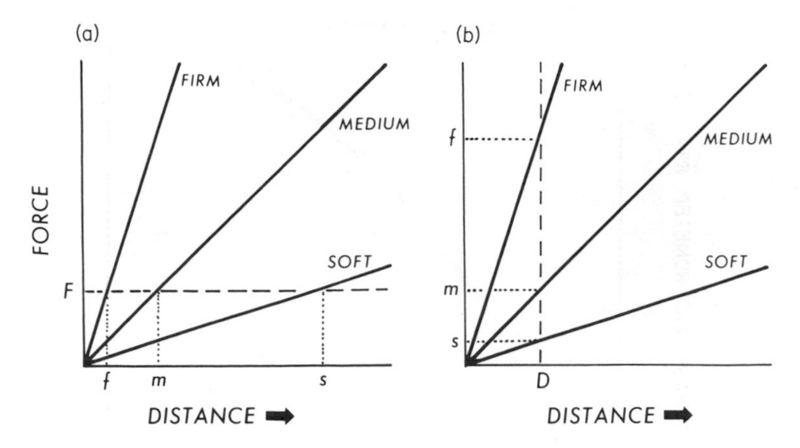

FIG. 29. Deformation of ideal firm, medium, and soft solids: (a) compression distance is measured at constant force F; (b) force required to achieve a standard compression is measured.

The Instron has been adapted for performing deformation testing (Bourne, 1967a), and equipment similar to the Instron could also be adapted in a like manner. Voisey *et al.* (1974) and Voisey and Buckley (1974) developed accessory equipment that can be used in an Ottawa Texture Measuring System (OTMS) or Instron for high-speed deformation testing coupled with digital readout of the data. The Penetrometer can also be adapted for deformation testing of foods that are not too rigid or too small (Bourne, 1973).

Figure 30 shows the most common types of force–deformation curves that are found on foods. The majority of products give an A type curve that is concave downward; this shape is typical of products such as marshmallows and the softer fruits and vegetables. The linear B type curve is found with rigid products such as firm green fruits and vegetables, hard candy, and eggs. This type obeys Hooke's law, which states that the deformation of a body is directly proportional to the force applied to it. Hooke (in 1678) enunciated this law on the basis of tension experiments with metal springs, but any body that obeys this law in tension or compression is called a *Hookean solid*. The C type curve, which is S shaped, is found with many breads and some cheeses.

It is obvious that the higher the applied force the greater will be the deformation, even when the relationship is nonlinear. From the geometry of Fig. 30 it can be seen that the deformation of a B type product is directly proportional to the applied force; hence, the deformation at two forces F_1 and F_2 are directly proportional to those forces; that is,

$$b_1/b_2 = F_1/F_2.$$

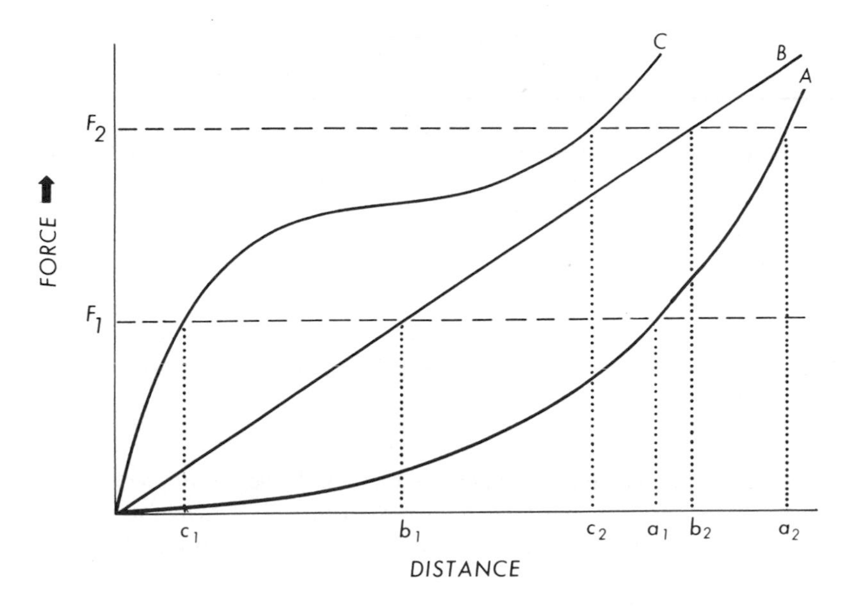

FIG. 30. Three characteristic types of force–deformation behavior.

Type A products are characterized by a rapid rate of increase in deformation at low force with the rate decreasing as the force increases; that is,

$$a_1/a_2 > F_1/F_2.$$

For C type products the deformation increases slowly at first and then more rapidly as the force is increased; hence,

$$c_1/c_2 < F_1/F_2.$$

Since the relationship between force and distance is linear in B type products the degree of force that is applied does not change the deformation ratio when comparing two samples. For A type products greater resolution between samples that are of similar quality is obtained with a small deforming force. This is demonstrated in Table 7, which shows the deformation of a fresh, soft marshmallow and a stale and firmer marshmallow. The amount of deformation increases as the deforming force increases for both marshmallows, but the ratio of the deformations declines with increasing force. Table 8 shows that a similar relationship exists for apples. Under 2-kg force the ratio of the deformations of a flaccid and firm apple was 3.2 while above 14 kg the ratios of the deformations became equal.

For foods that exhibit A type characteristics it is desirable to use as small a deforming force as practicable in order to achieve maximum resolution between

TABLE 7
DEFORMATION OF MARSHMALLOWS[a]

Deforming force (g)	Deformation (mm)		Deformation ratio (A/B)
	A. Soft	B. Firm	
20	0.80	0.022	3.64
100	2.32	0.92	2.52
1000	10.7	6.9	1.55
5000	18.7	14.9	1.25

[a]Data from Bourne (1967a); reprinted from *J. Food Sci.* **32,** 605, 1967. Copyright by Institute of Food Technologists.

samples. In order to maintain equivalent resolution, the degree of precision in the measurement of distance must be higher for firm foods than for soft foods because the change in the height of the firm food is less than for soft food.

Geometry of the Test Piece

Small irregularities in the surface of the test product have the potential to give large errors in deformation. This is demonstrated in Fig. 31, which shows the deformation curves on vertical cylinders cut from a single frankfurter and tested in the Instron. The first curve was obtained on a piece with plane, parallel ends;

TABLE 8
DEFORMATION OF ROME APPLES WHEN SQUEEZED[a]

Force range (kg)	Deformation of flaccid apple (mm)	Deformation of firm apple (mm)	Ratio $\frac{\text{Flaccid apple deformation}}{\text{Firm apple deformation}}$
0–2	2.7	0.85	3.2
2–4	1.0	0.4	2.5
4–6	0.7	0.3	2.3
6–8	0.55	0.3	1.8
8–10	0.45	0.3	1.5
10–12	0.4	0.3	1.3
12–14	0.4	0.3	1.3
14–16	0.3	0.3	1.0
16–18	0.3	0.3	1.0
18–20	0.3	0.3	1.0

[a]Data from Bourne (1967b); reprinted with permission of New York State Agricultural Experiment Station.

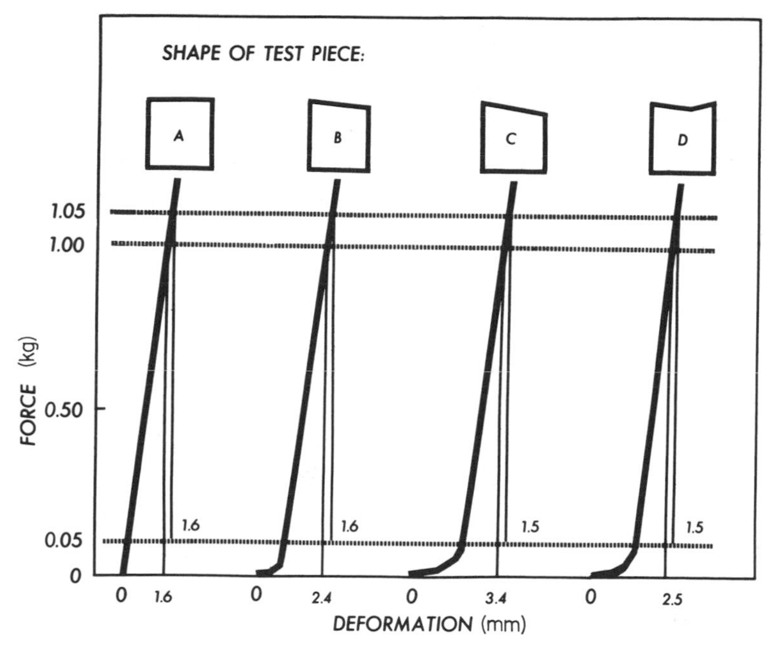

FIG. 31. Effect of irregularities in surface of test piece on deformation test. Note how the errors are overcome by starting the deformation a little above zero force. (From Bourne, 1967a; reprinted from *J. Food Sci.* **32,** 603, 1967. Copyright by Institute of Food Technologists.)

the next two curves were obtained on test pieces with plane ends that were not parallel; and the last curve was obtained on a test piece with one end curved. The effect of these irregularities in shape is to produce a "tail" at the low force end of the curve. These tails represent that portion of the deformation when less than the complete cross-sectional area of the test piece is being compressed. They can introduce large errors into the measurement. The test pieces A–D, all cut from the same frankfurter, show deformations of 1.6, 2.4, 3.4, and 2.5 mm under a 1-kg deforming force. This source of error can be overcome by measuring the deformation from some reference force level that is sufficiently above zero to eliminate the effect of the tails. In this case, the deformations of these four test pieces measured between 0.05 and 1.05 kg are 1.6, 1.6, 1.5, 1.5 mm, which is as uniform as one can expect. While every care should be taken to minimize the small irregularities at the contact surfaces, corrections for these irregularities can be made by beginning the deformation measurements slightly above zero force (Bourne, 1967a).

For those commodities that give a linear or near-linear force–deformation relationship (i.e., they obey Hooke's law), and also for other commodities that are stressed lightly, it has been shown that the deformation is directly proportional to the height of the samples if the cross-sectional area is uniform (e.g., rectangular or circular in cross section).

The equation for Young's modulus of elasticity (see p. 46) can be rearranged as follows:

$$\Delta L = FL/EA \tag{17}$$

Young's modulus of elasticity E is constant for a given sample; hence, if a constant force F is used, the deformation (ΔL) is directly proportional to the unstressed height L and inversely proportional to the area A. This equation has been confirmed on standard shapes cut from agar gels (Brinton and Bourne, 1972).

The deformation of horizontal cylinders of a commodity is inversely and linearly proportional to the length of the cylinder.

The effect of the changing diameter is more complex. According to Roark (1965), the change in diameter of a horizontal cylinder under compression is given by the equation

$$\Delta D = \frac{4}{3}P\frac{(1 - \nu^2)}{\pi E} + 4P\frac{(1 - \nu^2)}{(\pi E)} \log_e \frac{(DE)^{1/2}}{1.075P^{1/2}}, \tag{18}$$

where D, is the diameter; ΔD, the deformation; P, the force per unit length; ν, Poisson's ratio; E, Young's modulus; and b, 2.15 $(PD/E)^{1/2}$.

According to this equation, the deformation of an horizontal cylinder is the sum of the two terms. The first term $\frac{4}{3}P(1 - \nu^2)/\pi E$ is independent of the diameter and second term is a complex function of the diameter. Experimental data on cylinders of agar gels is in general agreement with Eq. (18) (Brinton and Bourne, 1972).

The effect of changing the diameter on the deformability of spheres is complex. Figure 32 shows the deformation of agar gel spheres between 50 and 200 g force. The 3% agar gel is softer than the 5% agar gel and gives a higher deformation. Both gels increase in deformation as the diameter increases and then decrease again, with the 5% gel showing a maximum at 2 cm diam and the 3% agar gel showing a maximum at 4 cm diam. Additional data show that when the deformation is measured between 10 and 50 g the peak deformation on the 3% agar gel occurs at the 2-cm-diam size. The deformation of spheres then is a complex function depending on the stiffness of the product, the diameter of the product, and the deforming force.

In testing approximately spherical commodities some workers have reported that the size of the commodity has little effect on deformation while others have reported that it increases or decreases the deformation. The above figure indicates that all these results are possible, depending on circumstances.

A number of agricultural engineers have applied classical engineering theory

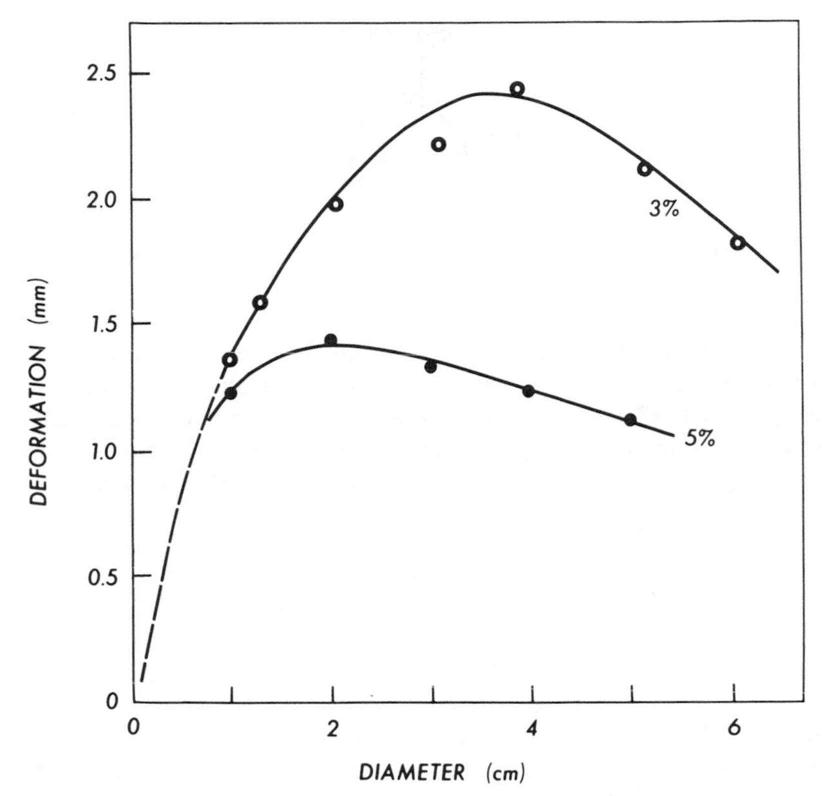

FIG. 32. Effect of diameter on deformation of 5 and 3% agar gel spheres illustrated by ● and ○, respectively. (From Brinton and Bourne, 1972; reprinted from *J. Texture Stud.* with permission from Food and Nutrition Press, Inc.)

to the uniaxial compression of a unit of food that is approximately spherical in shape. Arnold and Mohsenin (1971) give an excellent summary of this work. They state that, in general, the relationship between applied force and deformation of a sphere compressed between two extensive surfaces can be expected to obey Hertz theory:

$$E = \frac{0.338F(1 - \mu^2)}{D^{3/2}} \left[K_1 \left(\frac{1}{R_1} + \frac{1}{R_1'} \right)^{1/3} + K_2 \left(\frac{1}{R_2} + \frac{1}{R_2'} \right)^{1/2} \right]^{3/2}, \qquad (19)$$

where E is the modulus of elasticity; F, the force; D, the deformation; μ, Poisson's ratio; $R_1 R_1'$, R_2, R_2', the radii of curvature at the contact points; and K_1 and K_2 are constants.

Another method of measuring what is, in effect, deformation of a food is an acoustical method. The principle is shown in Fig. 33. A sound speaker which is

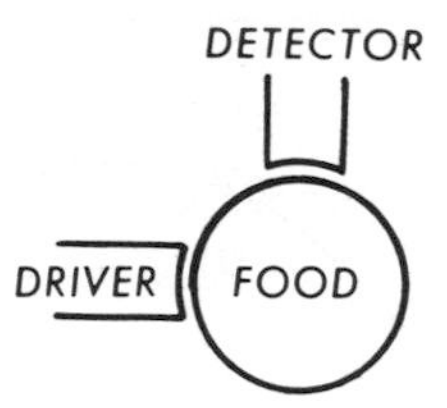

FIG. 33. Schematic of acoustic spectrometer method for measuring deformability.

the driver is placed in contact with a food and caused to produce sound (sonic waves) of constant amplitude which is transmitted into the food. The frequency of the vibration is gradually increased from a low to a high value. A microphone placed in contact with the food at a position 90° to the driver acts as a detector or receiver to pick up the vibrations within the food at each frequency. Electronic equipment plots the amplitude of vibration within the food as a function of the driving frequency. The method has been used for research purposes by Abbott *et al.* (1968a,b), Finney and Norris (1968), Finney *et al.* (1968, 1978), Finney (1970, 1971a,b,c), and Finney *et al.* (1978) and is reviewed by Finney (1972).

A typical curve is shown in Fig. 34. The amplitude–frequency relationship shows a series of peaks that occur at regular intervals. The first amplitude peak is called the "resonance frequency" because the natural period of vibration of food is the same as the driver at this frequency. Additional peaks are found at the first, second, and third harmonics, which are simple multiples of the resonance frequency and occur at exactly two times, three times, etc., the frequency of the resonance frequencies.

If the food is in the shape of a uniform cylinder, Young's modulus of elasticity can be calculated from the resonance frequency (assuming the food is elastic, isotropic, homogeneous, and continuous) by means of the following equation:

$$E = 4\rho f^2 L^2, \qquad (19)$$

where E is Young's modulus of elasticity (dynes per square centimeter); ρ, the density (grams per cubic centimeter); f, the fundamental longitudinal frequency in hertz (cycles per second); and L, the length of the cylindrical specimen in centimeters.

Since it is frequently inconvenient and sometimes impossible to cut a uniform cylinder of tissue, attempts have been made to measure the firmness of intact units that are approximately spherical in shape. For these cases Finney (1971a) defines a "stiffness coefficient" as f^2m, where f is the resonance frequency and m the mass of the article. Cooke (1972) and Cooke and Rand (1973) made a theoretical analysis of the deformation of spheres which indicated that $f^2m^{2/3}$ should be the mass independent indicator of the shear modulus rather than f^2m as used by Finney. Clark and Shackelford (1976) used the stiffness modulus $f^2m^{2/3}$ on peaches with limited success.

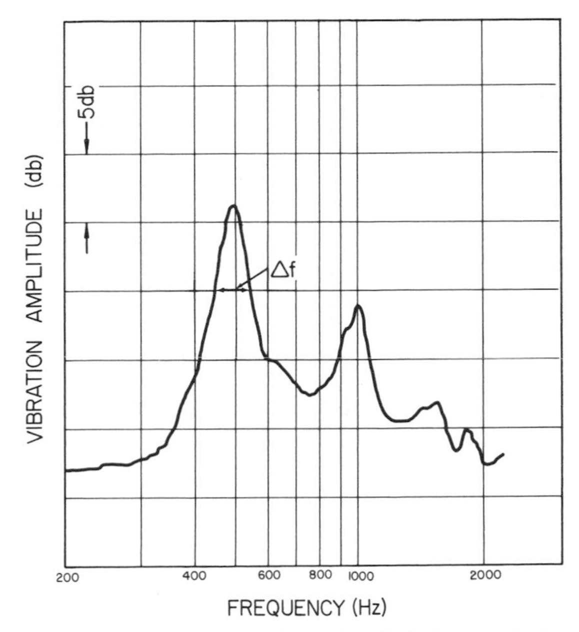

FIG. 34. A recorder curve showing amplitude of vibration of a fruit versus the frequency of the input vibration. (From Finney *et al.*, 1968; reprinted from *J. Food Sci.* **32,** 643, 1968. Copyright by Institute of Food Technologists.)

Time Aspects of Deformation

Suppose an article of food of uniform cross-sectional area is resting on a rigid surface with a weightless rigid plate resting on the upper side (Fig. 35). Suppose now that a weight is placed on the plate and that some mechanism is available to measure the change in height of the food under this constant compressing force.

Figure 36 illustrates what happens when the material is perfectly elastic. When the weight is placed on the food, there is an immediate deformation called "instantaneous elastic deformation" and no further change with time. When the weight is removed, the sample instanteously and completely recovers its original height.

Few foods are perfectly elastic. Most foods possess flow properties in addition to elasticity and are described as "viscoelastic." The behavior of a viscoelastic food under these conditions is demonstrated in Fig. 37. When the weight is placed on the food, there is an immediate compression of the food which is the instantaneous elastic deformation. This is followed by a prolonged, continuous but decelerating rate of deformation called "creep" or "retarded deformation." The deformation continuously increases with time and theoretically never stops; the slope of the line never becomes perfectly horizontal. When the weight is

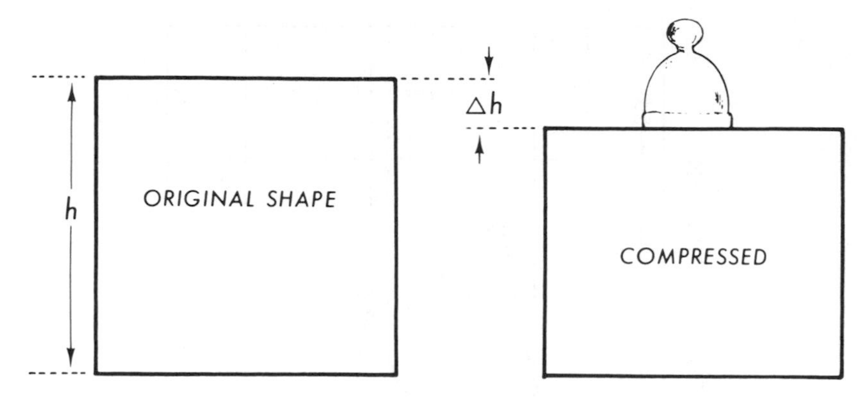

FIG. 35. Schematic of uniaxial deformation of a solid under a constant force.

removed, there is an instantaneous partial elastic recovery followed by further recovery with respect to time called "retarded recovery," or "creep recovery." Again, this line theoretically never becomes horizontal. With these products the commodity does not return to its original height; it is permanently and irreversibly compressed. This is known as irreversible or "permanent deformation," or "set."

The degree of viscoelasticity of food varies widely. A food that is mostly elastic and slightly viscoelastic will give a deformation–time response behavior similar to that shown in Fig. 36, while a highly viscoelastic product will exhibit behavior as shown in Fig. 37.

Creep and recovery are probably a minor part of the deformation that is normally sensed in the hand because of the short time duration of the squeeze.

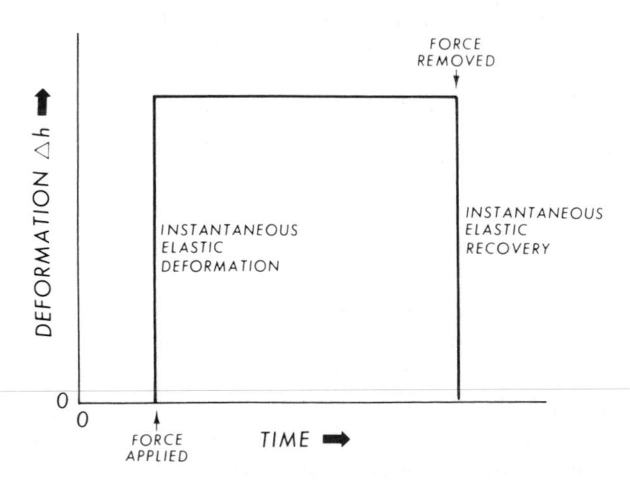

FIG. 36. Deformation–time relationship for an elastic body under constant stress.

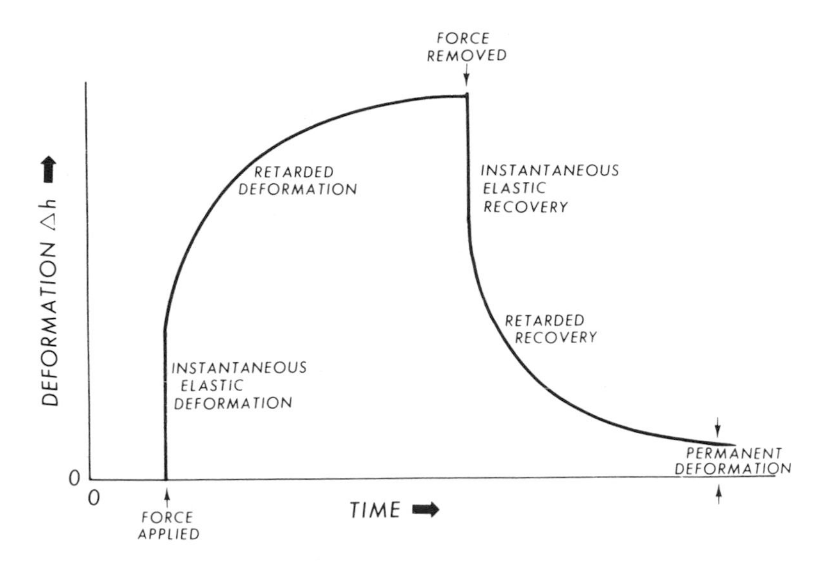

FIG. 37. Deformation–time relationship for a viscoelastic body under constant stress.

Occasionally it is important; for example, in bread doughs, which are highly viscoelastic. The irreversible deformation that results from viscoelasticity is found, for example, in grapefruit that have been tightly packed in a shipping carton; after being placed on the table where they are free to resume their spherical shape, they will retain flat compression faces for many days, demonstrating the permanent deformation that has occurred.

The above discussion relates to the change in height with respect to time under a constant deforming force. Another way of measuring the time aspects of deformation is to measure the change in force over a period of time at a constant level of deformation; that is, the product is compressed to a certain height and held at that compression while changes in force are measured. This is a test that is easily performed in the Instron and similar instruments (Bourne *et al.*, 1966). A typical curve for a viscoelastic solid is shown in Fig. 38. The force increases steeply and almost linearly from 0 to A when the commodity is compressed. At point A the compression is stopped and the product is held at a constant height. The force declines, rapidly at first and then more slowly as the product continues to deform under the force. This decay of stress under a constant strain is known as "stress relaxation." At point B the product is partially decompressed by raising the compression plate a short distance and stopping it again at C. As the produce is held with less compression the force will increase again as the product slowly recovers it original shape. This is known as "recovery." An elastic solid gives almost the same compression from 0 to A as the viscoelastic solid, but

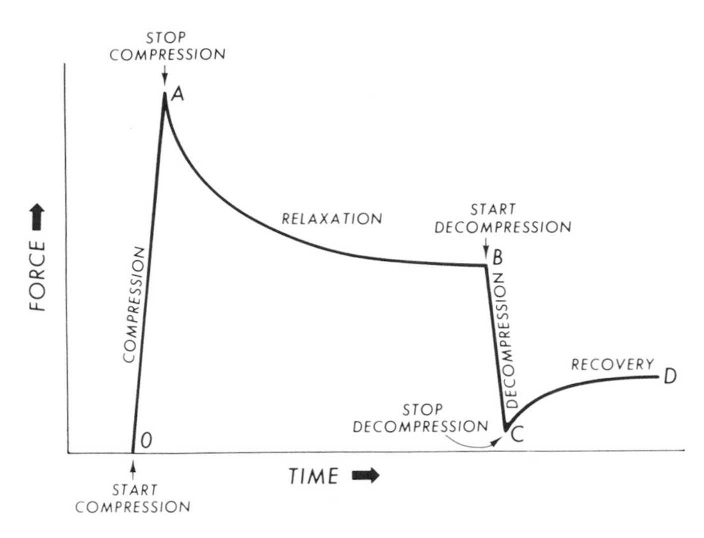

FIG. 38. Force–time relationship for a viscoelastic body under constant strain.

when compression is stopped at A the force does not change but gives a horizontal line until the solid is decompressed.

The relaxation time is the time required for the stress at constant strain to decrease to $1/e$ of it original value, where e is the base of natural logarithms (2.7183). Since $1/e = 0.3678$, the relaxation time is the time required for the force to decay to 36.8% of its original value. This can be measured on instruments such as the Instron where a constant compressive strain can be maintained and the change in force with respect to time is measured. In some cases the time to relax to $1/e$ is excessive, in which case some lower value is taken as an arbitrary relaxation time; for example, Rasper and deMan (1980) used an Instron to perform a tensile relaxation test on composite doughs and defined relaxation as the time required for 36.7% decay of the maximum load at the beginning of the relaxation period.

Area and Volume Measuring Instruments

There are probably no true area measuring instruments used in food texture measurements. Several consistometers such as the Adams Consistometer, the Cream Corn Meter, and the USDA Standard Consistency Tester (see p. 163) measure the flow of a fluid or semifluid food from a circular container out across a horizontal plate and the diameter is measured at two points at right angles. This is a distance measurement rather than area but is the closest that we have to area measuring instruments.

The volume of a loaf of bread made from a standard formula is measured by means of seed displacement using small seeds such as caraway seeds. The Succulometer (Kramer and Smith, 1946) measures the volume of juice expressed from sweet corn as an index of maturity. The volume of juice expressed from fish and meat has also been measured (Briskey *et al.*, 1959; Hamm, 1960; Dagbjartsson and Solberg, 1972; Karmas and Turk, 1976; Jauregui *et al.*, 1981). Khan and Elahi (1980) report that the maximum volume before collapsing of a batter prepared from wheat flour, baking powder, and water is highly correlated with volume and protein content of bread made from that flour.

Time Measuring Instruments

Kinematic viscometers, such as the Ostwald, measure the time for a standard volume of fluid to flow through a restricted opening. Falling-ball viscometers measure the time for a ball to fall a given distance through a liquid. These will be described more fully in the chapter on viscosity.

A few other time-measuring instruments are described in the literature. The Gardner Mobilometer measures the time required for a disk and shaft to fall a standard distance through a fluid (Gardner and VanHeuckeroth, 1927). The British Baking and Research Association Biscuit Hardness Tester measures the time required for a small circular saw rotating at 15 rpm to cut through a stack of biscuits (Wade, 1968) (see p. 166). Rogers and Sanders (1942) describe a method for measuring the firmness of cheese curd that is based on the time for a cutting head to move a standard distance through the curd under a constant force. Wilder (1947) describes a Fiberometer that measures the time required to cut across asparagus spears by a 0.041-in.-diam stainless-steel wire under a force of 3 lb.

The Falling Number Method is used by the baking industry to determine the α-amylase activity of wheat and rye flours is based on time measurement (Hagberg, 1960, 1961; Perten, 1964, 1967). It measures the time required by α-amylase to liquefy a gelatinized starch paste to a predetermined viscosity level. The viscosity level is that which allows the stirring rod to fall 70 mm in 1 sec under the influence of gravity. This technique has been adopted as a standard method by several countries and by the American Association of Cereal Chemists (No. 56–81A), the International Association of Cereal Chemists (No. 107), and the International Standardization Organization (ISO/DIS 3093). The apparatus is made by Falling Number AB, Box 32072, S126 11 Stockholm, Sweden.

Work, Energy, and Power Measuring Instruments

Instruments that use the principle of work and energy are not common. Energy and work both have the dimensions mass $\times$ length^2 $\times$ time^{-2} which is equivalent to force × distance. Instruments that plot out the force–distance relationship of a test on a strip chart can provide work functions by measuring the area under the force–distance curve or by any other technique that gives the force–distance integral. The area under the curve can be obtained by several methods, including (a) measuring the area with a planimeter (convert the area measurements into work units knowing the force and distance scales); (b) cutting out the paper under the curve, weighing it, and converting this into area knowing the weight of paper per unit area; and (c) counting the number of squares on the paper under the curve. Furthermore, several electronic methods are available to measure the area under the curves.

It is important to note that there are two work elements in a compression–decompression cycle: (1) The area under the curve during *compression* represents the work done on the food by the machine. (2) The area under the curve during *decompression* represents the work returned to the machine by the food as it recovers. Generally, the decompression work is small but in the interest of accuracy it is worth separating these two factors.

The Instron, the Food Technology Texture Test System, and the Ottawa Texture Measuring System each provide a force–distance curve and hence the area under the curve can be converted into true measurements of work. The compressing mechanism in the GF Texturometer moves with approximately sinusoidal speed while the chart moves at a linear speed; hence, the areas under these curves are not true measurements of work.

Power is the rate of doing work and has the dimensions mass × length^2 × time^{-3}. A few power measurements have been used. For example, Miyada and Tappel (1956) attached a Wattmeter to the electric motor that powered a meat grinder and, by taking a reading with the meat grinder running empty and then with meat being run through at a constant rate, they were able to obtain an index of the toughness of the meat. Kilborn and Dempster (1965) used a similar principle to measure the power required to knead dough.

Ratio Measuring Techniques

This is not a widely used technique. The numbers obtained are dimensionless since they are derived from the ratio of two measurements of the same variable.

The General Foods texturometer parameter of cohesiveness is a ratio because cohesiveness is defined as

$$\frac{\text{area under second bite curve}}{\text{area under first bite curve}}$$

Occasionally relative density correlates with texture. Relative density is the density of a product divided by the density of water and is a dimensionless number. For example, La Belle (1964) found that the texture of sour cherries and precooked dried beans correlated highly with density. Firm cherries have a low relative density because they do not pack down as much as soft cherries. Precooked dry beans with good texture have a high density because they are solid while those beans that have puffed up during the drying process (butterflied texture) have a much lower relative density. Scott Blair and Coppen (1940) measured the bulk density of cut cheese curd and used this as a textural index of the readiness to proceed with the next step in manufacturing. The specific gravity of batters has been associated with good volume and textural quality of baked goods (Funk *et al.*, 1969).

Hoseney *et al.* (1979) used a ratio method to measure the spreading properties of wheat doughs. A mixed bread dough is mechanically rounded and placed on a smooth horizontal plate in a fermentation cabinet at 30°C and 90% R.H. The dough is unrestrained and is free to spread out under the force of gravity. The width and height of the dough are measured with calipers every 15 min for 1 h. The spread ratio is calculated as width/height. The researchers obtained spread ratios ranging from 1 to 4.

Multiple Variable Instruments

With this class of instrument one variable is measured while several variables are uncontrolled. The variables may or may not be interrelated and may or may not be linear. It is usually impossible to establish true dimensional units and to convert the results from this type of instrument into a conventional system of measurement. Hence it is advisable to avoid using this type of instrument.

An example of this kind of instrument is the Durometer, which was designed to measure the stiffness of rubber but is occasionally used to measure the softness of foods. It consists of a spring-loaded probe that may be hemispherical or conical in shape that protrudes from an anvil. The probe is pressed onto the food until the anvil contacts the food, at which time a reading is taken from the dial. At this point the probe has penetrated partly into the food, and partly backward into the instrument by compressing the spring.

Figure 39 shows the relationships of the various parameters involved. The force exerted on the food by the probe has a direct linear relationship while the penetration of the probe into the food has an inverse linear relationship to the scale reading. The area and the perimeter have an inverse curvilinear relationship to the scale reading. It is impossible to convert a Durometer reading to any other system of measurement because the force, depth of penetration, area, and perimeter in contact with the food all vary in different ways. Table 9 shows the irregular manner in which the various parameters change with instrument readings.

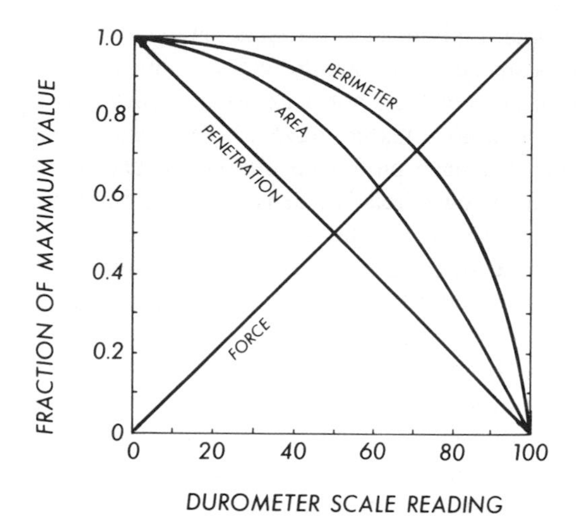

FIG. 39. Relationships between force, depth of penetration of the ball, area, and perimeter of the ball and the Durometer scale reading. (From Bourne and Mondy, 1967; reprinted from *Food Technol.* **21,** (10), 97, 1967. Copyright by Institute of Food Technologists.)

At high Durometer readings an increasing force is being applied to a diminishing area until, theoretically, at a scale reading of 100, the maximum force of the spring is being applied to the test material at a point source of zero area; there is zero penetration of the ball into the test material under an infinitely large force per unit area. A plot of force per unit area of the test material versus Durometer scale reading is shown in Fig. 40. This characteristic of the instrument explains why the Durometer is unreliable at readings above 80, even when used on rubber.

TABLE 9

RELATIONSHIP BETWEEN DUROMETER SCALE AND INSTRUMENT VARIABLES[a]

	Percent of maximum			
Durometer scale reading	Force	Area	Penetration depth	Perimeter
0	0	100	100	100
30	30	91	70	96
60	60	64	40	80
90	90	18	10	43
100	100	0	0	0

[a]From Bourne and Mondy (1967); reprinted from *Food Technol.* **21**(10), 1388, 1967. Copyright by Institute of Food Technologists.

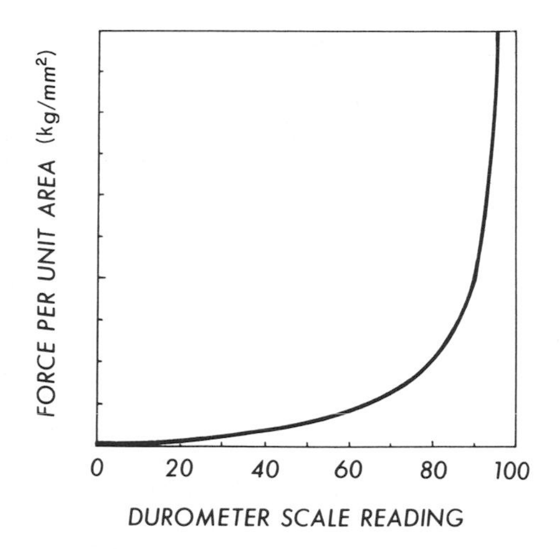

FIG. 40. Relationship between force per unit area on the sample and Durometer scale reading. (From Bourne and Mondy, 1967; reprinted from *Food Technol.* **21**(10), 98, 1967. Copyright by Institute of Food Technologists.)

Units of Measurement

Scientists and engineers have established a common worldwide standard system of units to replace the wide range of measuring units that have been used over the years. This system, called the Système International d'Unités (International System of Units) with the abbreviation SI, was adopted at an international conference in 1960. The SI is basically the metric system extended to give a uniform and rational set of units for all types of measurements. Of particular interest to texture technologists is that SI uses the newton as the standard unit of force replacing units of mass (e.g., the pound or kilogram) in expressing force. It is incorrect to use mass to express units of force because force has the dimensions mass $\times$ length $\times$ $(\text{time})^{-2}$. Changing to SI eliminates the disparity of using mass as a measure of force, and it also eliminates using the gravitational constant g to convert mass into force units. One newton force equals the force generated by gravity on 101.9716 g.

The SI system includes three classes of units: (1) base units, (2) supplementary units, and (3) derived units.

The base units and their definition are shown in Table 10. Supplementary units are defined angles and are of little interest in texture work. Derived units are expressed algebraically in terms of base units and/or supplementary units (Table 11). The prefixes that are used to form decimal multiples and submultiples of SI units are given in Table 12. The choice of the appropriate multiples of a SI unit is

TABLE 10
SI BASE UNITS

Quantity	Name	Symbol	Definition
Length	Meter	m	The length equal to 1,650,763.73 wavelengths of orange-red radiation in vacuum resulting from the transition between the levels $2P_{10}$ and $5d_5$ of the krypton-86 atom.
Mass	kilogram	kg	The mass equal to the mass of the international prototype of the kilogram.
Time	second	sec	The duration of 9,192,631,770 periods of the radiation corresponding to the transition between the two hyperfine levels of the ground state of the cesium-133 atom.
Thermodynamic temperature	kelvin	K	The fraction of 1/273.16 of the thermodynamic temperature of the triple point of water.
Amount of substance	mole	mole	The amount of substance of a system that contains as many elementary entities as there are atoms in 0.012 Kg of carbon-12.
Electric current	ampere	A	That constant electric current which, if maintained in two straight parallel conductors of infinite length, of negligible circular cross section, and placed 1 m apart in vacuum, would produce between these conductors a force equal to 2×10^{-7} N m^{-1} of length.
Luminous intensity	candela	cd	The luminous intensity, in the perpendicular direction, of a surface of 1/600,000 m^2 of a blackbody at the temperature of freezing platinum under a pressure of 101,325 N m^{-2}.

governed by convenience, the multiple chosen for a particular application being the one which will lead to numerical values in a practical range. The multiple is usually chosen so that the numerical value will be between 0.1 and 1000.

Working in SI units renders obsolete many of the old units of measurement. Some of these are listed in Table 13. Since these units have been in use for many years the researcher will often find them when reading the literature. These obsolete units should no longer be used in present work. However, two obsolete units are still being used by some researchers with some justification: (a) 1.000 centipoise viscosity equals 1.000 millipascal second; (b) 1.000 centistoke kinematic viscosity equals 1.000 square millimeter reciprocal second.

TABLE 11

SI DERIVED UNITS

Quantity	Name of SI unit	Symbol	Dimensions	Commonly used multiple	Conversion factor
Area	square meter	m^2	m^2	—	1 m^2 = 10.76391 ft^2
Volume	cubic meter	m^3	m^3	—	1 m^3 = 1000 liters
					1 liter = 1 dm^3 (cubic decimeter)
Frequency	hertz	Hz	s^{-1}	—	—
Force	newton	N	m kg s^{-2}	—	1 N = 101.9716 force
Pressure, stress	pascal (or newton per m^2)	Pa	N m^{-2} or m^{-1}kg s^{-2}	—	1 bar = 10^5 Pa
Dynamic viscosity	pascal second	Pa s	N S m^{-2} or m^{-1}kg s^{-1}	mPa s	1 centipoise = 1 mPa s
Kinematic viscosity	square meter per second	m^2s^{-1}	m^2s^{-1}	mm^2 s^{-1}	1 centistoke = 1 mm^2 s^{-1}
Work, energy, heat	joule	J	N m or m^2kg s^{-2}	—	—
Power	watt	W	m^2kg s^{-3}	—	1 W = 1 J/s

TABLE 12
SI MULTIPLYING FACTORS

	Prefix	
Factor	Name	Symbol
10^{12}	tera	T
10^{9}	giga	G
10^{6}	mega	M
10^{3}	kilo	k
10^{2}	hecto	h
10^{1}	deca	da
10^{-1}	deci	d
10^{-2}	centi	c
10^{-3}	milli	m
10^{-6}	micro	μ
10^{-9}	nano	n
10^{-12}	pico	p
10^{-15}	femto	f
10^{-18}	atto	a

Chemical Analysis

This indirect method is occasionally used as an index of textural qualities. Several examples are given below.

Alcohol insoluble solids (AIS) is a good index of the maturity and texture of fresh, raw green peas (Kertesz, 1935; Association of Official Analytical Chemists Method 32.006). This method comprises grinding fresh peas in 80% alcohol in a blender, filtering on a vacuum filter, washing well with 80% alcohol, and

TABLE 13
UNITS RENDERED OBSOLETE BY SI

Name	Symbol	Conversion to SI units
erg	erg	1 erg = 10^{-7} J
dyne	dyn	1 dyn = 10^{-5} N
poise	P	1 P = 0.1 Pa s
		1 centipoise = 1 mPa s
stoke	St	1 St = $10^{-4} m^2 s^{-1}$
		1 centistoke = 1 $mm^2 s^{-1}$
kilogram force	kgf	1 kgf = 9.80665 N
pounds force	lbf	1 lbf = 4.4482 N
calorie	cal	1 cal = 4.1868 J

drying and weighing the residue. The success of this test depends upon the fact that as the peas mature the total solids increase and sugar converts into starch. Since starch is insoluble in alcohol, there is a rapid change in AIS as the peas mature.

The pericarp content of sweet corn is a useful index of its maturity (Kramer *et al.*, 1949). Corn kernels (100 g) are ground in a blender with water, then washed well on a 30-mesh sieve. The starchy endosperm and germ are ground fine enough to pass through the sieve, but the pericarp is sufficiently tough that it is not chopped fine enough to pass through the screen. After washing thoroughly, the screen is dried and weighed to give the pericarp content, which normally ranges from 1.3 to 5.2%.

Moisture content is sometimes an index of maturity. An example of this is sweet corn where a commonly accepted standard is <68%, too tough, overmature; 68–70%, cream style for canning; 70–76% is whole kernel for canning; 76–78% is suitable for freezing; >78%, too immature.

A chemical index of the amount of collagen in meat performed by determining the hydroxyproline content has been proposed as an index of meat toughness. There is as yet no consensus as to the value of this determination as an index of meat toughness. The subject has been reviewed by Szczesniak and Torgeson (1965).

Miscellaneous Methods

The criteria for inclusion under this heading are (a) that it is an objective method, (b) that the measurement correlates well with texture, and (c) that the method does not fit into any of the categories described previously.

Optical methods. The cell fragility method is an optical method that is used to measure the toughness of fish (Love and Muslemuddin, 1972a,b; Love and Mackay, 1962). In this method a standard weight of fish is homogenized in a blender in a mixture of 2% trichloroacetic acid plus 1.2% formaldehyde for a standard time after which the optical density is measured. Tender fish grind into a fine state and give a high optical density while tough fish remain as fewer large particles and give a lower optical density. The cell fragility method is reported to give good results for nonfatty fish but less satisfactory results for fatty fish or fish in which advanced bacterial spoilage has occurred.

Sound. Drake (1963, 1965) analyzed the amplitude and duration of chewing sounds over a wide range of frequencies. This method shows some promise, particularly with noisy foods, but large variations between sounds generated by different individuals complicates the procedure.

Vickers and Bourne (1976a,b) postulated that the property of crispness is an acoustical sensation that is detected by the ear during the fracturing of crisp foods. Crisp foods produce a characteristic sound that has a broad range of frequencies with low notes predominating and irregular and uneven variations in loudness. The total amount of sound generated is an indicator of the degree of crispness. It is likely that acoustical methods will be developed in the future for texture analysis of crisp and crunchy foods.

Multiple Measuring Instruments

This class of instruments is characterized by the ability to measure several variables under controlled conditions. They consist of three essential parts:

1. A drive system that imparts motion to the test cell. The drive mechanism may be screw, hydraulic, or an eccentric and lever system.

The Instron is driven by twin screws, the Ottawa Texture Measuring System is driven by a single screw, the Food Technology Texture Test System is driven hydraulically, and the GF Texturometer is driven by an eccentric and lever system.

2. Test cells that hold the food and apply force to it. Different test cells can be used to puncture, compress gently, compress greatly, extrude, shear, snap, pull apart, etc.
3. A force measuring and recording system. The recording system plots the complete history of force changes throughout the test.

All recorders plot out a force–time relationship. Since the GF Texturometer operates in a sinusoidal speed pattern, only force, time, and functions of force and time can be read from the chart. The Ottawa Texture Measuring System and the Food Technology Texture Test System are driven at approximately constant speed; hence, the time axis can be used as a very good approximation of the distance of travel of the moving parts.

Both the crosshead and the chart in the Instron are driven synchronously, which means that the speed is kept in lockstep with the frequency of the alternating current supply. Since the ac line frequency is maintained very exactly by the electric generators at central power stations, the time axis on the Instron chart is an exact simple multiple of the distance traveled by the moving crosshead. The numerical value of the multiple depends on the gear trains that are used for the chart and the crosshead.

The advantage of the recording instruments is that the complete force history is plotted, giving all the changes that occur, including the rate of change (slopes), maximum force (peaks), area under the curve (work), and frequently other parameters of interest. The use of recorders tends to cause a loss of confidence in the old 1-point instruments. When a pointer moves over a dial and the maximum force reading is taken, there appears to be an element of certainty about the results which leads to a feeling of confidence in the instrument. When the same test is repeated in a recording instrument, the maximum force often seems to be an arbitrary point to use as an index of textural qualities; there is a loss of confidence in the accuracy of the test and the feeling of infallibility associated with some of these simple instruments is lost.

Speed of Recorder Pen

The recorders that are customarily used in the food industry measure variables that change slowly with time; for example, temperature, gas chromatography, and light spectrometry. Consequently, there is little question or concern about whether the recorder is faithfully plotting the measured variable. This fortunate state of affairs does not apply to texture measuring instruments where rapid changes in force often occur. Many food technologists innocently (and erroneously) place complete confidence in the graphs that are plotted on the charts of their recording texturometers.

Figure 41 shows a model that explains the problem. Suppose a force measuring instrument receives a full-scale force applied instanteously, held for 1 sec, and then removed instantaneously. The solid black line in the figure gives the correct representation of the change of force with time; however, no recorder will reproduce this line exactly because it requires a finite period of time for the pen to travel the width of the chart. This time is known as the *pen response time*. The

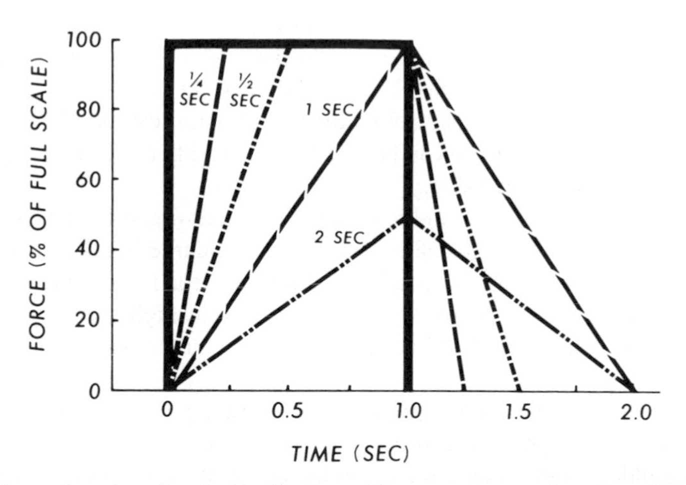

FIG. 41. Force–time plots given by ¼-, ½-, 1-, and 2-sec response recorders when a full-scale force is applied instantaneously, held for 1 sec, then removed. [Reprinted from "Rheology and Texture in Food Quality" (deMan, Voisey, Rasper, and Stanley, eds.), p. 248; with permission from AVI Publ. Co.]

TABLE 14

EFFECT OF RECORDER RESPONSE TIME ON MEASUREMENTS OBTAINED FROM FORCE–TIME CURVE[a]

Parameter	Pen response time (sec)				
	Instantaneous	¼	½	1	2
Maximum force	100	100	100	100	50
Time to reach maximum force (sec)	0	0.25	0.50	1.0	Not reached
Time maximum force is shown (sec)	1.0	0.75	0.50	Momentarily	Not shown
Time some force is shown (sec)	1.0	1.25	1.5	2.0	2.0
Total area under curve (relative values)	80	80	80	80	40
Area under load portion of curve (relative values)	80	70	60	40	20

[a]Full-scale load applied for 1 sec; from Bourne (1976); with permission from AVI Publ. Co.

dashed lines in Fig. 41 show the plots that will be made by recorders with pen response speeds of ¼, ½, 1, and 2 sec. The ¼-, ½-, and 1-sec response recorders will correctly give the peak force while the 2-sec response recorder will show only 50% of the correct peak force. Although the actual full-scale force was applied for exactly 1 sec, no recorder will show this correctly; the ¼-sec recorder shows the full-scale force having been in effect for ¾ sec, the ½-sec response recorder shows full-scale force for ½ sec, the 1-sec response recorder shows full-scale force momentarily, and the 2-sec recorder never reaches full-scale force. Table 14 summarizes the graphs shown in Fig. 41 and illustrates the fact that the errors caused by pen response can be substantial, and that the error increases as the pen response time increases.

Figure 42 shows the nature and magnitude of the errors that can be introduced by pen response speed in a real situation. The same curve has been traced in the four examples. In each case the solid line shows the true force–time relationship obtained by compressing a whole apple for 1 sec at high speed and then decompressing it, and the dashed lines show where the recorder deviates from the correct position.

The ¼-sec response recorder gives a faithful tracing of changes in force with time during the compression and correctly shows the initial slope, yield point, several shoulders, and maximum force at the end of the compression. There is an error during decompression because the force drops to zero almost instantaneously while the pen arrived at zero about ⅕ sec later. The ½-sec response recorder gives an incorrect initial slope, misses the abrupt drop in force after the yield

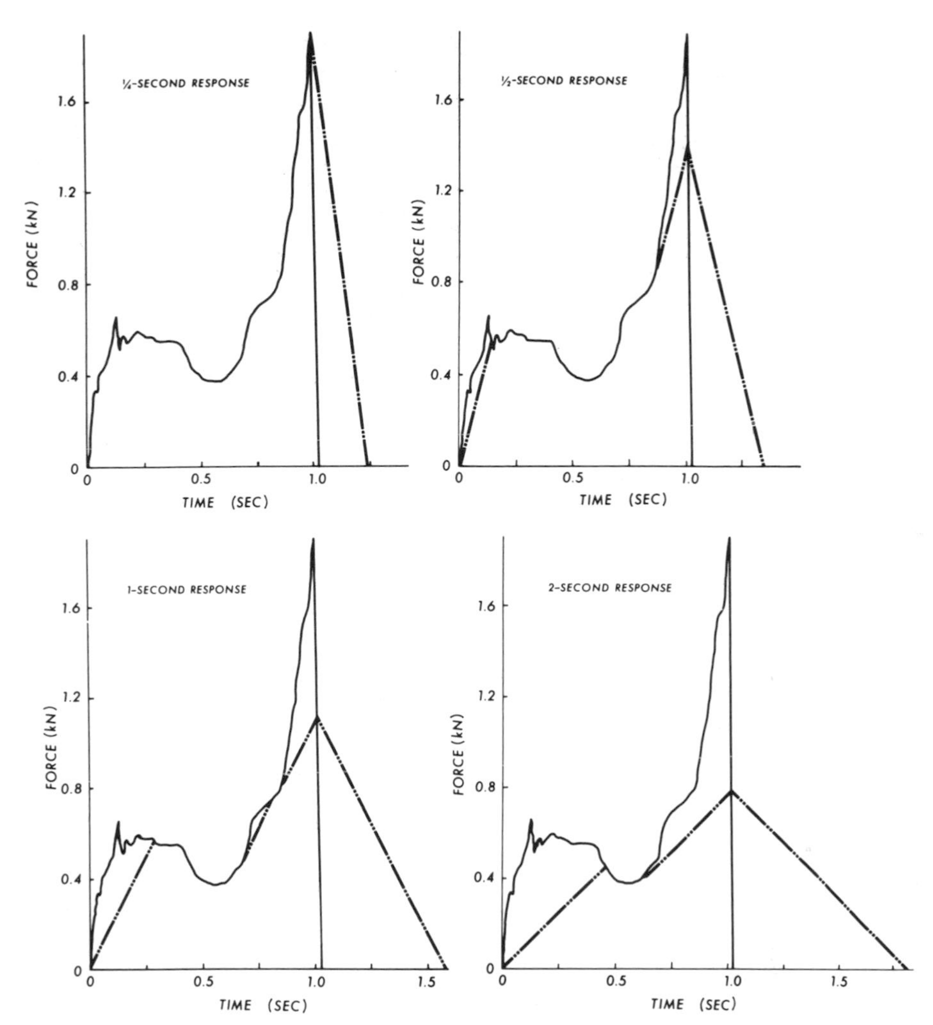

FIG. 42. Force–time plot for a whole apple that is rapidly compressed in the Instron for 1 sec and then decompressed. The solid line shows the true force, the dashed lines show errors made by ¼-, ½-, 1-, and 2-second response recorders. [Reprinted from "Rheology and Texture in Food Quality" (deMan, Voisey, Rasper, and Stanley, eds.), p. 250; with permission from AVI Publ. Co.]

point, misses the maximum force at the end of the compression by a large margin, and shows a larger decompression area than the ¼-second response recorder. It does provide an accurate record of the peaks and shoulders in the curve from about 0.2 to 0.8 sec.

The 1-sec response recorder shows an even greater error in the initial slope, maximum force, and decompression, and also fails to record the large shoulder at

0.8 sec. The 2-sec response recorder grossly misrepresents the record, aside from showing the trough at 0.5 sec for a short period. The evidence in this figure should shatter any feelings of infallibility of recorders and should make every texture technologist aware of the necessity of always being alert to the possibility of errors in the trace the pen makes on the chart.

There are three methods that can be used to overcome the problem of pen response errors.

1. Use a high-speed recorder. This is the preferred method. This writer recommends purchasing the highest-speed conventional recorder the laboratory can afford because it minimizes errors and allows the maximum range of speed of moving parts to be used.

Occasionally a high-speed recorder is a liability instead of an asset. Voisey (1971a) gave an example of this by attaching a recorder to a Mixograph to measure the changes in torque as a wheat dough was kneaded. Figure 43a shows the plot obtained with a 0.2-sec response recorder. This recorder shows the rapid fluctuations from moment to moment in such detail that it is difficult to see the trend over a period of some minutes. Figure 42b shows the same test repeated with a 12-sec response recorder. The momentary fluctuations are lost but the slow development of dough strength is shown more clearly. Also, the torque scale was reduced from 160 cm kg for the 0.2-sec recorder to 40 cm kg for the 12-sec recorder, spreading the development curve more fully across the chart. The point of this example is that the recorder's behavior in any instrument should not be taken for granted but should be selected to give the type and quality of plots that are best suited to the requirements of each particular experiment.

2. Use of a high full-scale force compresses the curve, thus giving the pen less distance to travel. Figure 44 demonstrates the use of this principle. Figure 44a shows the result of a high-speed compression test on a whole apple using full-scale force of 20 kN (kilonewton). The solid line gives the true force–time record and the dashed line shows the substantial deviations given by a 1-sec response recorder. Figure 44b shows the result of an identical test on another apple, but with the full-scale force increased to 50 kN. This curve has the same general shape but is compressed because the pen travels a smaller distance to record the same force: the recorder follows the change in force more closely because the pen had less distance to travel. The only errors occur in the maximum force peak and the decompression part of the curve. This procedure results in some loss of precision, but this is preferred to the substantial errors that occur when the force changes more quickly than the pen speed. This procedure cannot be used if the instrument is already working at its full-force capacity.

3. Run the test at a slower speed. This reduces the rate of change in force and gives the pen more time to keep up with the action. Figure 45 is the trace resulting from compression of an apple under the same conditions as Fig. 44a,

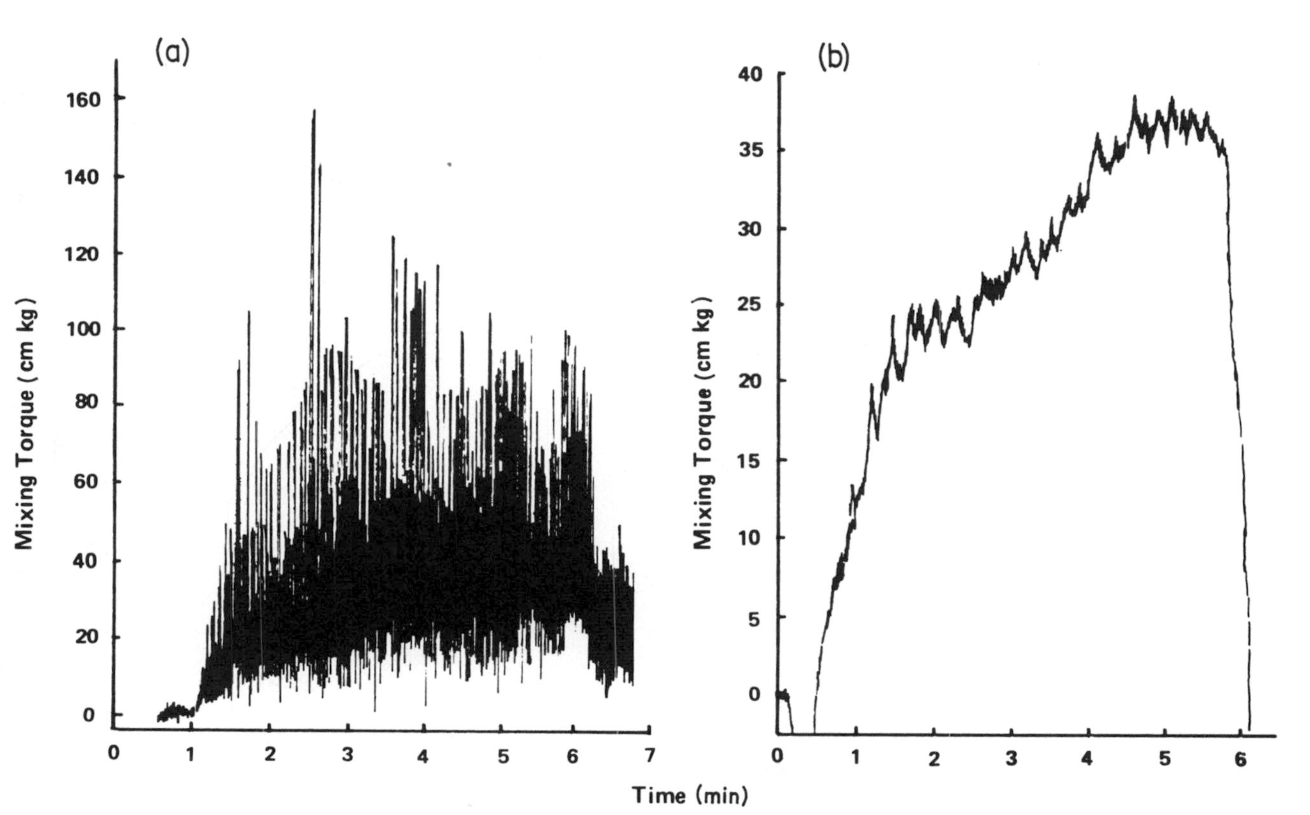

Fig. 43. Mixing torque versus time for the Mixograph: (a) 0.2-second response recorder; (b) 12-second response recorder. (Courtesy of P. W. Voisey; reprinted from *J. Texture Stud.* with permission from Food and Nutrition Press, Inc.)

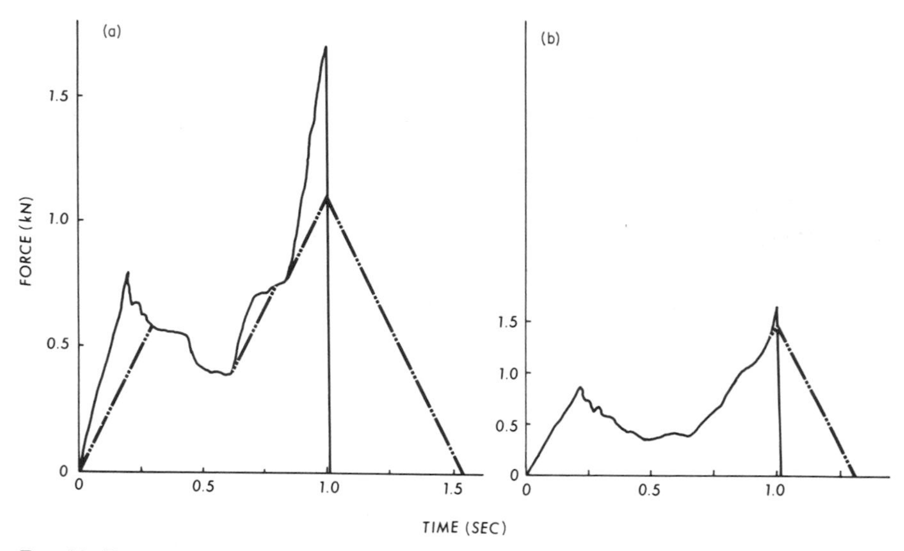

FIG. 44. Force–time plot for a whole apple that is rapidly compressed in the Instron for 1 sec and then decompressed. The solid line shows the true force; the dashed lines show the errors made by a 1-sec response recorder: (a) uses 20 kN full-scale force; (b) uses 50 kN full-scale force. [Reprinted from "Rheology and Texture in Food Quality" (deMan, Voisey, Rasper and Stanley, eds.), p. 253; with permission from AVI Publ. Co.]

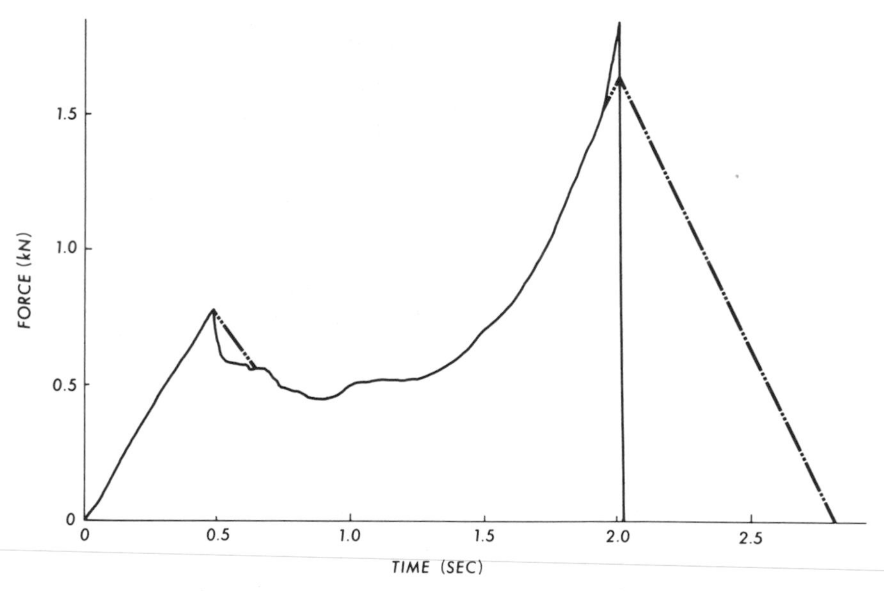

FIG. 45. Force–time plot for a whole apple that is compressed at half speed in the Instron for 2 sec and then decompressed. The solid line shows the true force and the dashed line shows the errors made by a 1-sec response recorder. [Reprinted from "Rheology and Texture in Food Quality" (deMan, Voisey, Rasper, and Stanley, eds.), p. 254; with permission from AVI Publ. Co.]

except that the compression speed was half of that which was used previously. Notice that the error, as shown by the dashed line, is much less.

One problem associated with this solution is that the rheological properties of some foods are time dependent; that is, the force–time relationship varies according to the speed at which the test is performed. For example, Shama and Sherman (1973a) showed that Gouda cheese can be made to appear firmer or softer than white Stilton cheese depending on the compression that is used.

It is impossible to establish a rule for maximum compression speed that can be used with a given recorder because the rigidity of the food affects the results. The important point is whether the rate of change of force occurs more or less rapidly than the ability of the pen to keep up with this rate of change. Figure 46 shows the curves that result from compressing a corn curl (a rigid food) and a cube of cream cheese (a nonrigid food) under identical conditions. The 1-sec recorder gives an accurate plot for the cheese, which exhibits a slow rate of change in force, but the same recorder gives an inaccurate trace for the corn curl, which undergoes rapid force changes. The critical factor is the *rate of change of force* in the specimen compared to the maximum rate of change that the recorder can accurately reproduce. Foods that give slow changes in force when compressed can tolerate slower response recorders or higher compression speeds than foods that give rapid changes in force.

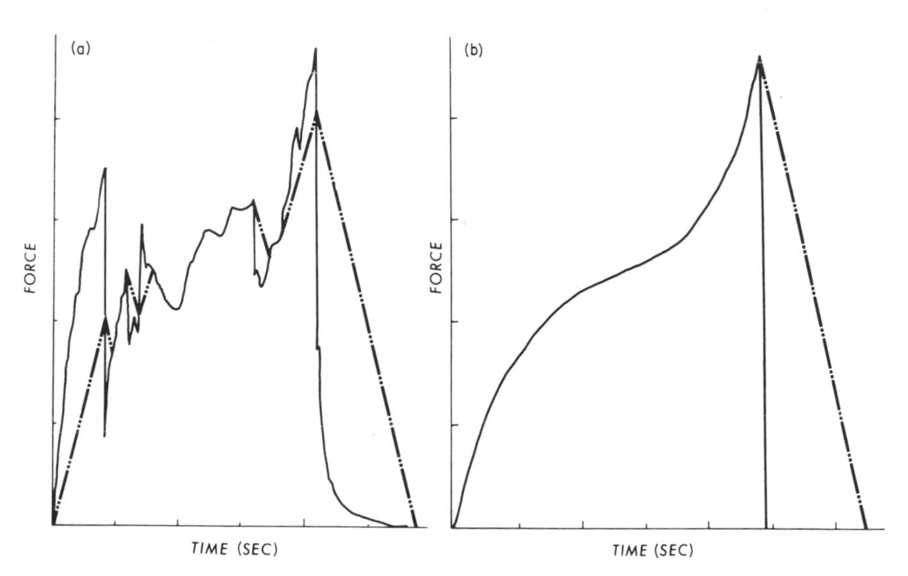

FIG. 46. Force–time plot for (a) a rigid food (corn curl) and (b) a soft food (cream cheese) compressed at the same rate in the Instron. The solid line shows the true force and the dashed lines show the errors made by a 1-sec response recorder. [Reprinted from "Rheology and Texture in Food Quality" (deMan, Voisey, Rasper, and Stanley, eds.), p. 255; with permission from AVI Publ. Co.]

The force–time plots produced by recording texturometers are generally curved. Straight-line portions are usually of short duration. Whenever one sees long straight lines of uniform slope on the chart one should suspect that perhaps pen response speed is being plotted rather than the true changes in force. After observing the straight-line portions and the true force–time plots in Figs. 42 and 46 one should become suspicious when extensive straight lines of standard slope are seen on their force–time plots.

Most recorder manufacturers will provide the stated pen response speed of their recorder. The pen response speed can be checked in the laboratory as follows: set the chart running at the maximum speed and then instantaneously apply a full-scale force to the load cell, leave it there for a second or two and then instantaneously remove it. The slope of the force–time line divided by the speed of travel of the chart will give the pen response time of that recorder.

Texture Profile Analysis (TPA)

A group at the General Foods Corporation Technical Center pioneered the test that compresses a bite-size piece of food two times in a reciprocating motion that imitates the action of the jaw, and extracted from the resulting force–time curve a number of textural parameters that correlate well with sensory evaluation of those parameters (Friedman *et al.*, 1963; Szczesniak *et al.*, 1963). The instrument devised especially for this purpose is the General Foods Texturometer. A typical GF Texturometer curve is shown in Fig. 47 (see also p. 168).

The height of the force peak on the first compression cycle (first bite) was

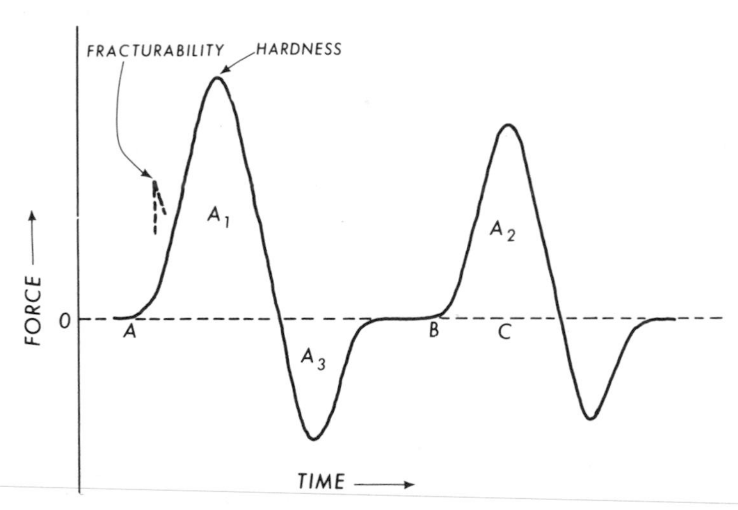

FIG. 47. A typical GF Texturometer curve (slightly altered) (Courtesy of Dr. A. S. Szczesniak with permission).

defined as hardness; in the figure, A is the beginning of the first compression and B is the beginning of the second compression. *Fracturability* (originally called brittleness) was defined as the force of the significant break in the curve on the first bite (shown as a dashed line in Fig. 46). The ratio of the positive force areas under the first and second compressions (A_2/A_1) was defined as *cohesiveness*. The negative force area of the first bite (A_3) represented the work necessary to pull the compressing plunger away from the sample and was defined as *adhesiveness*. The distance that the food recovered its height during the time that elapsed between the end of the first bite and the start of the second bite (BC) was defined as *springiness* (originally called elasticity). Two other parameters were derived by calculation from the measured parameters: *gumminess* was defined as the product of hardness × cohesiveness; *chewiness* was defined as the product of gumminess × springiness (which is hardness × cohesiveness × springiness). Szczesniak (1975b) gave an updated account of the development and changes in the technique since 1963.

The texture parameters identified by the General Foods group gave excellent correlations with sensory ratings (Szczesniak *et al.*, 1963). Figure 48 shows the

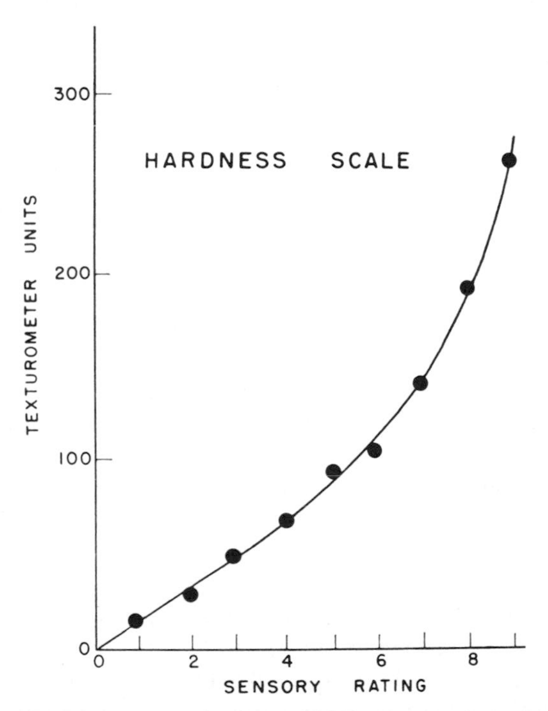

FIG. 48. Correlation between sensory evaluation and GF Texturometer for hardness of nine selected foods. (Courtesy of Dr. A. S. Szczesniak; reprinted from *J. Food Sci.* **28,** 401, 1963. Copyright by Institute of Food Technologists.)

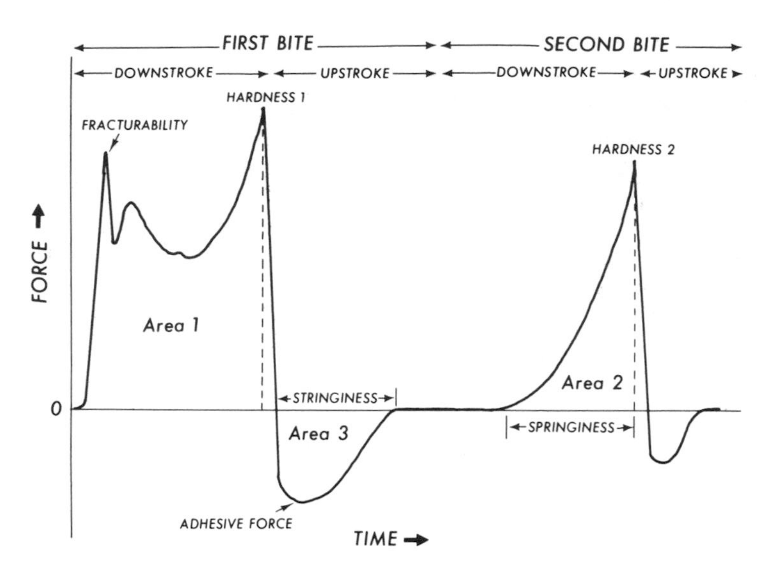

FIG. 49. A generalized Texture Profile Analysis curve obtained from the Instron Universal Testing Machine. (Reprinted from *Food Technol.* **32**(7), 63, 1978. Copyright by Institute of Food Technologists.)

correlation for the hardness scale. High correlations between sensory and instrument measurements were obtained for the other texture parameters.

The Instron has been adapted to perform a modified texture profile analysis (Bourne, 1968, 1974). A typical Instron TPA curve is shown in Fig. 49. Bourne closely followed the interpretation of Friedman *et al.* (1963) with one exception: instead of measuring the total areas under the curves to obtain cohesiveness, he measured the areas under the compression portion only and excluded the areas under the decompression portions.

A typical TPA curve obtained in the Instron differs in several major respects from that obtained by the GF Texturometer. This can be seen by comparing Figs. 47 and 49. The Intron curves show sharp peaks at the end of each compression while the GF Texturometer shows rounded peaks. These differences arise from differences in instrument construction and operation. The GF Texturometer is driven by means of an eccentric rotating at constant speed and imparting a sinusoidal speed to the compressing mechanism, while the Instron is driven at constant speed. The GF Texturometer decelerates as it approaches the end of the compression stroke, momentarily stops, and then slowly accelerates again as it makes the upward stroke. In contrast, the Instron approaches the end of the compression stroke at constant speed, abruptly reverses direction, and performs the upward stroke at constant speed. The constant speed of the Instron versus the continuously changing speed of the GF Texturometer largely accounts for the

TABLE 15

DIMENSIONAL ANALYSIS OF TPA PARAMETERS[a]

Mechanical parameter	Measured variable	Dimensions of measured variable
Hardness	Force	mlt^{-2}
Cohesiveness	Ratio	Dimensionless
Springiness	Distance	l
Adhesiveness	Work	ml^2t^{-2}
Fracturability (brittleness)	Force	mlt^{-2}
Chewiness	Work	ml^2t^{-2}
Gumminess	Force	mlt^{-2}

[a]From Bourne (1966a). Note: This table was incorrectly presented in the original publication. The correct table was published in *J. Food Sci.* **32,** 154, 1967. Copyright by Institute of Food Technologists.

sharp peaks in the Instron in contrast to the rounded peaks of the Texturometer.

Another difference is that the supporting platform of the GF Texturometer is flexible; it bends a little as the load is applied. The Instron is so rigid that bending of the instrument can be ignored. Yet another difference is that the compressing plate of the Texturometer moves in the arc of a circle, while in the Instron it moves rectilinearly. These three factors taken together account for the differences in the TPA curves obtained by the GF Texturometer and the Instron.

Since the Instron gives both a force–time and force–distance curve, the TPA parameters obtained from it can be given dimensions, which are listed in Table 15.

Henry *et al.* (1971) provided a more detailed analysis of the adhesiveness portion of texture profile curve for semisolid foods such as custard, puddings, and whipped toppings. In addition to measuring the area of the adhesiveness curve they measured its maximum force (symbolized by F_a to denote firmness under tension), the recovery in the adhesion portion between the first and second compressions (E_a to denote elastic recovery under tension), and the ratio of the two adhesion areas (C_a to denote cohesiveness under tension). They calculated gumminess under tension ($Ch_a = F_a \times C_a \times E_a$). They also measured the property of stringiness (or, inversely, shortness) as the distance the product was extended during decompression before breaking off (Henry and Katz, 1969). Their experiments showed that eight of these parameters accounted for more than 90% of the variation of four sensory factors.

Reports on the Texture Profile Analysis of a number of commodities have appeared in the literature with some variations on the main themes described above. Breene (1975) gave a complete review of this area.

CHAPTER 4

Practice of Objective Texture Measurement

Introduction

This chapter discusses the major commercially available texture measuring instruments that are available at the present time. Discontinued instruments are not discussed (with two exceptions). The discussion includes a brief description of the major features of each instrument and how it is used and calibrated. The principles on which these instruments operate were discussed in the previous chapter. The order in which the instruments are described will follow the same sequence as used in the previous chapter. The Appendix lists the suppliers of these instruments, to whom inquiries for further information and prices should be sent.

Force Measuring Instruments

Hand-Operated Puncture Testers

These testers are derived from the improved type of pressure tester developed by Magness and Taylor (1925). These are frequently called "pressure testers" but a better description would be to class them as "puncture testers." Four manufacturers make this class of instrument: Ballauf, Chatillon, Effi-Gi, and the UC Fruit Firmness Tester (developed by the University of California). Table 1 lists the specifications of these puncture testers and Fig. 1 shows some of the instruments. All these instruments use a spring to measure applied force with an indicator to show the maximum test force.

TABLE 1

SPECIFICATIONS OF HAND-OPERATED PUNCTURE TESTERS

Manufacturer	Force full scale × graduations	Plunger travel to full scale force (cm)	Punch		Instrument	
			Diam (in.)	Face	Length (cm)	Weight (g)
Ballauf Co.	30 × 1 lb	13	7/16, 5/16	Rounded	52	700
Ballauf Co.	10 × ½ lb	13	7/16, 5/16	Rounded	44	530
Chatillon 719-40 MRPFR	40 × ½ lb 18 kg × 200 g	10	7/16, 5/16	Flat	50	500
Chatillon 719-20 MRPFR	20 × ¼ lb	10	7/16, 5/16	Flat	50	450
Chatillon 719-10 MRPFR	10 lb × 2 oz 4.5 kg × 50 g	10	7/16, 5/16	Flat	50	420
Chatillon 719-5 MRPFR	5 lb × 1 oz 2.2 kg × 50 g	10	7/16, 5/16	Flat	50	400
Chatillon 516-1000 MRPFR	1000 × 10 g 2 lb × ½ oz	10	0.026, 0.032, 0.046, 0.058, 0.063	Flat	46	180
Chatillon 516-500 MRPFR	500 × 5 g 1 lb × ¼ oz	10	0.026, 0.032, 0.0468, 0.058, 0.063	Flat	44	180
Effi-Gi	12 × ¼ kg 5 × 0.1 kg	2	7/16, 5/16	Rounded	13	170
UC Tester	30 × ¼ lb		7/16, 5/16	Rounded		
(University of California)	10 × 0.1 lb		7/16, 5/16	Rounded		

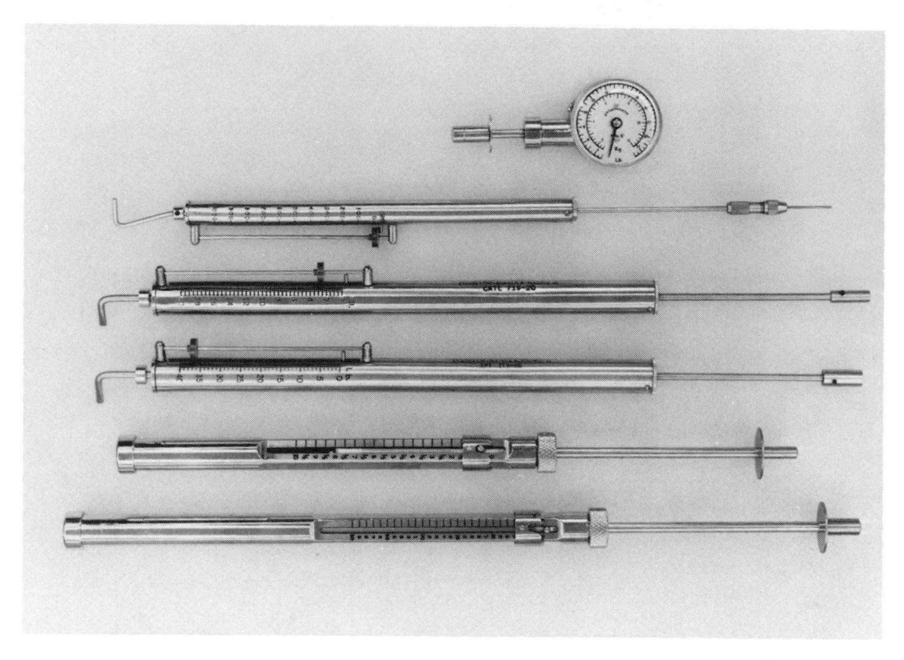

FIG. 1. Hand-operated puncture testers of the Magness–Taylor type. From the bottom up: 30-lb Ballauf with 7/16-in.-diam punch, 10-lb Ballauf with 5/16-in.-diam punch, 40-lb Chatillon with 7/16-in.-diam punch, 20-lb Chatillon with 5/16-in.-diam punch, 1000 g Chatillon with 0.058-in.-diam punch, and Effi-Gi with 7/16-in.-diam punch.

The Ballauf Company makes two testers, one with a 30-lb spring for firmer products and the other with a 10-lb spring. Two punches are provided: 7/16 in. and 5/16 in. diam. The punches have a rounded face and an inscribed line 5/16 in. back from the front end of the punch, indicating the depth to which it should be pressed into the test sample. A splash collar prevents juice from running back along the shaft.

Chatillon makes two series of testers. The 719 series covers force ranges of 5, 10, 20, and 40 lb and provides 7/16- and 5/16-in.-diam punches. These have a flat face, there is no inscribed line indicating how far the punch should penetrate into the test sample, and there is no splash collar. The Chatillon 516 series are smaller, lighter instruments with force ranges of 500 and 1000 g. A small chuck at the end of the shaft is used to hold one of the five punches that range from 0.026 to 0.063 in. diam.

The Effi-Gi is the smallest and lightest instrument and most convenient to handle. It has a dial force gauge and uses the same punches as the Ballauf; 5-kg and 12-kg force scales are available.

The UC tester uses the same punches as Ballauf but uses a Hunter spring force dial gauge. It is mounted in a stand and operated by a hand lever. A shallow cup in the base ensures that the fruit to be tested is always aligned correctly. This instrument is covered by United States Patent No. 3,470,737.

With the exception of the UC tester, the testers are held in one hand, the punch is placed against the sample to be tested (most commonly a fruit), and steadily increasing force is applied until the punch penetrates to the inscribed line. The operator has to decide how far to make the punch penetrate for those punches that have no inscribed lines. The maximum test force is recorded by a pointer on the force gauge. The pointer must be returned to zero after each test.

The manner in which these hand gauges are operated affects the readings that are obtained. Therefore, it is mandatory to use a standard method to operate this class of instrument. Table 2 lists the operating rules that have been devised in our laboratory.

The springs of these instruments should be calibrated regularly to ensure that they are giving the correct force readings. The Ballauf instrument is calibrated by placing the tester vertically on weight scales with the punch down, applying force steadily until a given force is registered on the scale, then releasing the force and taking the reading on the Ballauf instrument. Continue in this way increasing the force in increments of 10% of the range up to the force range of the instrument. The weight of the shaft assembly to which the punch is attached must be subtracted from the instrument reading for comparison with the spring balance reading. The shaft assembly in the 30-lb instrument weighs approximately 300 g (10 oz) and the shaft assembly in the 10-lb instrument weighs approximately 200 g (7 oz). The Ballauf Co. will recalibrate their instruments at the factory for a nominal fee.

The vertical calibration method described in the previous paragraph compares the instrument spring force plus the mass of the shaft assembly against a scale reading. There is, therefore, a positive constant error equal to the mass of the rod in the calibration readings that must be subtracted. When the instrument is used in the normal horizontal operating position, the mass of the shaft assembly no longer affects the reading.

The Effi-Gi is calibrated in the same way as the Ballauf. The zero is adjusted by adding or removing shims to the inside of the instrument.

The Chatillon pressure testers have a knurled knob at the end of barrel nearest the pressure tip. Rotating this knob adjusts the zero point. The zero reading should be checked while holding the instrument horizontal before each use and adjusted to the correct value. Once this adjustment has been made the knurled ring should not be moved. This instrument has a hook at the top end that can be used for measuring tensile forces. To calibrate, the instrument is held in a vertical position with the tip upward, the knurled ring is adjusted to account for the weight of the shaft to bring the indicator to the zero point, and then weights are hung on the hook to increase the force applied in suitable increments.

TABLE 2

OPERATING RULES FOR USE OF A HAND-OPERATED PUNCTURE TESTER ON FRUITS AND VEGETABLES[a]

Remove skin from test site (unless it has been shown the skin does not affect the result).

Hold food in one hand against a rigid vertical surface such as a wall, tree trunk, or heavy bench. Keep test surface perpendicular to the punch face.

Hold puncture tester in the other hand with the side of the hand resting on the hip and steadily "lean into" the tester with hip; this gives a more even rate of force application than pushing with the arm. Tester must be in a horizontal plane. (Erratic motions due to lack of firm support, and pushing with the hand only can cause jerky movement which may cause spuriously high readings. Since these are *maximum* force instruments a momentarily high force will be recorded as the test force.)

When penetration begins, the operator should pause momentarily in increasing the applied force. Penetration will continue at constant force and approximate constant speed with type B products (see Fig. 2, p. 124). In type C products the punch will accelerate into the flesh past the inscribed line. The penetration will stop for the type A products even though the force is maintained; the force is now increased in small increments with a pause between each increment in order to allow the punch to penetrate as far as possible at that force; continue until tip penetrates to the inscribed line.

Use a constant diameter punch for any series of tests. If necessary, change to another instrument with a different force spring to cover the force range but do not change the punch diameter during the experiment.

Do not test near the edge of the test sample. If the edge cracks or splits during a test the result should be rejected. This problem can be overcome by testing farther away from the edge or using a smaller diameter punch.

It is customary to use at least 20–30 fruits per test to obtain representative data. Two readings are usually made on each fruit: on opposite sides midway between the stem and the blossom ends, and away from the suture line.

---

[a]From Bourne (1979b); with permission from Academic Press Inc. (London) Ltd.

Voisey (1977a) studied the static and dynamic calibration of the Ballauf and Effi-Gi testers by mounting them in the Instron and found that the primary source of differences between instrument readings was systematic calibration errors. These can be corrected. Voisey (1977a) recommended dynamic calibration because this simulates the actual operation of the instrument. He also suggested that differences between operators could be reduced by training operators to achieve a constant rate of force increase.

These instruments are widely used in the horticultural industry for measuring the firmness of fruits and some vegetables, but they could quite well be used for a number of other foods. Ballauf units are operated by holding one hand over the smooth end of the tester. The Chatillon instrument must be held by the body because of the tension hook protruding from the top. This poses no problem for the lower force ranges but for the 40-lb instrument the body must be grasped very tightly to supply sufficient force to prevent the body from slipping through the hand. A complete set of Chatillon instruments is an economical method for obtaining a wide range of forces and a wide range of tip diameters. The Effi-Gi is the most compact of the instruments and can be carried in a pocket. The rounded dial fits comfortably between the forefinger and the thumb. The spring on the Ballauf and Chatillon testers needs to be compressed a long distance to reach full-scale force, while on the Effi-Gi the distance is much less (Table 1). This makes a difference when performing a large number of tests because the amount of work (force × distance) required to operate the Effi-Gi on similar foods is about one fifth that for the other testers that cover the same force range.

A person using these instruments for the first time may be perplexed by the different ways in which these puncture testers handle. Sometimes the punch can be pushed into the commodity smoothly and gently, making it easy to control the depth of penetration. At other times the punch tends to penetrate with less control, and with some foods it suddenly plunges in past the inscribed line. The punches from these instruments have been mounted in an Instron testing machine, which automatically records the complete history of the force changes that occur during tests (Bourne, 1965a; Voisey, 1977a). These studies have thrown considerable light on the performance of the hand puncture testers and on how they should be used. Figure 2 shows the three basic types of curves that are obtained with horticultural products. In each case there is an initial rapid rise in force over a short distance as the punch moves onto the sample. During this stage the sample is deforming under the applied force; there is no puncturing of the tissues. This stage ends abruptly when the punch begins to penetrate into the food, which event is represented by the sudden change in slope called the "yield point" or sometimes "bioyield point." The yield point occurs when the punch begins to penetrate into the food, causing irreversible crushing. The third phase of the puncture test, namely, the direction of the force change after the yield point and during penetration of the punch into the food, separates the puncture curves into three types: type A, the force continues to increase after the yield point; type B, the force is approximately constant after the yield point; type C, the force decreases after the yield point.

With type A products (typified by freshly harvested apples) the hand tester must be pushed with increasing force after the yield point to make the punch penetrate to the required depth. Each increment in force causes an increment in penetration and no further penetration occurs until the force is increased again. It is easy to control the depth of penetration of the punch tip into a type A commodity.

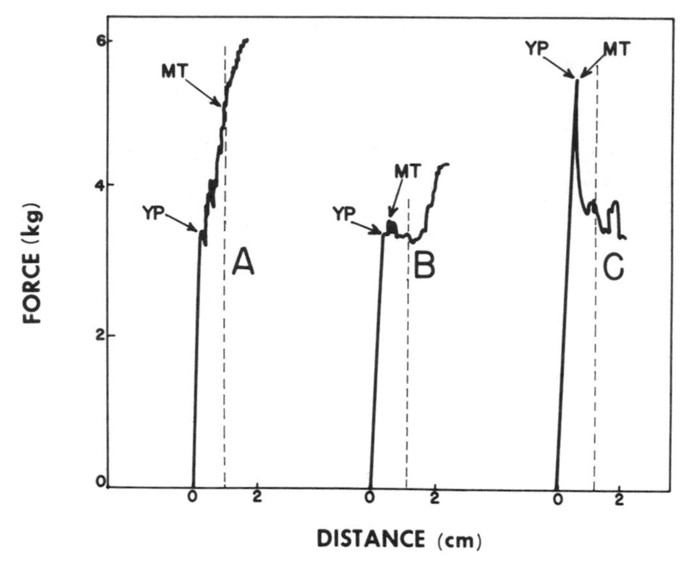

FIG. 2. Characteristic force–distance curves obtained by mounting the 5⁄16-in.-diam punch of the Ballauf tester in the Instron. *YP* is the yield point and *MT* the force reading that would have been obtained on a hand-operated puncture tester. The dotted vertical lines represent the depth of penetration to the 5⁄16-in. inscribed line on the pressure tip. (From Bourne, 1965a; reprinted from *Food Technol.* **19,** 414, 1965. Copyright by Institute of Food Technologists.)

With type B products (typified by ripe pears and peaches, and apples that have been held in cold storage a long time) the hand tester must be pushed until the yield point is reached, the punch then continues to penetrate the tissue with no further increase in force because penetration occurs at approximately constant force. It is fairly easy to control the depth of penetration into a type B product but not as easily as with a type A product. When the punch penetrates beyond the inscribed line, the force reading can still be used because the puncture force is almost independent of the depth of penetration.

For type C products (most raw vegetables exhibit this type of behavior), the hand tester must be pushed until the yield point is reached, when the punch plunges into the tissue very rapidly until it is stopped by the splash collar. Figure 3 shows typical force–distance plots for vegetables. It is very difficult to control the depth of penetration into type C products. When the yield point has been reached, the spring continues to push the punch with yield point force, although the resistance to penetration has become much lower. Consequently the punch accelerates so quickly that even an experienced operator cannot prevent the tip from penetrating into the food past the inscribed line.

Because it is impossible to stop the penetration at the inscribed line on the punch in a type C product, many operators consider the puncture test to be an

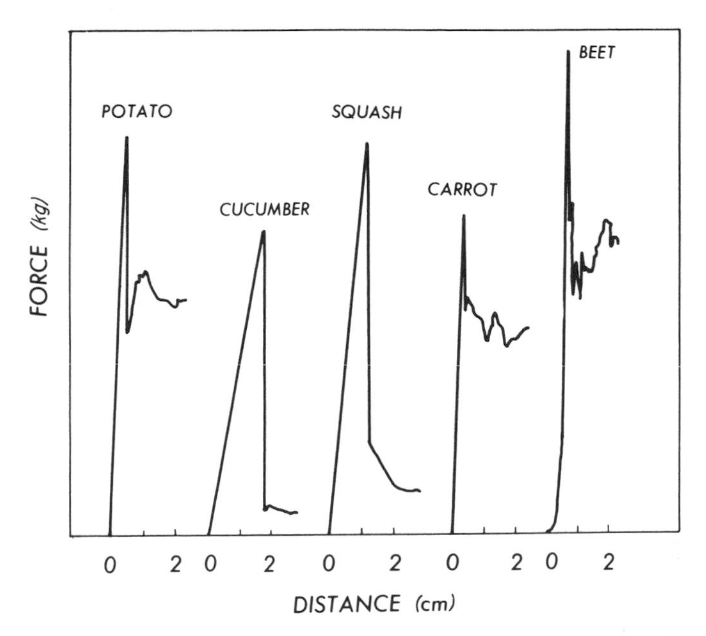

FIG. 3. Characteristic force–distance curves from puncture tests on raw vegetables. Note that each one is a type C curve. (From Bourne, 1975c; with permission from D. Reidel Publ. Co.)

unsatisfactory test for vegetables, but this is an erroneous opinion. The yield point for a type C product is the maximum force that is encountered during the test. Since the hand tester uses a maximum force reading dial, it will read the yield point (maximum force) correctly, even though the penetration goes beyond the inscribed line, provided that the test is performed correctly. Therefore, the hand puncture test can be a useful test on raw vegetables and other products that behave in this manner. Table 3 shows data in which the puncture test was performed using the same punches in the Effi-Gi puncture tester and in the Instron; the hand tester gives the correct measurement of the yield point of vegetables (within the limits of experimental error), although the punch penetrated up to the splash collar in every test.

There has been considerable discussion as to whether the skin should or should not be removed at the puncture test site. Figure 4, which is a schematic representation derived from thousands of tests with the Instron, clarifies this point. The left-hand column of graphs shows the three different shapes of force–distance curves that are obtained on horticultural products with the skin removed (as discussed above; see Fig. 2).

Whenever skin is present it must be ruptured by the punch before any substantial penetration of the punch into the food can occur; the force to rupture the skin is therefore included in the penetration force and this appears as a peak in the

TABLE 3
COMPARISON OF PUNCTURE TEST BY HAND TESTER AND INSTRON[a]

Commodity	Hand tester[b] mean force (kg)	Instron[c] mean force (kg)
Irish potatoes	10.76	10.86
Summer squash	9.78	9.50
Beets	12.23	12.59

[a]From Bourne (1975c); with permission for D. Reidel Publ. Co. 5⁄16-in.-diam punch; mean of 25 punches.
[b]Effi-Gi tester, 5⁄16-in.-diam tip, operated by hand, read dial force.
[c]5⁄16-in.-diam tip mounted in Instron, read off yield point.

yield point force which is superimposed upon a regular type A or type B curve and an increase in the height of the peak of a type C force–distance curve. The height of the superimposed peak depends upon the toughness of the skin. In type B and type C products, where the flesh yield point force is the measured quantity, the increase in reading caused by the skin is always reflected in a higher force reading. In some cases (e.g., strawberries) the skin is so soft that the increase is negligible. In the case of a type A commodity with soft skin the force to rupture the skin is less than the force required to penetrate 5⁄16 in. If the skin is of medium toughness, it generally ruptures at about the same force required as at the 5⁄16-in. penetration and only a small increase in force reading results. Finally, if the skin is quite tough, the force to rupture the skin is well beyond the normal Magness–Taylor force and the force reading will be noticeably increased. Figure 4 explains why the presence of skin sometimes causes an increase in puncture force and at other times it does not.

Since the strength of the skin is not necessarily related to the firmness of the underlying flesh, it is evident that the skin should be removed if a true measurement of flesh firmness is required, unless it has been established that the skin is so tender that it causes a negligible increase or that the product exhibits type A characteristics and does not have a tough skin.

Equation (5) in Chapter 3 (p. 60) explains the relationship between punch diameter and puncture force and shows that both area and perimeter of the punch are important. This equation demonstrates that the puncture force depends on two different properties of the test material and on both the area and perimeter of the punch. It explains why it is difficult to convert data obtained with one punch diameter to data obtained with another punch diameter. For this reason it is mandatory to standardize the punch diameter in any one set of experiments. It is quite acceptable to change the strength of the spring as one moves into a higher or lower force range, but punch diameter should not be changed.

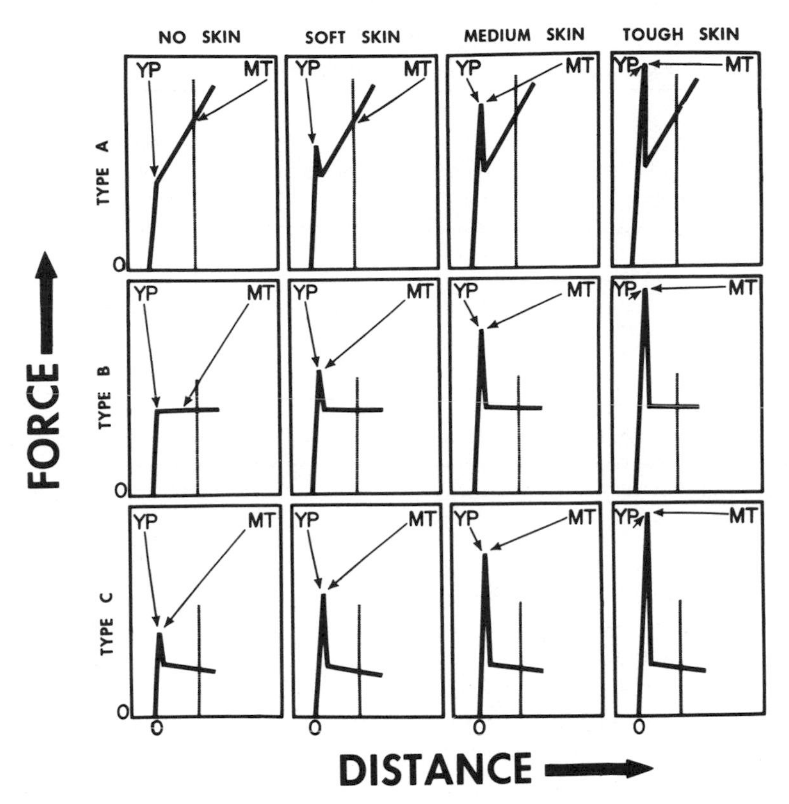

FIG. 4. Schematic representation of force–distance curves obtained when puncturing horticultural products with and without skin. The left-hand column shows the three types of curves obtained with skin removed at the test site. The next three columns show the effect of soft, medium, and tough skin on each type of curve. *YP* is the yield point and *MT* the force reading that would have been obtained on a hand-operated puncture tester. Note that the only effect of the skin is to increase the yield point. The vertical lines indicate 5/16-in. penetration point. (From Bourne, 1965b; with permission from New York State Agricultural Experiment Station.)

In general, the 7/16-in.-diam punches are used on most fruits because the force required will be less than 30 lb. The 5/16-in.-diam punch is used on very hard fruits and raw vegetables when the force would exceed 30 lb with the larger punch diameter. The small diameter punches of the Chatillon 516 series are frequently used on commodities such as sweet corn, green peas, and strawberries. Haller (1941) gives a good discussion of fruit puncture testers and their practical applications and typical results for puncture tests on apples, pears, plums, and peaches. Table 4 gives typical puncture force figures for some apple varieties, and Table 5 gives typical figures for several fruits.

The VanDoorn tester is a hand-held instrument that is used for measuring the

TABLE 4

RANGE OF FIRMNESS READINGS OF SOME APPLE VARIETIES BY PUNCTURE TEST[a]

Variety	Hard	Firm	Firm ripe	Ripe	Prime eating	Overripe upper limit
Ben Davis	24–17.5	18–14.5	15–12	13.5–8	13–9	9
Delicious	20–16.5	17.5–14	15–11	12–8	12–8	8
Grimes golden	27–18	18.5–15	16–12.5	13.5–9	12–8	8
Jonathan	21–16	16.5–13.5	14–10.5	12–8	12–8	8
Rome beauty	23–18	19–13	16–12.5	13.5–9	13–9	9
Stayman winesap	21–16	16.5–13	14–11	12–7	12–8	8
Wealthy	20–16	17–13	14–10	11–6	—	—
Yellow transparent	22–16	17–13	14–10	11–6	—	—
York imperial	24–18	19–16	17–14	15–9	13–10	10

[a]lb force with 7/16-in.-diam Magness–Taylor tip. Data from Haller (1941).

firmness of butter. A circular flat-faced punch with a cross-sectional area of 4 cm^2 (2.26 cm diam) and 1 cm thick is pressed into the butter. The sides of the punch are concave to minimize friction of the butter along the sides. The punch is pressed 1 cm into the butter in 30 sec and a spring scale on the instrument records the maximum force encountered on a 16-kg scale. This device is relatively insensitive to variations in the rate of penetration, gives good correlation with judgments of professional butter graders, and gives reproducible figures that differentiate between different butters. Since the firmness of butter is highly dependent upon temperature, this test must be performed at a constant temperature (Kruisheer and den Herder, 1938; Kruisheer, 1939; Mulder, 1953; Prentice, 1972).

Mechanical and Motorized Puncture Testers

Bloom Gelometer. The Bloom Gelometer (Bloom, 1925) is a puncture test designed to measure the strength of gelatins and gelatin jellies. It consists of a hopper full of lead shot that flows through a tube onto a pan, thus providing the force necessary to make a plunger penetrate into a standard jelly. Borker *et al.* (1966) reviewed the early history of gelatin gel testing and discussed the necessity for frequent maintenance of alignment and adjustment of the Bloom gelometer.

The 1980 edition of *Official Methods of Analysis* (published by the Association of Official Analytical Chemists) gives the details for the preparation of a standard gel. For gelatin it is method 23.007, and for gelatin dessert powders, method 23.013. The standard method for determining the jelly strength of glue is described by DeBeaukelaer *et al.* (1930). The standard jar containing the stan-

TABLE 5
FIRMNESS OF VARIOUS FRUITS BY PUNCTURE TEST[a]

Fruit	Variety	Color Stage	Firmness (lb)
Apricots	Royal	Yellowish green	14.5
		Greenish yellow	10.0
		Greenish yellow to yellow	7.1
		Yellow to orange	4.1
Plums	Beauty	Green to straw tip	13.2
		Straw to slight pink tip	9.0
		Straw to red tip	6.1
		½ to ¾ red	4.9
	Climax	Green to faint straw tip	25.1
		Straw to greenish yellow	20.7
		Greenish yellow to red tip	15.5
		¼ to ¾ red	8.9
Peaches	Elberta	Yellowish green, slight blush	17.6
		Cream to light yellow, slight blush	12.4
		Full yellow, ⅓ to ½ red	3.7
	Phillips cling	Greenish yellow to yellow	12.0
		Yellow, ¼ to ½ red	8.8
		Golden yellow, ¼ to ¾ red	8.4
Pears	Bartlett	Original green	29.2
		Original green to light green	26.9
		Light green to yellowish green	21.0
		Yellowish green	15.2
	Beurre hardy	Original green	12.3
		Light green	10.8
		Light green to yellowish green	8.6

[a]Data from Allen (1932); with permission from Division of Agricultural Sciences, University of California.

[b]Magness–Taylor tester: 7/16-in.-diam tip for apricots and plums; 5/16-in.-diam tip for peaches and pears.

dard jelly is placed in the Bloom Gelometer and adjusted until the flat face of the probe is just resting on the surface. A ½-in.-diameter punch is used for gelatin and a 1-in.-diameter punch is used for gelatin desserts. A lever is tripped allowing lead shot to flow from the hopper into a lightweight aluminum dish on the scale supported by the punch pan at the rate of 200 ± 5 g per 5 sec. When the plunger has penetrated 4 mm into the jelly (which usually occurs suddenly), an electrical contact shuts off the flow of shot. The shot is weighed and the weight

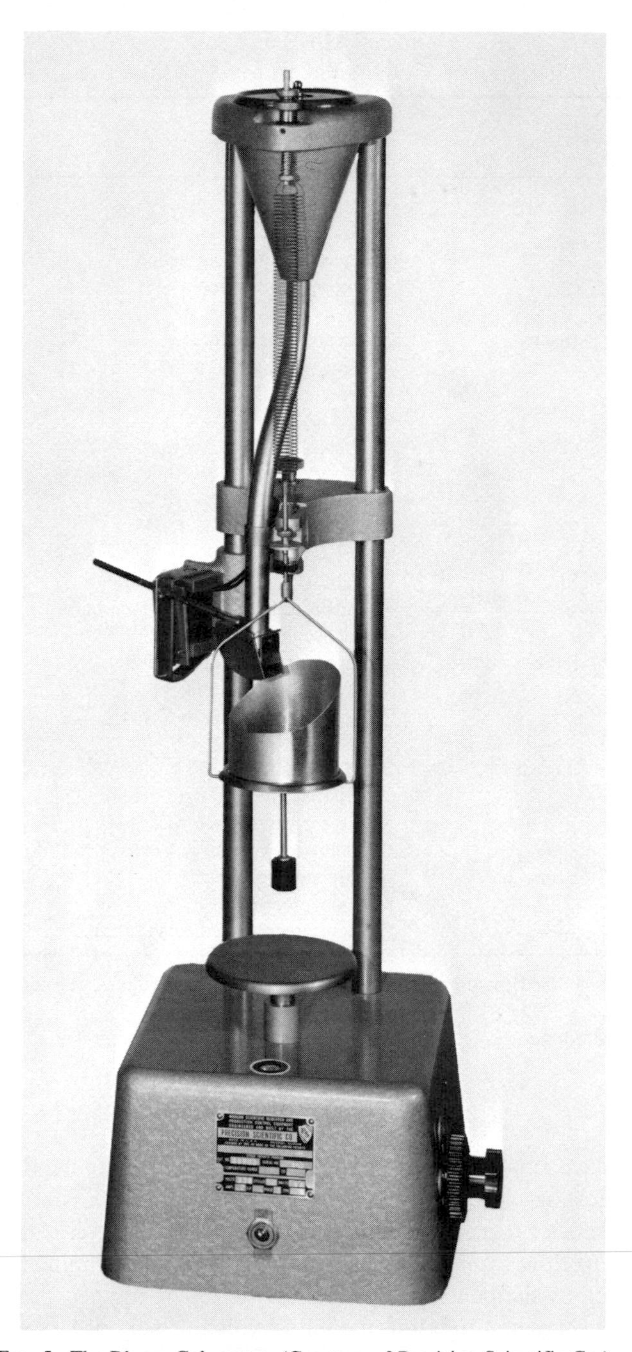

FIG. 5. The Bloom Gelometer. (Courtesy of Precision Scientific Co.)

of shot in grams is expressed as the Bloom of that gel. The Bloom Gelometer is 18 × 19 × 63 cm high and weighs 13 kg (see Fig. 5).

DeBeaukelaer *et al.* (1945) showed that the flow rate of 200 g per 5 sec causes errors in soft jellies because the lead shot runs out too fast, and suggested that for soft jellies the flow rate should be reduced to 40–50 g per 5 sec. Borker and Sloman (1969) also found that slowing the flow rate of shot to 45 g per 5 sec gave more precise results and recommended that this flow rate be incorporated as an official standard. As noted above, the 1980 official standard continues to use the 200 g per 5 sec rate of shot flow.

The Stevens LFRA Texture Analyzer

This instrument, developed by the Leatherhead Food Research Association (LFRA) in England, was designed to perform the standard Bloom test plus a number of other tests. The instrument stands about 50 cm high, 24 cm wide, and 23 cm deep, and weighs about 12 kg (see Fig. 6). It replaces the Boucher Electronic Jelly Tester, which is no longer manufactured.

The standard probe is a ½-in.-diam flat-faced straight-sided acrylic punch that

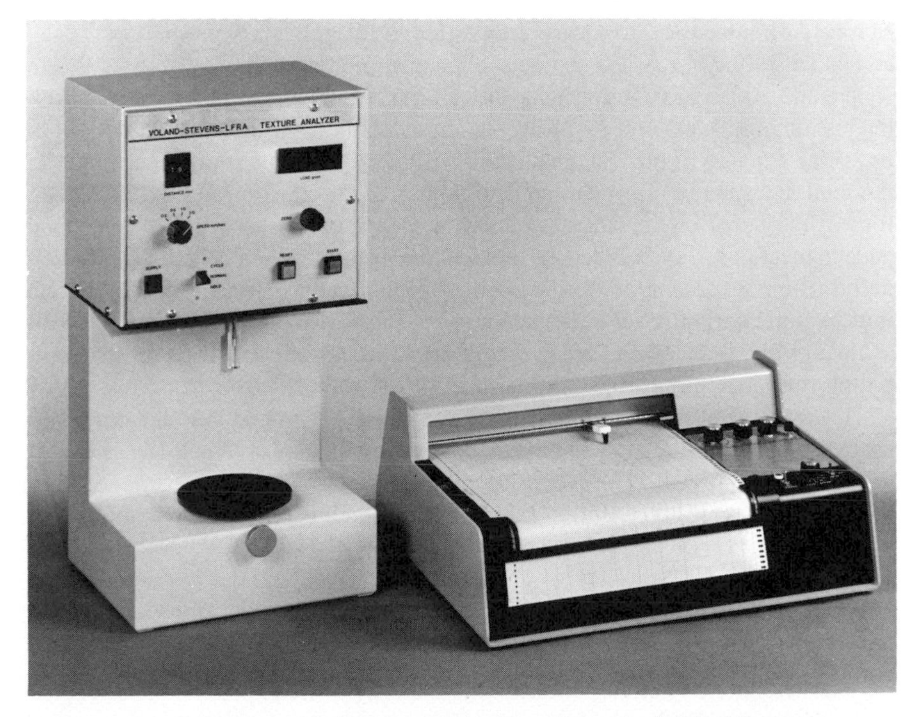

FIG. 6. The Stevens LFRA Texture Analyzer. (Courtesy of Voland Corp.)

has the same dimensions as the punch used for the Bloom test. Punches of other diameters and punches in the form of a needle, ball, or blade are also available. Four speeds of punch travel are available: 12, 30, 60, and 120 mm min^{-1}. The maximum stroke of the punch is 15 cm. The penetration distance is adjustable from 1 to 29 mm in 1-mm steps. In operation, the test sample is placed beneath the punch and the motor activated. The punch moves downward at the maximum speed until a force of 5 g is registered, when it automatically steps down to the selected set speed and travels at this speed for the selected distance. At the end of the stroke it returns to its original position at maximum speed. An electronic load cell in the base of the instrument senses the force and registers it on a digital readout, which shows the maximum force obtained in the test. The instrument has a capacity of 1000 g force and reads within 1 g. It can be adapted to a 100-g force capacity and a reading within 0.1 g for very soft products.

A recorder is an optional accessory giving force–distance plots of the puncture tests. This instrument is a useful general-purpose puncture tester for soft products. It is used on meat pastes, foams, various gels, and some fats.

The Marine Colloids Gel Tester

This instrument was evolved from the Cherry-Burrell Curd Firmness Tester by Marine Colloids, Inc., to measure the strength of gels made from carrageenan and other refined seaweed extracts. The instrument is approximately 35 cm square and 70 cm high and weighs about 12 kg. Two punches are provided with the instrument, 0.845 and 0.431 in. diam. A Chatillon spring scale of 500, 1000, or 2000 g capacity is provided with the instrument. For routine work the punch is driven downward by a synchronous motor at 18 cm min^{-1}, and the maximum force is measured on the Chatillon scale. A second model uses an electronic load cell connected to a recorder to replace the Chatillon scale and give force–deformation records. For this work the speed is adjustable from 4 to 18 cm min^{-1} but is normally set to run at 5 cm min^{-1}. A slip clutch prevents overload of the driving mechanism. When the plunger reaches the end of its set stroke, it returns automatically to the starting position for the next test.

This instrument is used to measure the firmness of various gels and other soft foods by puncture. A curd knife can be provided as an optional accessory for testing the firmness of cheese curd. With this accessory the instrument becomes identical with the Cherry Burrell Curd Tester, which is no longer manufactured.

Maturometer

This instrument, designed to measure the maturity of fresh green peas, was developed in Australia and is extensively used in that country. Its intended purposes were to objectively measure the maturity of fresh peas and to select the optimal harvest time during the growth of the crop that gives the yield and quality required to meet the production objectives of the processor. Its manufacture is covered by Australian Patent No. 143,316. It consists of 143 ⅛-in.-diam flat-face punches set in an array of 11 rows by 13 rows with individual punches spaced on 7⁄16-in. centers. A metal plate containing 143 matching countersunk holes is positioned underneath the punches.

A pea is lodged in each recess. When the plate of peas is driven upward by a motor, the peas are punctured simultaneously by the pins and the maximum force is measured on a force scale at the top of the instrument (Lynch and Mitchell, 1950, 1952; Mitchell *et al.*, 1961). A matching perforated plate mounted over the metal plate that holds the peas prevents the peas from sticking to the punches during the return stroke. The instrument has a force capacity of 440 lb in 5-lb graduations.

Based on extensive field testing and sensory evaluation, it was found for Australian conditions that peas harvested at a maturometer reading of 250 lb gave the maximum yield of highest quality peas for canning. A somewhat lower figure is needed for the maximum yield of best-quality peas for freezing. In using the Maturometer Index (MI) as a basis of payment for quality the following ranges are recommended for field run ungraded peas:

Grade 1 (canning) consists of peas in the range of 230–270 MI
Grade 2 (canning) consists of peas in the range of 190–230 and 270–320 MI.
Grade 1 (freezing) consists of all peas up to 200 MI.

The MI of peas in the field increases by an average of 20 lb per day. By testing field samples daily it is possible to predict when the figure of 250 lb will be reached. This enables a pea processor to know several days in advance when to harvest a field and the number of fields that will be harvested on a given day.

Casimir *et al.* (1971), using a single-punch Maturometer, found a simple correlation coefficient r ranging from 0.96 to 0.99 between puncture force and alcohol insoluble solids of individual peas. Casimir *et al.* (1967) showed that high speed of operation of the pea viner caused some bruising and tenderization of the peas resulting in lower Maturometer readings.

Christel Texture Meter

This instrument (Christel, 1938) consists of a set of 25 flat-faced 3⁄16-in.-diam punches that are held in a metal plate above a metal cup 2 in. internal diam and 1 ¾ in. deep. A removable metal cover containing a set of holes that match the array of punches above it rests on top of the cup. The food is placed in the cup, the set of punches is driven down by a hand-operated gear and rack assembly, and the force is registered on an hydraulic pressure gauge with a force capacity of 100 or 300 lb. This device was first developed for measuring maturity of green peas but was displaced from this use with the advent of the FMC Tenderometer.

It is a low-cost device that can be taken out into the field, and has proven to be useful for routine quality control measurements of firmness on a number of commodities. Unfortunately, the Seifert Manufacturing Co. of Wisconsin, which used to supply the instrument, is no longer in existence, but a number of these instruments are still being used in various laboratories.

Armour Tenderometer

This instrument consists of an array of 10 ⅛-in.-diam stainless-steel probes, 3 in. long, with the last inch tapered to a point (Hansen, 1972). The instrument and its operation are covered by the basic United States Patent No. 3,593,572 and by several later patents. Morrow and Mohsenin (1976) analyzed the mechanics of a multiple conical probe system of this type. In operation the array of 10 needles is manually pressed 2 in. into the fifteenth rib eye of a beef carcass. The maximum force during penetration is recorded on a portable strain gauge force transducer fitted with a peak force indicator. Measurements of the maximum force on cold rib eyes in the chill room on the day after slaughter were found to correlate well with subjective panel tenderness scores on the same meat cooked after 1 week of aging. Huffman (1974) found the Armour Tenderometer to be superior to USDA quality grade or marbling as a means of placing cattle into homogeneous tenderness groups.

This instrument is not available commercially. Armour and Company use the Tenderometer to select beef for its TesTender® merchandising program in which beef is sold to the consumer on a guaranteed-tender basis without the use of enzymes or other artificial tenderizing methods. The company also licenses the Tenderometer as part of TenderChek® program to other companies who use it to select beef for tenderness.

Voisey (1976) pointed out that the electronic peak force detector of the Armour Tenderometer will record transient peak forces because peak force detectors have a rapid response rate. Hence, operators should be trained to penetrate the sample at a constant speed and to avoid jerky jabbing motions that are likely to give spuriously high results. Voisey (1976) also pointed out that a measurement of the energy used to penetrate the sample (i.e., the integral of force × penetration) would minimize the effects of short-term force impulses and other incorrect operator techniques and would probably be more reliable although it would increase the sophistication of the instrument.

Some researchers have found low correlations between the Armour Tenderometer readings and sensory panels (Carpenter *et al.*, 1972; Dikeman *et al.*, 1972) while others find poor correlation (Parrish *et al.*, 1973; Henrickson *et al.*, 1972; Campion *et al.*, 1975; Harris, 1975). Some of these low correlations may have resulted partly from incorrect operator technique (Voisey, 1976). Nevertheless, this Tenderometer was granted the Industrial Achievement Award by the Institute of Food Technologists in 1973.

Other Puncture Testers

Most of the multimeasuring instruments can be set up to do a wide variety of puncture tests. Because of the expense, their use is generally restricted to research purposes. Nevertheless, it is worth noting that most of the multimeasuring instruments can be used as puncture testers.

Compression–Extrusion Testers

It was pointed out in the previous chapter that the extrusion principle test cells usually involve complex combinations of compression, extrusion, shear, friction, and perhaps other effects. For the sake of brevity, the word "extrusion" will be used to describe this type of test, but the reader should remember that this class of test cell usually involves more than extrusion.

FMC Pea Tenderometer

This instrument was developed by the Food Machinery Corporation as an objective means for measuring the quality and maturity of fresh green peas (Martin, 1937; Martin *et al.*, 1938). A motor-driven grid of 19 stainless steel blades ⅛ in. thick and spaced ⅛ in. apart are rotated through a second reaction grid of 18 similar blades. The peas placed in the cavity between the two grids are cut and extruded through the slits between the blades. This is commonly known as a shearing device, but it is evident that most of the action on peas is extrusion. The reaction grid is mounted in bearings and is free to rotate, but its rotation is resisted by a weighted pendulum hanging from the second grid which swings out of the vertical as the reaction grid rotates. The force exerted during extrusion of the peas is reflected in the angular movement of the pendulum and is recorded by a pointer that moves across a sinusoidal scale. The pointer records the maximum force encountered in each test. The machine is rugged, self-contained, easy to clean, and can stand a lot of abuse in a processing plant or at a pea vining station.

Although it is widely used by the pea processing industry as an index of quality and price to be paid for the peas, it has some serious drawbacks, notably the problem of calibration. If the blades become dented or warped, a friction component is introduced. Voisey and Nonnecke (1971) performed a detailed appraisal of the Pea Tenderometer and found serious differences among different Tenderometers being used in industry. The problem of standardizing this instrument has also been discussed by Bourne (1972). Unilever Research in England have devised a standardization procedure they claim maintains agreement to within ±1.5 Tenderometer units between all instruments in their continental European and British factories (Pearson and Raynor, 1975). However, Voisey (1975) still considers the Tenderometer to have serious deficiencies. Despite these problems, this instrument continues to be a widely used method for measuring the quality of peas in the industry.

The Food Machinery Corporation discontinued the manufacture of the Pea Tenderometer about 1978, but it maintains spare parts in stock. A number of small-capacity pea processing plants have ceased operation over the last two decades and consequently some used Tenderometers are available for purchase. The FMC Tenderometer will probably continue to be used for many more years, although it seems that it will eventually become obsolete.

Texture Press

This versatile and well-known instrument was developed at the University of Maryland (Kramer *et al.,* 1951; Decker *et al.,* 1957). Although it is commonly known as the "Kramer Shear Press," the name of the instrument has undergone several changes. The instrument was first manufactured by the Bridge Food Machinery Co. of Philadelphia, and later made by the Lee Corporation of Washington, D.C., and called the Lee Comptroller and later the Lee–Kramer Shear Press. Later, the rights to manufacture the instrument were acquired by Allo Precision Metals Engineering, Inc., of Rockville, Maryland, and was called the Allo–Kramer Shear Press. Presently, it is manufactured by the Food Technology Corporation of Rockland, Maryland, and is known as the Food Technology Corporation Texture Test System, abbreviated to FTC Texture Test System.

At this time we will discuss only the basic machine fitted with the presently available force measurement devices, that is, Digital Texture gauge and Texturecorder. A number of accessories, including a Texture integrator, special modes of operation, and various test cells, can be attached to this instrument converting it into a multiple measuring instrument. This mode of operation is discussed on p. 175.

The basic machine, known as the "Texture Press," is 64 cm wide, 66 cm deep, 86 cm high, and weighs about 86 kg (see Fig. 7). This is a robust machine that is designed for hard reliable work under wet food processing plant conditions. The system is driven hydraulically. An electrically driven oil pump powers the ram to which the moving parts are attached. A switch controls the up and down motion of the ram. The working space for the test cells is 4 × 4½ in.

In the older models the force was measured by a proving ring placed between the test cell and the bottom of the ram. In the newer models the force is measured by means of a force transducer placed between the bottom of the ram and the test cell that is electrically connected to either a direct-reading digital texture gauge or a strip chart texturecorder, both of which are calibrated in pounds/force. The force transducers have long-term stability; once calibrated they hold their performance for extended periods of time unless overloaded or abused. Force trans-

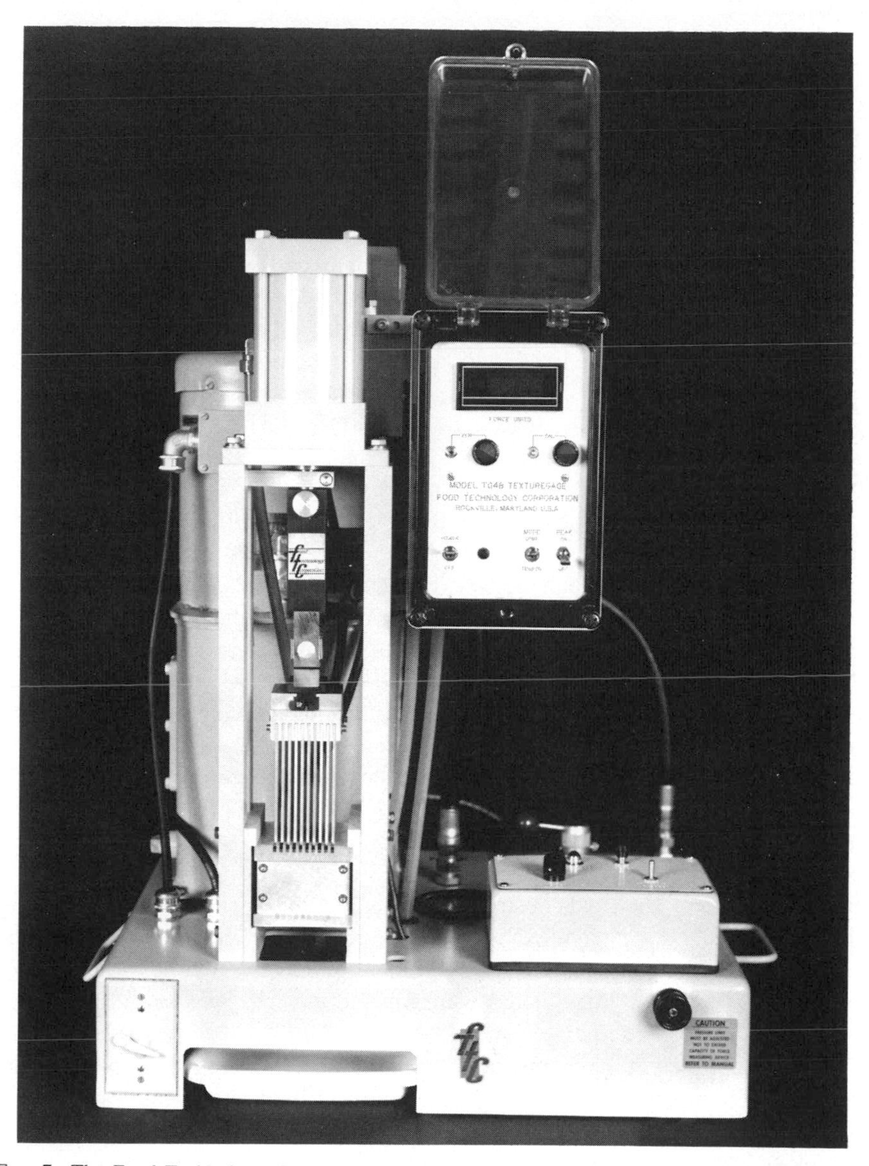

FIG. 7. The Food Technology Corporation Texture Press (''Kramer Shear Press''). (Courtesy of Food Technology Corporation.)

ducers should be returned to the manufacturer periodically for recalibration and inspection.

Six force transducers ranging from 0–30 to 0–3000 lb capacity are available. A special force transducer is available for use with fresh peas, which is calibrated directly in Pea Tenderometer units, and covers the range of 0–500 equivalent Tenderometer units. One Tenderometer unit is equivalent to approximately 6.2 lb/force. The digital texture gauge can be used as a maximum force measuring instrument by utilizing the peak holding switch on the front of the texture gauge. With the switch in the peak position, the digital meter will read the peak maximum force and hold this reading until manually reset with the zero switch.

The standard test cell of the Texture Press consists of a metal box with internal dimensions 2⅝ × 2⅞ × 2½ in. high (6.6 × 7.3 × 6.4 cm). A set of ⅛-in.-wide bars spaced ⅛ in. apart are fixed in the bottom of the box. Guide ridges from the ends of these bars rise vertically up the sides of the box. A set of ten blades, each ⅛ in. thick and 2¾ in. wide, spaced ⅛ in. apart, is attached to the press ram. A metal lid containing a set of bars that match the bars in the bottom fits over the box. In operation, the food is placed in the test cell, the lid is positioned, and the test cell is placed in the machine such that the slits formed by the bars in the lid are aligned with the blades on the ram. When the ram is activated, the set of blades is forced down through the box, first compresssing and then extruding the material. Some of the material extrudes upward between the moving blades, and the remainder extrudes downward through the bars in the bottom of the test cell. The moving blades are propelled down until they pass between the bars in the bottom of the test cell. When the ram is reversed, the moving blades ascend and return to their original position. As they ascend, the bars of the stationary test cell lid scrape off into the cell the food lodged between the moving blades.

The first standard test cell was fabricated in stainless steel, and the moving blades were a rigid welded unit with the bottom faces of the blades flat and parallel. This type of cell was manufactured by the Lee Corporation and Allo Precision Metals. The rigid construction posed a number of problems, including that of friction between the fixed bars and moving blades which can cause serious errors in measurement, particularly with soft products (Bourne, 1972). Also there could be friction if the blades were burred, twisted, bent, or in some other way moved out of strict alignment.

The standard test cell manufactured by the Food Technology Corporation was changed to aluminum alloy, which makes the cell lighter (from almost 6 to 2.4 kg) and easier to handle. In the new design the moving blades are not welded but are pinned together, leaving a small amount of free play of the blades in the attachment connected to the ram. The blades in the new test cell design self-align with the slots in the box. The clearances between the moving blades and the slots in the stationary box have been increased slightly, which reduces the problem of friction between the parts.

Voisey (1977b) found that some friction still occurred with the aluminum test cell and that the amount of friction varied greatly from cell to cell. He considered that these errors may be acceptable for samples that require a high force, but noted that errors could become large for samples that require a low force.

The bottom faces of the moving blades are slanted in alternate directions, which eliminates the sudden peak force that sometimes occurred when the flat and parallel blades first engaged the stationary bars at the bottom of the box.

The new design test cell has reduced some of the problems of the old cell and is preferred for general use. Occasionally the old cell may have an advantage over the new cell. For example, Ross and Porter (1968, 1969, 1971, 1976) used the old design test cell to study the texture of french fries. They were able to obtain good results with the old model cell with squared-off ends on the moving blades, but their results cannot be duplicated with the new type cell with the slanted blades.

The relationship between the weight of material in the cell and the maximum force during the compression stroke was studied by Szczesniak *et al.* (1970), and is shown in Fig. 8. For two products (white bread and sponge cake) a linear relationship is found between sample weight and maximum force over a limited range of sample weight. DeMan and Kamel (1981) also found a linear relationship between maximum force and sample weight for cooked poultry meat. The relationship for the other foods was nonlinear, tending toward constant force–weight relationship at high fill weights. Some products (e.g., raw apples and cooked dry beans) never reach a linear relationship. Many products attain a constant force independent of sample weight before the cell is filled (e.g., canned beets, peas, carrots, lima beans; frozen peas and lima beans; and raw snap beans and bananas). Thus, for most foodstuffs the force per sample weight is not constant but decreases as the sample weight increases. On these grounds it is advisable to use a constant weight of sample in the test cell unless tests show that there is a linear relationship between sample weight and maximum force for that food. Many researchers report Texture Press data as pounds force per gram weight of product. Figure 8 shows that this procedure is likely to introduce errors, and it should be discontinued.

The speed of travel of the hydraulic ram is infinitely variable from 0 to 20 in. min^{-1} by adjusting a flow control valve located in the oil supply pipeline to the ram. Ram speed is usually expressed as seconds to travel its full stroke of 3½ in. This procedure poses the problem of using a reciprocal scale, that is, the higher the number in seconds the slower the speed. The formula for converting seconds to travel full stroke length to inches per minute, assuming constant ram speed, is

$$(3.5/\text{sec}) \times 60 = \text{inches per minute}$$

Figure 9 converts time for full stroke to the inches per minute and millimeters per minute for the most widely used range of speeds.

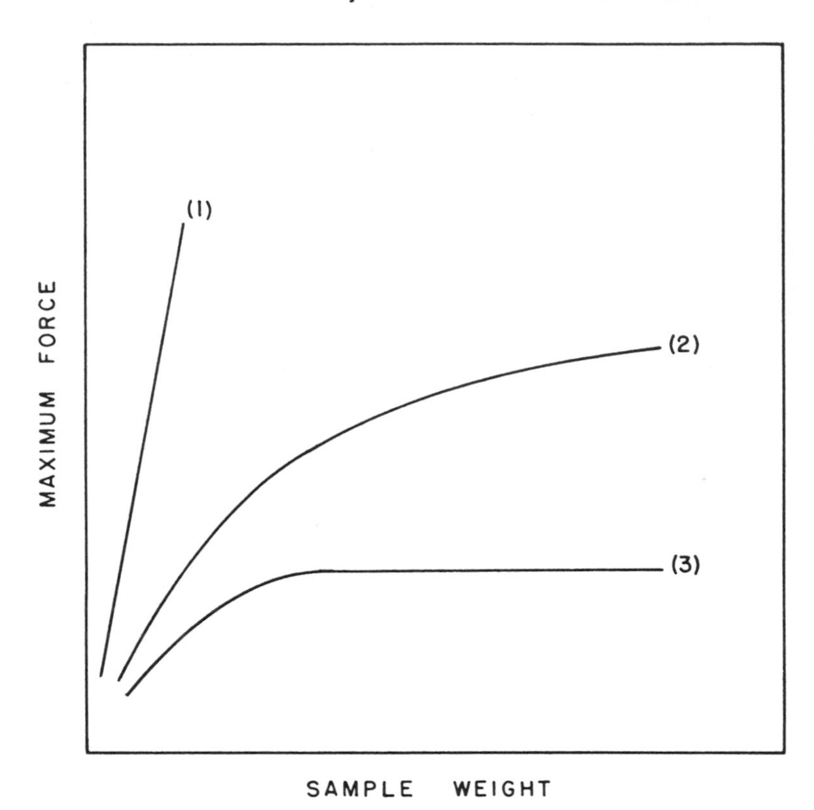

FIG. 8. Typical maximum force versus sample weight relationship for standard Texture Press test cell. Behavior (1) is exemplified by white bread and cake, (2) by raw apples and cooked dry beans, and (3) by canned or frozen vegetables. (Courtesy of Dr. A. S. Szczesniak. Reprinted from *J. Texture Stud.;* with permission from Food and Nutrition Press.)

The viscosity, which affects the rate of flow of the hydraulic oil, depends on the oil temperature. Hence, at a given setting of the control valve the speed of the ram will change with changing oil temperature. Therefore, the ram speed should be checked after the instrument has been running for some time to compensate for the effect of the heating of the oil. This is particularly important for those commodities that are strain-rate sensitive and for very slow ram speeds. Ang *et al.* (1960) used the Texture Press at a very slow rate of 0.46 in. min^{-1} and found that after 2 hr of operation the oil had heated to 165°F, and the speed of travel of the ram had changed. In order to overcome this problem they placed a thermostatically controlled electric immersion heater in the oil bath to preheat the oil to 170°F before testing began. This is the only recorded instance where the temperature of the oil bath needed to be controlled in order to maintain adequate control of the speed of the ram. Voisey (1972) in a study of the Texture Press

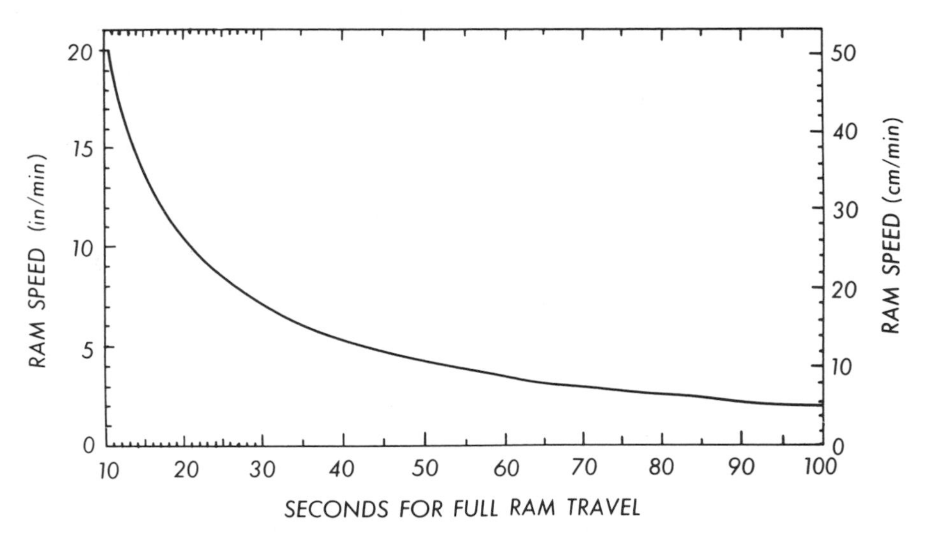

FIG. 9. Relationship between ram speed of Texture Press and time for ram to travel full stroke length of 3½ in.

discusses the problem of speed control and concluded that the early models gave inadequate control of ram speed. Later models incorporate an improved flow control valve in the hydraulic control circuit.

The Food Technology Corporation provide another extrusion cell that can be operated in several modes. It consists of a cast-iron cylinder that mounts in the machine frame. A circular piston is attached to the ram. In one mode the piston is a close fit in the cylinder, and all food placed in the cylinder is pushed out before it. There is some friction between the piston and the walls of the cylinder. A grid of metal bars or a flat plate containing a single orifice is inserted in the base of the cylinder, and the food is extruded through the grid or the orifice plate. In the second mode, a piston with a smaller diameter is used and a solid plate is placed in the bottom of the cylinder; in this configuration it acts as a back extrusion cell because the food is extruded upward between the walls of the cylinder and the sides of the piston, moving in the opposite direction to the motion of the piston. The annulus width in this back extrusion cell is ⅛ in. (3.2 mm), and there is no friction between the piston and the cylinder.

The Ottawa Pea Tenderometer

The Ottawa Pea Tenderometer is a special version of the Ottawa Texture Measuring System (Voisey, 1971b, 1974; Voisey *et al.*, 1972; Voisey and Nonnecke, 1973a,b) that was adapted specifically for measuring the maturity of fresh peas (Voisey and Nonnecke, 1972a,b, 1973a,b,c). The standard test cell is constructed of ½-in.-thick aluminum plate and is square in cross section. The internal cross-sectional area of the cell is 30 cm^2 (55 mm along the edge), and it stands about 13 cm high. A rectangular plunger made of ½-in.-thick aluminum plate is attached to a 1-in.-diam shaft. The plunger has a clearance of 0.275 mm from the wall on each side to eliminate friction. The peas are extruded through a replaceable wire grid that slips into the bottom of this cell. The grid consists of nine wires 2.36 mm diam with a gap of 3.3 mm between the wires.

The plunger is driven down into the test cell at 18.2 cm min^{-1} by a synchronous motor connected through a gear box to a single vertical screw that moves the crosshead. An electronic load cell placed between the crosshead and the plunger is connected to a signal-conditioning-amplifying unit, which is connected by a module to a peak force detection unit whose output is displayed on a digital panel meter in a waterproof case at the side of the press. Comparative studies have shown that the Ottawa Pea Tenderometer readings correlate highly with the Food Technology Corporation Texture Press and the FMC Pea Tenderometer. It offers the advantages of easy calibration, high precision, no friction between moving parts, and test cells that are separate from the operating mechanism.

Vettori Manghi Tenderometro

This Italian-built instrument is similar to the FTC Texture Press in that it uses an array of metal blades that move down and through slots formed by a set of stationary bars (Andreotti and Agosti, 1965). The instrument is constructed of stainless steel and is driven by a hand-powered crank handle. The capacity of the test cell is 166 ml, which is approximately one third the 450-ml capacity of the FMC Pea Tenderometer. An hydraulic gauge measures maximum force in Tenderometer units on a 0–250 scale. It is used mostly for measuring the maturity of fresh peas. It can be easily disassembled for cleaning.

Pabst Texture Tester

The extrusion testers described above are designed to take a sample weight of the order of 100–200 g. With foods that come in small-unit sizes (e.g., peas), a large number of units are tested simultaneously. While this is useful for obtaining a measure of the average quality of the food, it gives no information on the distribution of textures within the sample.

The Pabst Texture Tester was designed to individually test bite-size pieces of foods such as peas (Szczesniak *et al.*, 1974). The standard test cell consists of eight moving blades that mesh with eight stationary bars. The blades are 1/16 in. thick, 1 in. high, and the entire blade assembly is 1 in. wide. The clearance between the blades and the bars is 0.002 in. The maximum-size sample that can be placed in this cell is 1 $in.^3$. The unit has the dimensions 35 × 82 × 37 cm high and weighs about 35 kg. It requires a water supply of not less than 50 psi at a flow rate of 1 gal min^{-1} and a connection from a 1-in. flexible drain line into an open drain to carry away the wash water.

Unlike the other extrusion testers that work in a vertical plane the Pabst tester operates in a plane inclined about 20° from horizontal. The moving blades are driven by an hydraulic servo that is powered by water pressure. The stationary set of bars is mounted in an antifriction suspension system to which the force gauge is attached. A null-seeking force balance system driven by a servo motor provides the balance force to maintain the stationary bar assembly in its positon. The output goes to a recorder pen that operates on a 2-in.-wide strip chart recorder that uses standard thermowriting paper.

In operation, a bite-size piece of food is dropped through a 1 × 1-in. cavity into the test cell. Then the moving set of blades is activated and extrudes and shears the sample through the stationary bars. When the extrusion is complete, the blades withdraw and the test chamber is automatically flushed with water to clean it for the next test. As a result of the self-cleaning cycle, many food products can be processed through the test cell on automatic cycle, resulting in a minimum of the operator's time. The rinse water used to scavenge the food debris from the test cell may contain a detergent that is dispensed from an optional metering accessory to facilitate cleaning.

The instrument can accommodate a full-scale force range from about 200 g to 300 kg. The chart paper can be driven at speeds of 17.5, 35, or 70 mm min^{-1}. In normal use only the peak force is read from the chart, but additional information can be obtained manually from the chart if needed. The relationship between sample weight and peak force is complex and not well understood. Hence the texture tester is limited as a research tool where accuracy and well-defined test conditions are desired, but it is useful in practical situations where reasonable accuracy coupled with automation and speed are required.

Because of the rapidity of measurements and the fact that the tests are made on small individual food pieces, the Pabst Pressure Tester lends itself to easy generation of data needed to construct force distribution curves for the sample. These distribution patterns characterize the test material in a manner not possible when average force values are used.

An optional accessory is an automatic feeding system, which consists of a vibratory feeder that organizes the food units into a single-file system up a spiral ramp that empties into the test cell. A light–photocell combination monitors the passage of a food unit into the test chamber, stops the feeder, and initiates the test cycle. On completion of the unit test, the feeder is restarted to supply the next unit of food. This automatic feed system works well with products that are nearly spherical (e.g., peas, dry beans, peanuts).

Optional test cells of different geometry are available to meet different test requirements. One of these is a heavy-duty shear cell that uses five 3⁄32-in.-thick blades and is recommended for use with hard foods. Another one is a puncture assembly consisting of a moving punch and a stationary anvil. The punch may have a flat or rounded end. A single-blade cell is also available.

An optional readout unit consists of a set of 10 electromechanical counters that divide the force amplitude scale into 10 equal segments. Each register enters one count for each peak force value that occurred in testing a series of samples within its increment of scale. On completion of a test series the total in each register is noted, which gives a profile of the variability within the sample.

Instron

The Instron Corporation provide a back extrusion cell as an optional accessory to their instrument. This cell to 4 in. internal diam and 4 in. high, and the plunger is designed to give a 4 mm-wide annular gap. The Instron Corporation provide as an optional accessory a jig that will hold all the standard working parts of the Food Technology Corporation test cells, including the standard Texture Press cell and their line of extrusion cells. The company also provides adaptors that will accept the Ottawa Testure Measuring System test cells.

Shear Testing

Warner–Bratzler Shear

The test cell of this apparatus consists of a stainless-steel blade 0.040 in. thick in which a hole, consisting of an equilateral triangle circumscribed around a 1-in.-diam circle, is cut and the edges rounded off to a radius of 0.02 in. (Warner, 1928; Bratzler, 1932, 1949). In some publications the Warner–Bratzler Shear is misrepresented as having a rectangular-shaped hole in the blade. This confusion probably results from the fact that the first experimental model of the Warner–Bratzler Shear used a blade with a square hole. Also, some researchers are presently experimenting with square blades.

Two sharp-edged borers that resemble cork borers are provided with the instrument and are used to cut a ½- or 1-in.-diam sample of meat. This sample is placed through the hole and two metal anvils, one on each side of the blade, move down, forcing the meat into the V of the triangle until it is cut through. A 50-lb capacity spring force gauge with a maximum pointer measures the maximum force encountered during this cutting action. The principle of this test has been described on p. 71. Although commercially available, the Warner– Bratzler Shear has not been patented (see Fig. 10).

This instrument measures approximately 23 × 30 × 56 cm high and weighs 14 kg. The anvil moves downward at 23 cm min^{-1}. However, the actual shearing

FIG. 10. The Warner–Bratzler Shear: the cylinder of wood inserted in the triangular blade represents a piece of meat.

rate is less than 23 cm min^{-1} because the spring in the gauge is highly extensible and the blade and the meat move downward to some extent as the force increases.

A number of studies have been performed to compare Warner–Bratzler Shear figures with subjective estimates of tenderness of meat. Szczesniak and Torgeson (1965) thoroughly reviewed the subject of meat tenderness and its measurement. In summarizing 38 studies on beef, four on pork, and nine on poultry these authors list correlation coefficients (*r*) between the Warner–Bratzler shear and some method of sensory testing ranging from −0.001 to −0.942. Of the 51 papers listed in this review, 41 reported good agreement or better, and the remainder indicated that correlation was borderline to poor. Szczesniak and Torgeson (1965) commented on the high degree of variability in the correlation between Warner–Bratzler Shear and sensory testing and point out that many factors come into play, one of which is the reliability of the taste panel that is used. At the present time there is no other device that consistently gives better correlations, although the FTC Texture Press is about as good (Szczesniak and Torgeson, 1965).

Although its reliability has often been questioned, the Warner–Bratzler Shear is the most widely used device in the United States for measuring toughness of meat. One serious difficulty with this test is the great variability of meat. Meat toughness varies markedly from animal to animal, from muscle to muscle within an animal, and also from point to point in the same muscle. Meat to be tested should be sheared across the muscle fibers; hence, samples should be cut parallel to the fibers (Hostetler and Ritchey, 1964). The boring tool should be sharpened regularly.

When cutting the sample core, it is necessary to use a technique that will give a standard diameter sample because the shear force is affected by the diameter of the test sample (see p. 73). A steady, moderate pressure should be maintained on

TABLE 6

EFFECT OF SAMPLE CUTTING METHOD ON WARNER–BRATZLER SHEAR TEST[a]

Cutting tool diam (cm)	Hand-bored samples		Machine-bored samples	
	Mean diam (cm)	Shear force (lb)	Mean diam (cm)	Shear force (lb)
2.54	2.41	18.4	2.48	19.6
1.90	1.79	11.6	1.88	12.1
1.27	1.21	7.5	1.25	8.4

[a]Data from Kastner and Henrickson (1969). Sample material was porcine longissimus dorsi muscle heated to an internal temperature of 72°C in deep fat at 140°C and chilled for 24 h in a 4°C cooler before cutting.

the cutting tool as it is twisted in order to obtain a uniform diameter along the length of the sample. High pressure will give an hourglass-shaped meat core that is thinner in the center than at the ends. Uneven pressure will give a core with uneven diameter along its length. Kastner and Henrickson (1969) recommend mounting the borer in a drill press because they found that samples cut with the aid of a drill press were more uniform in diameter, closer to the diameter of the borer, and slightly larger than the handcut samples (Table 6). These authors also found that more uniform cores were obtained when cooked pork was held at 4°C for 24 h before cutting the samples.

The degree of cooking has a great effect on the toughness of meat; hence, it is necessary to have all meat cooked to the same degree of doneness in any one study. A higher final internal temperature (or degree of doneness) results in a higher shear reading. The range of shear readings usually varies from about 5 to 25 lb, depending upon the size of the sample, doneness of the meat, and toughness of the meat.

Torsion Devices

Most of the torsion measuring instruments are used to measure viscosity of fluids, which will be discussed in Chapter 5 where the subject of viscosity is covered. Three instruments that use the principle of torque and are used on semisolid foods are described below.

Farinograph

This is a basic testing instrument that is used in flour mills, bakeries, and cereal research laboratories to determine the baking quality and moisture-absorbing capacity of flour and the handling properties of bread dough (Munz and Brabender, 1940; Locken *et al.*, 1960; Brabender, 1965). The instrument works by mixing wheat flour, water, and sometimes other ingredients in a small mixing bowl that has two Z-shaped paddles that rotate on a horizontal axis. The torque required to mix the resulting dough and how this changes during mixing provide a quantitative measure of rheological properties of the dough that correlate well with the way it handles in the bakery. The method is highly empirical and requires strict control of the conditions.

The basic instrument occupies approximately 120 × 120 cm of bench space and is about 90 cm high. Three models are available: Model FA2 is powered by a two-speed 0.5-hp electric dynamometer motor that drives the paddles at either 63 or 31.5 rpm. Model FAH is also driven by a two-speed 0.5-hp dynamometer motor that drives the paddles at either 63 or 126 rpm. Model DO-V153 (Do-Corder) is powered by a 0.8-hp dc dynamometer motor that drives the paddles at any speed between 20 and 210 rpm by means of an infinitely variable speed control. The mixing bowl is made of stainless steel with a jacket through which water is circulated from a temperature bath to maintain constant temperature. Two sizes of mixing bowls are available: 50 and 300 g capacity. The capacity refers to the amount of flour that is used. The actual capacity of the bowl is about 50% more than the weight of the flour. A pair of sigma-shaped blades are standard for mixing flour. A pair of delta-shaped blades can be supplied and are used to study ingredients such as shortenings that cause a change in the consistency.

The blades of the Farinograph mixer/measuring head are driven by a motor that is suspended to swing freely between precision bearings to form a dynamometer. As the mixer blades encounter a resistance torque from the test material, the dynamometer reacts by swinging in the opposite direction of the shaft rotation. The reaction torque acts through the lever system of an analytical scale and is simultaneously recorded on a strip chart recorder. The baking and milling industry commonly express their results in Brabender units. One Brabender unit is one meter-gram torque.

The Brabender instruments can be calibrated with weights if needed. Most users of the Brabender instruments have a company serviceman routinely check each machine yearly. Check (standard) flour samples are routinely distributed by the American Association of Cereal Chemists, 3340 Pilot Knob Road, St. Paul, Minnesota 55121, to compare instruments.

Resistograph

This instrument was introduced in 1972. It works on the same principle as the Farinograph and is incorporated in one machine, the Farino/Resistograph. A 500-mg sensitivity range is used instead of the 1000-mg sensitivity range, and airfoil-shaped blades replace the sigma blades of the Farinograph. The Resistograph uses the constant dough weight method; 160 g of dough is charged into the R-100 mixer.

Mixograph

This instrument is a recording dough mixer that performs substantially the same functions as the Farinograph, using a small sample of flour (30 g). A smaller assembly that accepts a 10-g sample of flour is also available (Finney and Shogren, 1972). The cup contains three pins and four contrarotating pins in the mixing head that knead the dough (Swanson and Working, 1933; Larmour *et al.*, 1939). In contrast to the Farinograph, where the mechanical dynamometer measures the reaction of the motor, the Mixograph has the mechanical dynamometer attached to the mixing bowl and measures its reaction as the dough is formed and kneaded. A pen attached to the arm records the movement on strip chart.

The resistance offered by the dough to four vertical pins revolving around three stationary pins in the mixing bowl creates a torque in the bowl that is proportional to the shear strength and elasticity of the dough. The Mixograph is a standard physical dough tester (American Association of Cereal Chemists, 1969). The Mixograph is 80 × 80 × 45 cm high, weighs 50 kg, and uses 15-cm-wide chart paper.

Plint Cheese Curd Torsiometer

This instrument, developed in cooperation with the United Kingdom National Institute for Research in Dairying, is intended to monitor the process of setting of curd in a cheese vat and to indicate when the curd is ready for cutting (Burnett and Scott Blair, 1963, 1964). The instrument is clamped to the side of the cheese vat. A four-bladed rotor attached to an extended shaft that positions it in the contents of the cheese vat is oscillated sinusoidally through an amplitude of ±7.5° with a frequency of 2 cycles per minute by means of a small synchronous motor. The drive from the measuring head to the rotor shaft is taken through calibrated springs and a lever. When the measuring head is moving clockwise, any resistance offered to the motion of the rotor by the curd causes the rotor, shaft, and lever to rotate slightly relative to the measuring head. The lever carries a silver contact that bears against a second contact when it rotates sufficiently, closing an electrical circuit that causes a red light to flash on, and a bell to ring. The position of the second contact relative to the stationary lever is adjusted by a micrometer so that the torque necessary to ring the alarm can be varied.

For routine production, the micrometer is set to a point established by a prior test that corresponds to the desired degree of setting of the curd. When the curd is ready for cutting, the warning light turns on and the bell sounds.

A second method of operation is to monitor the setting process. The micrometer is first set to close the electrical contact at a low torque. When the bell rings, the time is noted, and the micrometer is reset to a slightly higher torque value. When the bell rings again, the time is noted and the micrometer is reset again. In this way one obtains a series of torque (expressed in micrometer reading) versus time readings that can be plotted or subjected to mathematical analysis.

Bending

Structograph

This instrument operates on the triple-beam principle (see p. 80). The sample rests on two parallel support bars that are attached to an elevator platform that is raised at constant speed to contact a sensor bar mounted above the sample and equidistant between and parallel to the two lower knife edges. A strip chart recorder gives a force–time plot.

This instrument is useful for measuring the force to snap brittle foods. It can also be used to measure bending deformation from the slope of the force–distance plot on the chart. For nonbrittle foods, a sharp-edged upper knife or a pointed cone can be used to measure the force to cut through or penetrate the product. This instrument has a variable stroke length up to 70 mm, and a rate of travel variable from 8 to 320 mm min^{-1}. Samples up to 80 mm wide can be accommodated. The strip chart recorder is 180 mm wide and the standard chart speed is 10 mm min^{-1}, but this can be varied by changing gears inside the instrument case. The instrument is approximately 50 cm wide, 28 cm deep, 51 cm high, and weighs 18 kg. The force ranges available are 0–500, 0–1000, and 0–2000 g.

Tensile Testers

Brabender Extensograph

This instrument is used in conjunction with the Farinograph to evaluate the rheological properties of bread dough in laboratories associated with the flour milling and bread baking industry. It consists of three parts: (1) the dough-forming devices, which round and roll the dough to standard dimensions, (2) a temperature-controlled fermentation cabinet to allow the dough to relax, and (3) the mechanism that stretches the dough and reads the changes in force with extension.

Three parameters are obtained from the Extensograph curve: (1) the energy, which is measured as the area under the curve, (2) resistance to extension, which is the force at 50 mm stretching measured in EU force units (Extensograph units), and (3) extensibility, which is the length of the curve, measured in millimeters.

Instron

The Instron Universal Testing Machine is designed to perform a wide range of tensile tests. The company manufactures a number of accessories for holding samples, but most of these were not designed specifically for food use. Nevertheless, the Instron is a very suitable instrument for tensile tests.

FTC Texture Test System

This company now provides a tension test cell (model TT-1) comprising a pair of serrated gripping jaws 1-in. wide. It also has available a thin-slice tensile test cell (model ST) with a horizontal work table for tensile tests on products such as sliced bologna, cheese, and bread. This is modeled on the accessory developed by Gillett *et al.* (1978).

Distance Measuring Instruments

Bostwick Consistometer

This simple instrument consists of a level stainless-steel trough that is rectangular in cross section and comprises two compartments. The first compartment is 5 × 5 × 3.8 cm high, and it is separated from the second compartment by means of a spring-loaded gate. The second compartment, which is contiguous with the first compartment, is a trough 5 cm wide, 24 cm long, and about 2.5 cm high. The floor of this compartment has a series of parallel lines drawn across it at 0.5-cm intervals beginning at the gate and extending to the far end. It weighs about 800 g. See Fig. 11.

In operation, the gate is pressed shut and locked in place by means of a trigger. The first compartment is filled with the material whose consistency is to be tested. This is usually a comminuted fruit or vegetable such as applesauce, carrot puree and other baby foods, or tomato catsup. The consistometer is leveled and the trigger is pressed, releasing the gate, which springs up out of the way. The fluid material is then free to flow under the force of gravity from the first compartment into the second compartment. The distance it has flowed from the gate after 30 sec is measured in centimeters as the Bostwick Consistometer reading.

When the moving front edge of the flowing product is curved, the distance to the forward edge of the curve is taken. In some products syneresis occurs; in these cases the clear liquid is generally ignored, and the reading is taken at the front edge of the puree. The width of the clear serum is sometimes also measured in those products in which considerable syneresis occurs.

The United States standard for tomato catsup stipulates that grade A and grade B quality should be of good consistency, and flow not more than 9 cm in 30 sec at 20°C in a Bostwick Consistometer. Grade C tomato catsup must have a "fairly good consistency" and flow not more than 14 cm in 30 sec at 20°C in a Bostwick Consistometer.

Rutgers (1958) reported that this instrument is suitable for nonthixotropic purees and thick porridges but not for starch-thickened milk puddings. Bookwalter *et al.* (1968) found the Consistometer to be suitable for processed cornmeals and their protein-enriched blends. Rao and Bourne (1977) found that the Bostwick Consistometer was suitable for fruit and vegetable purees but not suitable for nonpureed foods because they adhered to the gate. It is not suitable for high

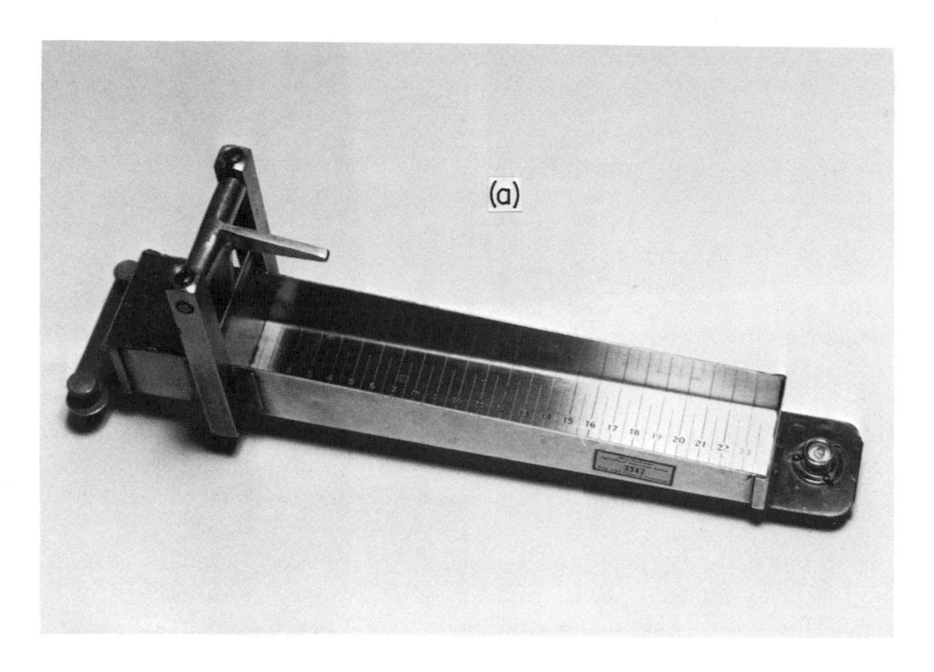

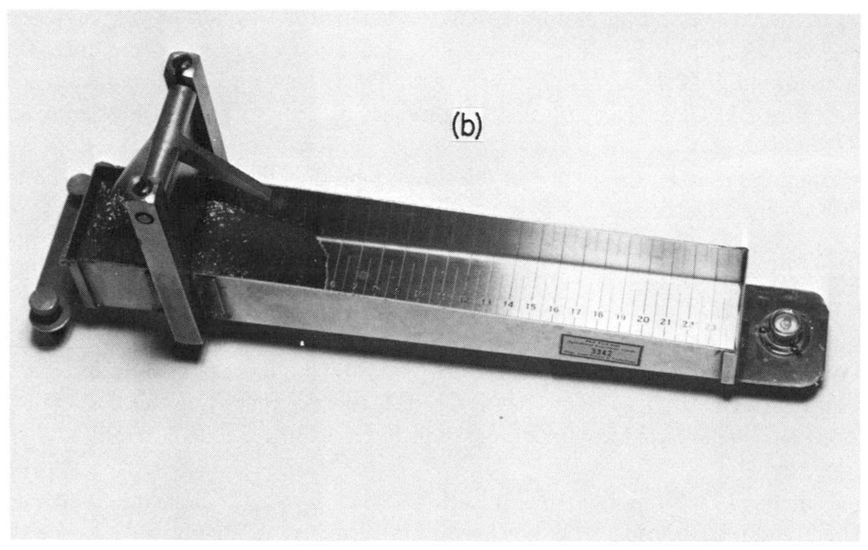

FIG. 11. The Bostwick Consistometer: (a) sample is in first compartment with gate closed; (b) gate is open and sample has flowed along the second compartment.

solids tomato paste because the paste does not flow far enough in 30 sec to give measurable differences between samples.

The results from this instrument cannot be converted into fundamental rheological parameters because surface tension, wetting power, and possibly other factors other than viscosity are also involved. Nevertheless, it is a useful, rapid quality control tool for products that have a yield point but are not too stiff.

The Hilker–Guthrie Plummet

This simple device was developed to measure the consistency or "body" of cultured cream, but it can also be used on other products that have a similar consistency (Hilker, 1947; Guthrie, 1952, 1963). The plummet consists of a hollow aluminum tube ½ in. diam and 4½ in. long weighing about 15 g. The lower end tapers to ⅛ in. diam and is closed off. A series of inscribed lines numbered 1–10 are etched into the tube at ⅜-in. intervals beginning from the top. The plummet is mounted in a stand vertically over the commodity to be tested and with the lower tip exactly 12 in. from the surface of the product. It is released and allowed to fall freely into the product. The depth of penetration into the commodity is read off the scale after 5 sec. It is customary to take the mean of three tests. See Fig. 12.

Hilker (1947) gives the following figures for relating the plummet reading to the viscosity of cultured cream: very thin, 0–2; thin, 2–4; medium, 4–6; good, 6–7.5; slightly heavy, 7.5–8.5; heavy, 8.5–10; very heavy, greater than 10.

Ridgelimeter

This little device was developed for judging the grade of fruit pectins and the stiffness of pectin jellies (Cox and Higby, 1944; Anonymous, 1959). The instrument is essentially a height-measuring gauge. A pectin jelly is made under standard conditions specified by The Institute of Food Technologists Committee on Pectin Standardization (Anonymous, 1959). To make a standard jelly, one assumes a jelly grade and uses (650/assumed grade) grams of pectin to make the jelly. The jellies are poured into tapered glass tumblers that are 1.75 in. i.d. at the bottom, 2.5 in. i.d. at the top, and internal height exactly 3.125 in. Masking tape is applied to the top of the jar to protrude at least ½ in. above the jar. See Fig. 13.

The boiling jelly is poured into the jar until it is ½ in. above the top of the jar, the excess being retained by the tape. After standing for 20–24 h at 25 ± 3°C the tape is removed, a wire cutter is moved across the top of the jar to remove the excess jelly, and the jelly is carefully tipped out onto a small square of plate glass that is furnished with the instrument. The pointer of the dial is moved down close to the surface of the jelly. After exactly 2 min the pointer is moved until it just

FIG. 12. The Hilker–Guthrie Plummet: the plummet has been released from the holder and allowed to fall freely into the cultured cream; two spare plummets are shown on the right.

contacts the jelly. The scale gives the percent sag to the nearest 0.1%. A jelly of "standard firmness" has a sag of 23.5%. The true grade of the test is obtained from the formula

$$\text{true grade} = \text{assumed grade } (2.0 - \%\ \text{sag}/23.5).$$

FIG. 13. The Ridgelimeter.

If a more precise calculation is needed, a conversion curve given by Cox and Higby (1944) may be used.

This is a simple but effective instrument and is the standard test used by the industry for establishing the grade of pectins and fruit jellies.

Penetrometer

This useful instrument was first developed for measuring the firmness or the yield point of materials such as petroleum jelly and bitumen, but it is widely used for measuring the firmness or yield point of butter, margarine, and other solid fats.

The Penetrometer manufactured by Precision Scientific will now be described (see Fig. 14). Penetrometers that are very similar are made by several other manufacturers. The Penetrometer consists of a vertical rod 3/16 in. diam and weighing 47.5 g that can be locked in position and then released to fall freely under the force of gravity. At the lower end is a small chuck that can hold various cones and needles. A 4-in.-diam dial gauge that is connected to a depth gauge is

FIG. 14. The Penetrometer: different types of cones and needles are shown resting on the white cloth.

used to measure manually the distance the rod falls after release to within 0.1 mm. The dial is graduated from 0 to 380 in ⅒-mm increments. Penetration measurements can be made to a total depth of 62 mm because the dial pointer can make approximately 1⅔ revolutions.

This assembly is attached to a vertical shaft by means of a rack and pinion to adjust the height above the sample. The vertical shaft is held in a heavy cast aluminum base that has a built-in spirit level and two leveling feet. A trigger normally holds the rod at its highest point in a locked position. When the trigger is pressed, the rod is free to drop. An optional addition is a solenoid trigger assembly controlled by an electrical timer that when switched on releases the trigger and then locks it again after 5 sec.

To operate, a suitable cone or needle is placed in the chuck, the trigger is released, and the rod and cone assembly is lifted up until it reaches the upper stop where it is locked. The needle should then read 0. If it is not 0, a small adjusting knob beneath the instrument is turned to bring the needle to 0. The material to be tested is positioned beneath the cone, and the whole cone and dial assembly is lowered by means of the rack and pinion until the point of the cone almost touches the surface of the fat. The rack and pinion assembly is then locked. The final adjustment to bring the point of the cone exactly into contact with the surface of the sample is made by means of a micrometer adjusting screw. The cone and rod assembly is then released and allowed to sink into the food under the force of gravity for 5 sec when it is locked again. Weights are provided that can be added to the top of the rod to increase the force on the cone. Then the depth gauge is pressed down gently until it reaches a stop on the rod. The dial reading indicates the depth of penetration directly in tenths of a millimeter. When the dial reading has been recorded, the cone and rod is returned to the 0 position and the instrument is ready to make the next reading.

There is discussion in the literature as to the exact meaning of the Penetrometer reading and whether the reading measures the yield point of the fat, the consistency, or some combination of these. Although this matter is not completely settled, the Penetrometer test is a useful test for solid fats. It is common to make measurements on fats at several temperatures in order to determine the range of temperatures over which the fat is workable (i.e., its plastic range).

Maleki and Siebel (1972) used a Penetrometer to measure the deformability of bread and reported a correlation coefficient (*r*) of 0.88 between Penetrometer units and sensory score of softness. These authors used 5-cm-thick slices of bread, a deforming weight of 203 g for 5 sec. Fresh bread gave a mean deformation of 13.5 mm while 5-day-old bread gave a mean deformation of 5.3 mm. Underwood and Keller (1948) used the Penetrometer to measure the consistency of tomato paste.

The Penetrometer can be adapted to measure the deformation of many foods by using a flat disk in place of the cone (see Fig. 15) (Bourne, 1973). A small flat disk 5 mm thick and 50 mm diam is cut from a piece of hard plastic and a ⅛-in.-diam brass rod inserted in the center of one side of the disk normal to the plane of the disk. The article of food (e.g., a tomato) is placed in position in the Penetrometer and the disk is brought down close to (but not touching) its surface by means of the rack and pinion. The Penetrometer is turned on and the disk and rod

FIG. 15. Adaptation of the Penetrometer for deformation testing. Note the disk that replaces the conventional cone, and the weight on top of the shaft. (From Bourne, 1973; reprinted from *J. Food Sci.* **38,** 720, 1973. Copyright by Institute of Food Technologists.)

are allowed to drop freely for 5 sec and then locked. The distance the rod and disk have fallen is measured on the dial gauge. A selected weight is then placed on the upper end of the rod, and the weighted disk and rod assembly is allowed to drop freely again for 5 sec, then locked, and a second reading is taken from the dial gauge. The difference between the two dial readings gives the deformation of the article in units of 0.1 mm for a force change equal to the difference in weight between the rod and disk assembly and the added weight.

In conventional Penetrometer testing with a cone, it is critical that the point of the cone be placed exactly at the surface of the food. In the deformation test the initial placement is not critical, since this is a test by difference. The 64-g weight of the unloaded disk and rod assembly is sufficient to give the small preliminary compression that eliminates those errors that might be caused by intrinsic irregularities in the surface of the food piece (Bourne, 1967a). This technique has the advantage that the Penetrometer is a relatively inexpensive instrument that can easily be adapted for measuring the deformability of foods that are reasonably soft and of reasonable size.

Figure 16 shows how this technique measured the change in deformation of two tomatoes as they ripen. Since this is a nondestructive test, the same tomato can be tested repeatedly, eliminating the problem of sample to sample variation, provided the force applied is sufficient to cause no irreversible change at the test site.

The SURDD Hardness Tester

This instrument was developed by the Southern Utilization Research and Development Division (SURDD) of the United States Department of Agriculture (USDA) and is designed to determine the hardness and softening characteristics of fats and waxes. The test is based on the Brinell tester that is used to measure the hardness of metals (see p. 83).

The instrument consists of a vertical rod that is free to move within a stand. The lower end of the rod holds a steel ball having a selected diameter from 0.125

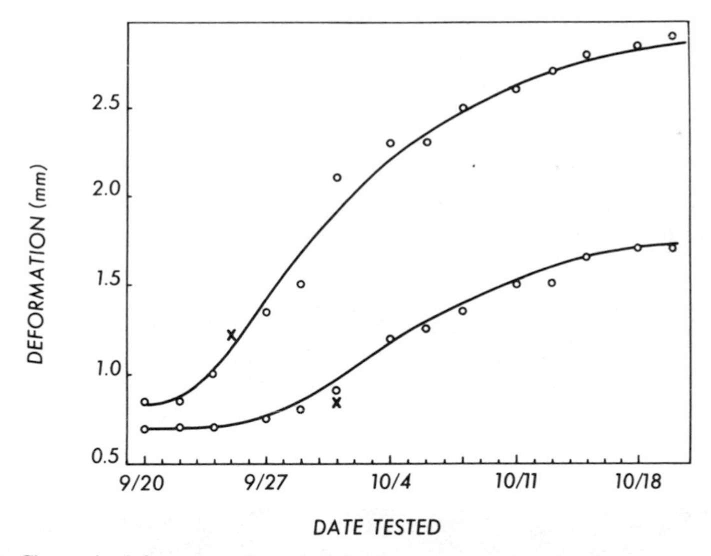

FIG. 16. Change in deformation of two individual tomatoes during ripening as measured by the Penetrometer. Lower curve is a firm variety and upper curve is a soft variety. The "X" on each curve denotes the day when the first sign of pink color appeared near the blossom end of the fruit. (From Bourne, 1973.)

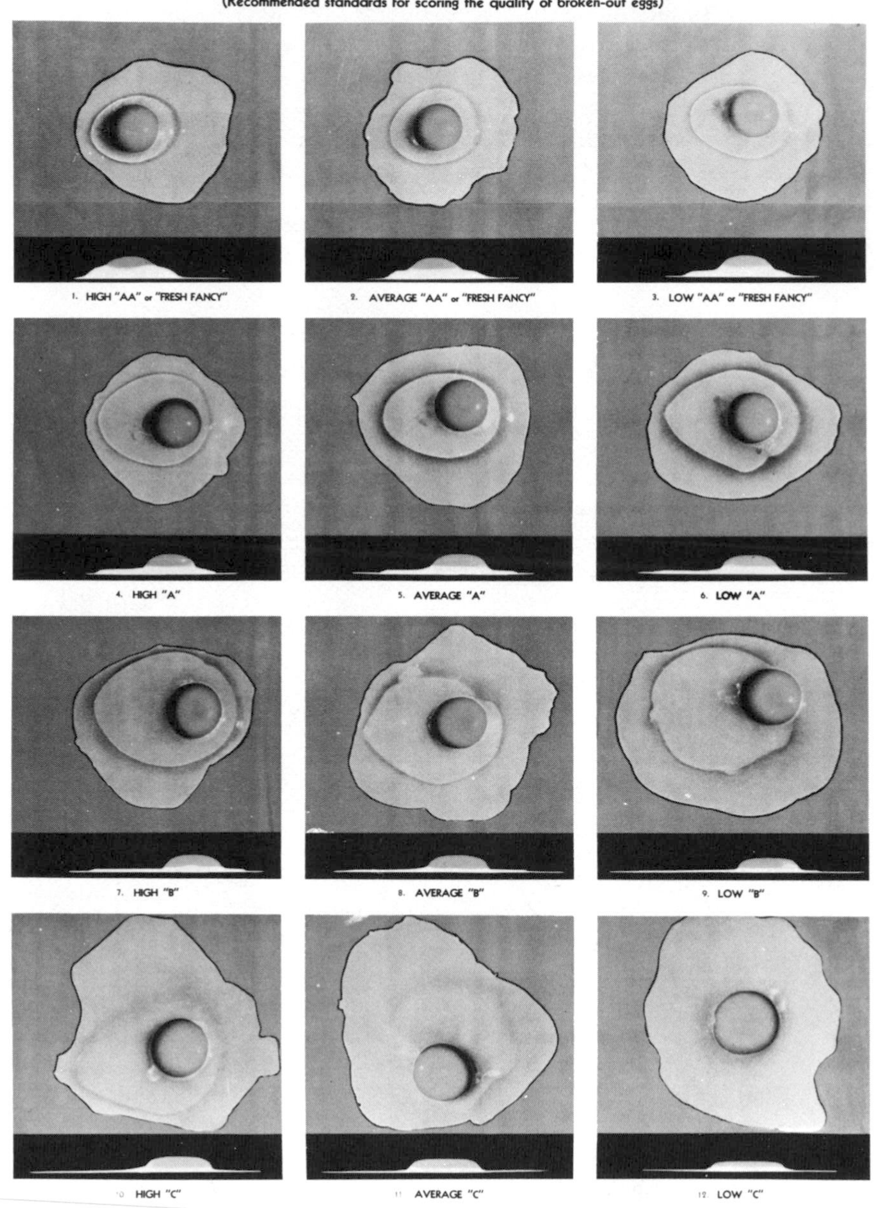

FIG. 17. Relationship between albumen height and egg quality (USDA photo).

to 0.500 in. The upper end of the rod holds a small platform on which weights may be placed. The force is applied by raising the test sample support platform upward until all the weight of the ball rests on the sample. In operation, a sample of the fat is placed beneath the ball, a suitable weight ranging from 0.2 to 6 kg is placed on the plate, and the ball is allowed to penetrate into the fat under this force for 1 min. The diameter of the ball and the weight are selected so that the diameter of the impression made in the fat is about one third the diameter of the ball. The diameter of the impression in the fat is measured by means of a cathetometer or a magnifying glass with a built-in scale. This measurement can be made to within 0.02 mm.

The Haugh Meter

The Haugh meter is used to measure the quality of eggs. It consists of a tripod stand through the center of which a pin is located that can be moved up and down by means of a screw. It is a small instrument, weighing less than 1 kg. The principle of the measurement is based upon the fact that high thick albumen (the egg white) indicates good quality, fresh eggs. Figure 17 shows in top view and silhouette the relationship between albumen height and egg quality.

In operation, an egg is weighed, the shell is broken gently, and the egg is spread out on a horizontal glass plate. The Haugh meter (see Fig. 18) is placed such that the center pin is over the thick white about 10 mm out from the edge of the yolk. The screw is turned until the face of the pin just touches ("kisses") the albumen. The gauge measures the height of the albumen above the plate. Haugh (1937) established that the log of the albumen height is directly proportional to the egg quality. The basic equation is

$$\text{Haugh units} = 100 \log H,$$

where H is the albumen height in millimeters. The factor 100 is used to remove the decimal.

A correction for the weight of the egg is needed because large eggs will have higher albumen than small eggs of equal quality. The equation correcting for the weight of the egg is

$$\text{Haugh units} = 100 \log\left[H - \frac{\sqrt{G}(30W^{0.37} - 100)}{100} + 1.9\right],$$

where G is 32.2 and W, the weight of the egg in grams. The dial gauge mounted on the tripod is usually calibrated directly in Haugh units by means of a scale that compensates for variations in egg weight.

Since egg quality is largely determined by heredity, the Haugh meter is used to identify and breed hens that lay top quality eggs with high albumen. Detailed specifications concerning egg quality have been published, and the Haugh units

FIG. 18. The Haugh meter (USDA photo).

of the eggs are one index of that quality (USDA Handbook No. 75, ''Egg Grading Manual''). In order for eggs to be graded AA or Fancy, they need to maintain a moving average of 72 Haugh units or higher. For eggs that are labeled Grade A, the flock must maintain a moving average of 60 Haugh units or higher.

The Baker Compressimeter

This instrument measures the force to press a metal plate onto a 12-mm-thick slice of bread until it has been compressed to 9 mm (25% compression). A small motor gradually applies a force by winding up on a spool a cord attached to the compressing lever. The force and degree of compression are measured on two scales, and it is possible to measure forces and deformations at other than 25% compression. The instrument is also used to measure the softness of buns, rolls, cakes, and other leavened baked goods.

The Baker Compressimeter is a standard method for measuring the staleness of bread (Cereal Laboratory Methods No. 74–10, published by the American Association of Cereal Chemists). The relationship between the actual Compressimeter reading and degree of staleness and quality of the bread is a matter of discussion among cereal chemists, and some judgment is needed to interpret the results (Platt and Powers, 1940; Bice and Geddes, 1949; Thompson and Meisner, 1950; Crossland and Favor, 1950; Edelman *et al.*, 1950; Bechtel *et al.*, 1953).

Willhoft (1970) developed an empirical equation describing the staling of bread over a period of six days:

$$F_t = A(t/t_i)^B,$$

where F_t is the firmness at time t, measured by an objective test; t_i, the time of initial measurement; A, a constant which is equal to firmness at time equal to unity; and B, a constant which is equal to the rate of firming.

He showed that a plot of log F_t versus log t is rectilinear with slope equal to the rate of firming constant B. The same author reviewed the theory and mechanism of bread staling and associated changes in textural properties (Willhoft, 1973).

Adams Consistometer and Tuc Cream Corn Meter

These instruments measure the distance a semifluid food flows across a plate in a standard time (Adams and Birdsall, 1946). They should be known as the Grawemeyer and Pfund Consistometer if priority is recognized (Grawemeyer and Pfund, 1943).

The Adams consistometer consists of a 12-mm-thick sheet of clear hard plastic (Plexiglas) approximately 37 cm square that is leveled by means of adjustable legs. A series of concentric circles are inscribed on the underside of the sheet at ¼-in. intervals. The frustum of a stainless-steel cone that is 3 in. diam at the lower face, 2 in. diam at the upper face, and 5 in. high is placed in the center of the plate. The Tuc Cream Corn Meter is very similar to the Adams Consistometer.

In operation, the cone is placed in position and filled to the top with the product. The cone is gently lifted up and the product is allowed to flow out in two dimensions across the plate. After a standard time the diameter of the product is measured along two axes at right angles to each other. The USDA specification for standard quality cream-style corn stipulates that the average diameter should not be greater than 30.5 cm after 30 sec flow. The product is substandard if the diameter is greater than the 30.5 cm.

USDA Consistometer

This is similar in principle to the Adams Consistometer. The USDA flow sheet #1 consists of a thin flexible plastic sheet over which the product flows. The receptacle holding the food is a Perspex cylinder 3 in. i.d. and 3¼ in. high. The distance of flow is measured from the outer edge of the cylinder in centimeters (see Fig. 19). This contrasts with the Adams Consistometer, which measures inches diameter and includes the diameter of the cone.

FIG. 19. The USDA Consistometer: (a) cylinder is filled with applesauce; (b) cylinder is removed, allowing sauce to flow out.

The USDA standards for canned applesauce measures the distance of flow after 1 min on this sheet. The flow value is taken as the average of the readings at four quandrants of the flow sheet. The readings are taken at the edge of the applesauce and do not include any free serum that exudes from the sauce. The amount of free serum that exudes may also be measured. For grade A applesauce, regular style, the flow should be not greater than 6.5 cm and for chunky style not more than 7.5 cm. For grade B applesauce the flow should not exceed 8.5 cm for the regular style and 9.5 cm for the chunky style.

Volume Measuring Instruments

Loaf Volume Meter

This apparatus consists of a metal box connected through a rectangular chute to a hopper containing rapeseed. A loaf of bread is placed in the box, which is closed, a slide in the chute is pulled out, and the rapeseed is allowed to fill the box. A calibrated scale on a Pyralin face of the volumeter column gives the direct reading of the volume of the bread in cubic centimeters. This device is widely used in the baking industry to measure loaf volume, which is one index of quality of the loaf (Cathcart and Cole, 1938; Funk *et al.*, 1969).

The standard volumeter consisting of a box 5⅝ × 11⅝ in. is designed for the 1-lb loaf of bread and can read volumes between 1675 and 3000 cc. Other sizes available are the "Pup size," measuring 400–1000 cm^3 "micro" size, measuring 100–270 cm^3, "half-pound" size, measuring 900–1500 cm^3, "1½-lb" size, measuring 2475–3800 cm^3, and a "round cake" size, designed for cakes with measuring volumes of 500–1600 cm^3. A "dummy" loaf of standard size is provided with each volumeter to calibrate the rapeseed level in the hopper.

The Succulometer

This instrument is designed to measure the volume of juice that can be pressed from fresh sweet corn and is used as an index of maturity and quality (Meyer, 1929; Sayre and Morris, 1931, 1932; Kramer and Smith, 1946). In operation, a 100-g sample of cut corn is placed in the sample chamber and pressure is applied through a hydraulic ram that is pumped by hand. A pressure of 500 lb is maintained for 3 min and the juice that flows out is collected in a 25-ml graduated cylinder. The volume of juice decreases as the corn becomes more mature. Kramer and Smith (1946) relate the Succulometer values to quality of sweet corn as follows: fancy quality, more than 22 ml of juice; extra standard, 19–22 ml; standard, 12–18 ml; substandard, less than 12 ml.

The Food Technology Corporation Texture Press is provided with an optional accessory in the form of a succulometer cell, which allows the succulometer test to be performed in this apparatus.

Time Measuring Instruments

The kinematic viscometers and the Stormer Viscometer are time measuring instruments. These will be discussed in Chapter 5.

The BBIRA Biscuit Texture Meter

The British Baking Industry Research Association Biscuit Tester is designed to measure the hardness of biscuits (cookies and crackers) (Wade, 1968). A stack of biscuits is placed in the sample holder and pressed with constant force against a small circular saw blade that is rotating at 15 rpm. The time taken to make the saw cut through the stack is recorded by a counter, which stops when the operation is complete. A brush is positioned behind the saw to clean the teeth. The unit is housed in a fiberglass cover with a door to enable the operator to gain access to the working elements. The door has a Perspex window in it to enable the operator to watch the saw cut operation. This instrument is essentially a comparator; standards must be established experimentally for each type of cookie or cracker.

Miscellaneous Methods

Torry Brown Homogenizer

This instrument is an homogenizer that has been especially designed to measure the toughness of fish by the "cell fragility method" (Love and Mackay, 1962; Love and Muslemuddin, 1972a,b; Whittle, 1973, 1975). A 200-mg sample of fish tissue is dropped into the homogenizer cup with 20 ml of a solution consisting of 2% trichloracetic acid and 1.2% formaldehyde cooled to below 6°C. The paddle in the cup is rotated at 8750 rpm for 30 sec. Muscle from fresh fish breaks up into small particles forming a cloudy soup that has a high optical density while muscle from tough fish is shredded into large pieces and has a low optical density. For rough work the optical density of the homogenate can be viewed by eye. For more accurate work the optical density can be read in an absorptiometer.

This method has been shown to give a good index of the deterioration in the texture of frozen fish. Fatty fish and fish in an advanced stage of bacterial decay give spurious readings because of the higher optical density imparted to the homogenized liquid by the bacterial cells or fat globules (Love and Muslemuddin, 1972a,b).

Corn Breakage Tester

This instrument is designed to measure the susceptibility of dry corn to breakage during commercial handling (Miller *et al.*, 1981a,b). A 100-g sample of cleaned 12⁄64-in. mesh-sized dry corn is placed in a stainless-steel cup with internal dimensions 3⅝ in. diam × 3½ in. high in which a stainless-steel blade rotates at a nominal speed of 1725 rpm. After 4 min the sample is transferred to a sieve with 12⁄64-in. round holes and shaken for 30 strokes. The breakage is that fraction of corn that passes through the sieve. This test has been reported as effective for measuring the breakage susceptibility of dry soybeans (Miller *et al.*, 1981c).

The model CK2-M Stein Corn Breakage Tester weighs 16 kg, stands 55 cm high, 38 cm deep, and 43 cm wide, and is driven by a ⅓-hp electric motor.

Multiple Measuring Instruments

GF Texturometer

This Texturometer simulates the motions of mastication by means of a mechanical chewing arrangement. It was developed by the central research group of the General Foods Corporation (Friedman *et al.*, 1963; Szczesniak *et al.*, 1963). The commercial instrument is manufactured in Japan. Originally the GF Texturometer was designed to be part of the total Texture Profile Method (Szczesniak, 1963b; Brandt *et al.*, 1963), but in fact the Texturometer is generally used as a test instrument in its own right, and it is not necessary to use it in conjunction with a sensory panel.

A 1⁄16-hp electric motor drives an eccentric that is linked to the activating arm of the instrument. Because it is driven by an eccentric, the arm moves through the arc of a circle with the speed varying approximately in a sine wave function. To this arm can be attached a variety of plungers made of Lucite, aluminum, or nickel, ranging in diameter from 3 to 50 mm. Most of the plungers have flat faces, but a tooth-shaped punch and a V-shaped punch are included among the plunger accessories. See Fig. 20.

Positioned underneath the moving arm to which the plungers are attached is a platform on which the food sample is placed. This platform rests on a cantilever beam to which strain gauges are attached. Bending of the beam under the application of a force is sensed by the strain gauges and recorded electronically on a strip chart recorder. This instrument is designed for testing "bite-size" pieces of

FIG. 20. The commercial GF Texturometer: the test is performed at the right and the force–time history is recorded on the horizontal chart at the left. (Courtesy of Zenken Co., Ltd.)

food. The standard piece is a ½-in. cube, but particulate foods (e.g., peanuts) are normally tested as whole units.

The 1963 paper by Friedman *et al.* (1963) describes the prototype instrument that is still in use at the General Foods Technical Center Laboratory. The first model of the commercial Texturometer was manufactured by Tominaga and Company Ltd., Japan. The latest version (model GTX-2-IN) is manufactured by Zenken and Company Ltd. of Tokyo, Japan.

In operation, the sample of food is placed on the platform, the chart and drive motors are switched on, and the food is compressed under the descending plunger two times, imitating the first two bites that would occur if the food were placed in the mouth. The original commercial model had two mastication speeds: 12 and 24 bites min^{-1}. The latest commercial model also has two mastication speeds: 6 and 12 bites min^{-1} The normal plunger stroke is approximately 35 mm. The *mean* speed at 12 chews min^{-1} is approximately 840 mm min^{-1}; the actual speed at any moment will vary widely from this figure because of the sinusoidal speed pattern that is imparted to the plungers by the eccentric drive. The mechanics of the movement of the eccentric and compressing arm have been analyzed by Brennan *et al.* (1975).

The standard cantilever beam on which the platform rests has a maximum load capacity of 40 kg. A preloaded spring in the compressing arm begins to flex when a force of about 45 kg is reached, which protects the instrument from overload. It is possible to order a beam with a larger cross section that has a force capacity of 60 kg for use with hard products and a beam with a small cross section that has a maximum force capacity of 3 kg for use with soft products such as whipped cream. The deflection of the standard platform and compressing assembly is approximately 0.01 mm N^{-1} applied force.

The strip chart on the horizontal recorder is 250 mm wide (10 in.). The response time of the recorder pen is ⅓ sec. Two chart speeds are available: 1500 and 750 mm min^{-1}.

A Texturometer is normally calibrated by compressing a piece of sponge rubber. It can also be calibrated by placing weights on the platform, provided they are centered on the platform because the response is sensitive to the position of the weight. One Texturometer unit is equivalent to approximately 105 g force ($\simeq$1.03 N), but the actual value can vary to some extent depending on the voltage output of the recorder batteries.

The commercial instrument is mounted on casters with a small cupboard to hold chart paper and accessories in the lower area. The instrument is 1.2 m high, 0.8 m long, 0.5 m wide, and weighs 135 kg. It is powered by ac 120 or 220 V at 50 or 60 Hz. An important feature of the GF Tenderometer is the analysis of the force–time curve that is traced on the chart. This analysis was discussed in the previous chapter (see p. 114).

Since the articulator of this Texturometer moves with a variable speed and the

TABLE 7

SUMMARY OF GF TEXTUROMETER CONDITIONS USED WITH DIFFERENT FOOD PRODUCTS

Foodstuff	Plunger type and diam (mm)	Clearance (in.)	Power supply (V)	Chews (min^{-1})	Sample presentation
Bread	Lucite, 17	0.042	10	12	¼ slice, crust removed
Cheese	Brass, 17 or 10	0.057	1	12	1-in. square × ½ in. high
Chocolate	Brass, 7	0.021	0.25–1	12	¾–1 in. cubes
Cold cereals	Lucite, 17	0.083	0.5–2	6 or 12	Small cup[b], filled
Coconut shreds or flakes	Lucite, 17	0.073	1	12	Small cup[b], filled
Cookies	Lucite, 17	0.083	1	12 or 24	2 pieces stacked
Crackers	Lucite, 17	0.083	1	12 or 24	2 pieces stacked
Dough	Brass, 50	0.311	2–5	12	1 in. ball
Eggs, scrambled	Brass, 50	0.083	10	12	Small cup[b], filled
Frankfurters	Brass, 17	0.050	1	12	½-in.-long cylinders
Gels	Lucite, 17	0.120	12	12	Small cup[b], filled
Macaroni	Lucite, 17	0.081	5	12	Small cup[b], filled
Meat	Brass, 17	0.057	0.5–1.0	24	1 in. square × ½ in. high
Muffins	Lucite, 17	0.083	1	24	
Mushrooms	Brass, 7	0.042	5	12	5-mm-thick slices
Nuts	Aluminum, 13	0.0625	0.51–1	6	One piece
Peas	Lucite, ridged 25	0.083	1	24	Small cup[b], filled
Potatoes, french fried	Tooth-shaped brass	0.057	5	12	One strip parallel to the plunger arm
Potatoes, mashed	Lucite, ridged 25	0.083	5–10	12	Small cup[b], filled
Puddings	Brass, 50	0.555	10	12	Large cup[c], filled
Sour cream	Brass, 50	0.375	10	12	Large cup[c], filled
Strawberries, fresh	Brass, 17	0.042	5	12	1 cm high, cut surface down
Whipped toppings	Brass, 50	0.375	3–10	12	Large cup[c], filled

From Szczesniak and Hall (1975); reprinted from *J. Texture Stud.* with permission from Food and Nutrition Press.

[b]Small cup is 2 in. diam × ⅞ in. high.

[c]Large cup is 3¼ in. diam × ¾ in. high.

chart moves at constant speed the plots given on the chart are force–time plots; hence, only functions of force, time, and force–time integrals can be obtained from the chart. It is possible to convert these Texturometer curves to a force–distance curve by graphical means (Brennan *et al.*, 1975), but this is a time-consuming procedure.

Table 7 summarizes the Texturometer conditions used with different food products. The shapes of the Texturometer curves vary widely, depending upon the textural properties of the product. Some textural parameters are absent in certain foods. For example, it is unlikely that the same commodity will exhibit the properties of both crispness and adhesiveness. The observed textural parameters for several foods are shown in Table 8.

The Texturometer can be used to characterize textural changes caused by changing temperature. Figure 21 shows how the textural properties of different formulations of whipped toppings change with rising temperature. Other applications of the GF Texturometer are discussed by Szczesniak and Hall (1975) and Tanaka (1975).

The newest model of the Texturometer offers as an option a built-in electrical digital integrator that automatically displays the areas under three portions of the curve.

Fukushima (1968) proposed a new parameter "induction property" defined as the ratio of hardness measured at 1.5 mm clearance to hardness measured at 5 mm clearance. The induction property showed good correlation with sensory evaluation of texture of *kamaboko* (fish sausage).

TABLE 8

OBSERVED PARAMETERS FOR FOODS TESTED IN THE GF TEXTUROMETER[a]

Food	Textural parameters
Cooked pudding	Hardness, cohesiveness, adhesiveness
Whipped topping	Hardness, cohesiveness, adhesiveness, gumminess
Wheat dough	Hardness, cohesiveness, adhesiveness
Cooked rice (one kernel)	Hardness, cohesiveness, adhesiveness
Cold breakfast cereal	Hardness, area under first peak
Bread crumb	Hardness
Strawberries	Fracturability, hardness, cohesiveness, adhesiveness
Frankfurters	Hardness, cohesiveness, springiness, oiliness
French fries	Hardness, crispness, cohesiveness, mealiness
Chocolate	Fracturability, hardness, cohesiveness, adhesiveness
Coconut	Hardness, cohesiveness, area under first bite
Cooked macaroni	Hardness, cohesiveness, gumminess
Mushrooms, frozen and thawed	Hardness, cohesiveness, adhesiveness
Cheese	Hardness, cohesiveness, adhesiveness

[a]Extracted from data of Szczesniak and Hall, 1975.

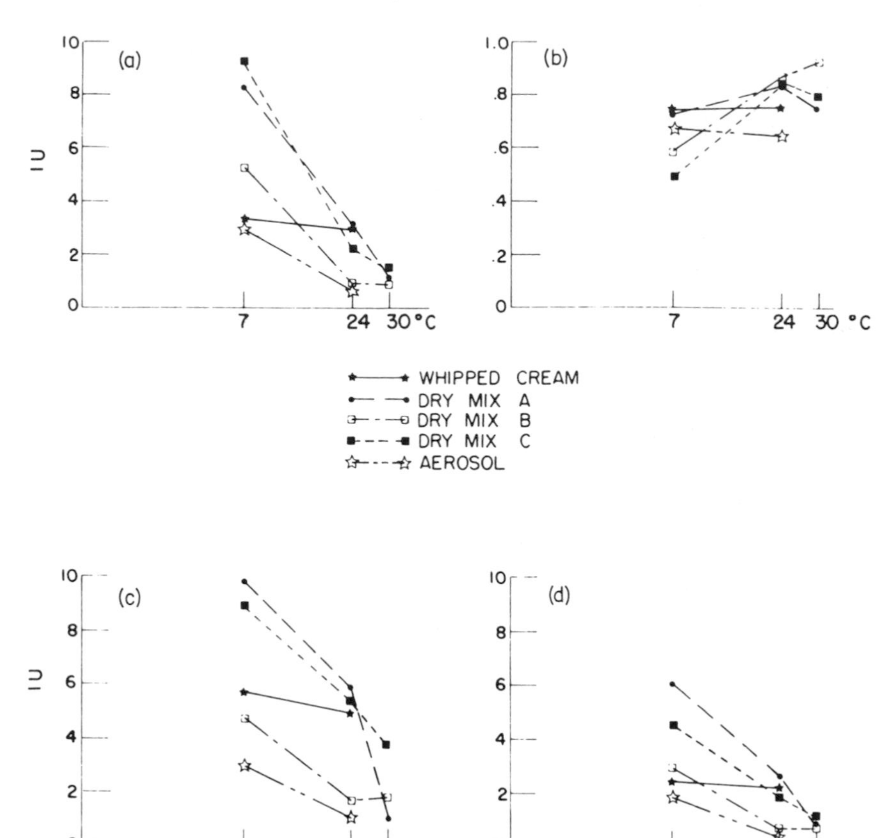

FIG. 21. Characterization of textural properties of whipped toppings as a function of temperature with the GF Texturometer where IU is instrumental units: (a) hardness; (b) cohesiveness; (c) adhesiveness; (d) gumminess. (From Szczesniak, 1975a; courtesy of Dr. A. S. Szczesniak. Reprinted from *J. Texture Stud.;* with permission from Food and Nutrition Press.)

Okabe (1971) classified GF Texturometer curves into five different categories based on the characteristics of the peaks obtained in three successive chews: (1) elastic, (2) recoverable retarded elastic, (3) unrecoverable retarded elastic, (4) plastic flow, and (5) brittle fracture (Fig. 22). This classification attempts to derive rheological meaning from the curves.

The GF Texturometer is used in Japan to measure the textural quality of cooked rice. The palatability of cooked rice for Japanese was found to be governed primarily by hardness, stickiness, and the ratio of stickiness to hardness (Okabe, 1979). Rice of high hardness can be palatable if stickiness is also high.

Type	Perfect elasticity	Retarded elasticity		Plastic flow	Brittle and fracturable
		Recoverable	Irrecoverable		
Typical mastication curve	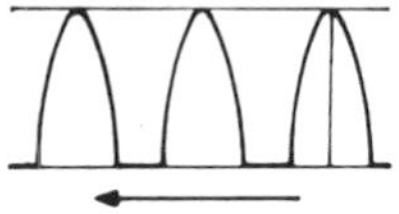	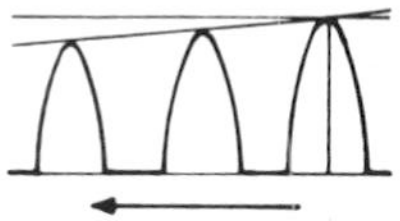	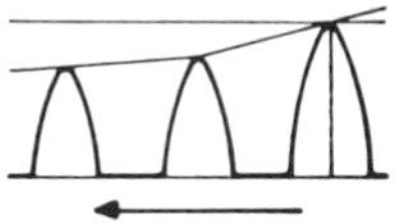	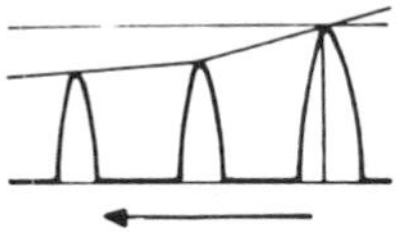	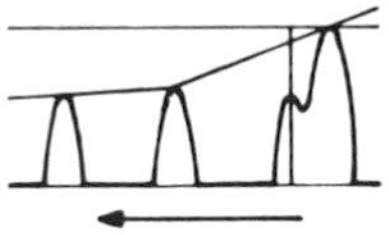
Characteristics of consecutive three mastication curve peaks	The shape and the height of three peaks are the same.	Height of three peaks decreases linearly.	Height of the second peak decreases pronouncedly.	Second and third peaks become thin and small in shape.	First peak is pronouncedly different in shape and height.
Characteristics of the first mastication curve peak	Peak of the curve appears at the same time with the lowest position of plunger. Peak is symmetric.	Peak of the curve appears at the same time with the lowest position of plunger. Peak is symmetric.	Peak of the curve appears at the same time with the lowest position of plunger. Peak is symmetric.	Peak of the curve appears at the same time with the lowest position of plunger. Peak is nonsymmetric.	First peak appears earlier than the lowest position of plunger. Several subpeaks are seen.

FIG. 22. Classification of GF Texturometer curves. (From Okabe, 1971. Reprinted from *J. Texture Stud.;* with permission from Food and Nutrition Press.)

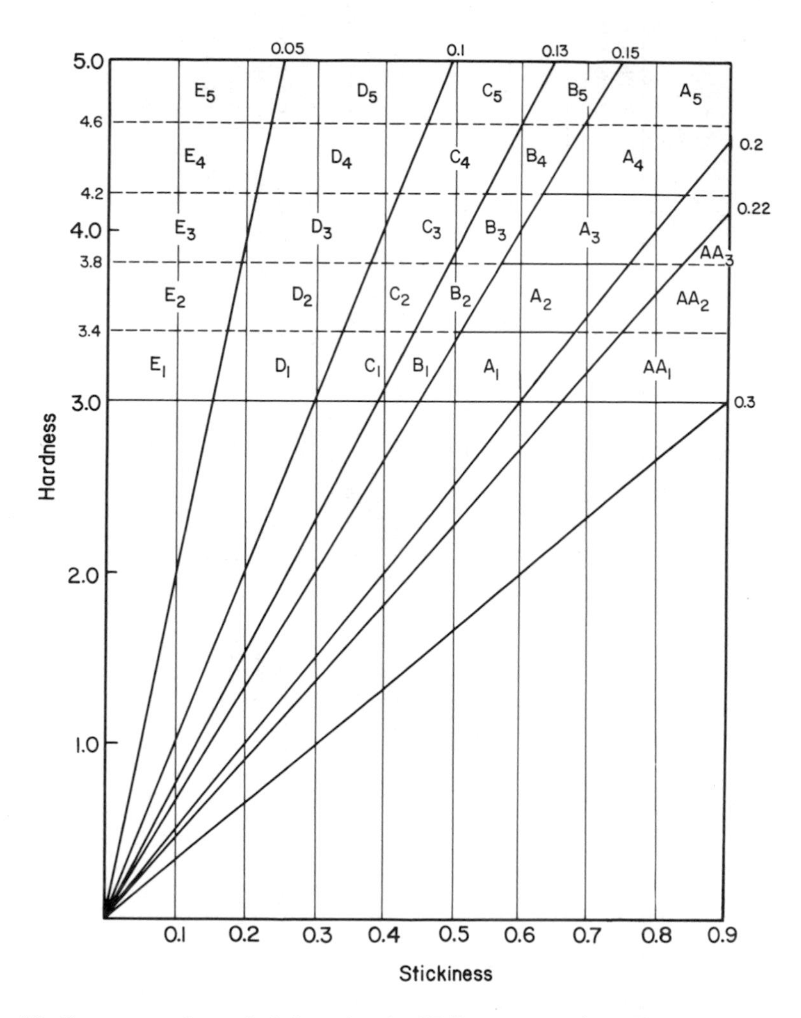

FIG. 23. Texturogram for cooked rice using the GF Texturometer (from Okabe, 1979). Zones of acceptability are based on hardness (vertical scale 0–5), stickiness (horizontal scale 0–0.9), and stickiness/hardness ratio (sloping lines, scale 0.05–0.3): A, excellent; B, good; C, acceptable; D, poor; and E, unacceptable. Each acceptability zone is also ranked in terms of hardness: for example, A_1 is low hardness, A_3 moderate hardness, and A_5 high hardness for excellent quality rice, while E_1–E_5 covers the hardness range for unaccepable quality rice. The AA zone comprises glutinous rice and some special new rice varieties. (Reprinted from *J. Texture Stud.;* with permission of Food and Nutrition Press.)

Figure 23 indicates zones of acceptability of rice as a function of hardness, stickiness, and thė harness/stickiness ratio. This diagram has been used in Japan to characterize factors affecting the palatability of different varieties, and also storage and processing factors that affect rice quality.

FTC Texture Test System

This apparatus is basically the Food Technology Corporation Texture Press, which was described on p. 136 but with a number of additions that give it expanded utilization, more flexibility in operation, and make it attractive for many research purposes. The major feature of the Texture Test System is the addition of a recorder that traces a complete force–time history of a test and replaces the maximum force reading dial of the Texture Press. Some additional accessòries will also be described below.

The hydraulic drive mechanism and other characteristics of the ram press have been described previously (see p. 136). The force sensor in this model is the strain gauge type force transducer, which provides a continuous signal that is directly proportional to the force applied and is transferred electronically to the recorder. The recorder chart is 125 mm wide (5 in.). The recorder paper is driven forward by direct mechanical linkage to the ram; hence, the chart speed is a direct function of the ram speed. The actual chart speed depends on the gearing system that is used. Under normal conditions there is only one chart speed. A synchronous electric motor drive is offered as an option that allows the chart to be driven independently of the ram.

The original recorder used a pen with a response time of approximately 1 sec. The latest model, the TR-5 Texturecorder, uses an inkless thermal-writing recorder with a response time of 0.03 sec. A useful addition is the model TG-4B digital Texture gauge, which displays either transient force or maximum force on an electronic solid-state numerical display. In normal operation the chart moves forward as the hydraulic ram moves down and stops when the ram begins its upstroke to return to its original position.

Some additional options available in the model T-2100 texture test systems are the following:

1. Automatic ram cycling using solid-state electronic controls that cause the ram to cycle between preset distances enabling the instrument to be used for Texture Profile Analysis (TPA) and similar cyclic tests;

2. A number of different test cells, including (i) a single-blade shear cell with a flat cutting face; (ii) a meat shear cell with a V-shaped cutting face that is similar to the Warner–Bratzler Shear although it does not have the same dimensions as the standard Warner–Bratzler Shear blade; (iii) compression test accessories that can be used for gentle or extreme compression or cycling compression; (iv) a Succulometer test cell for pressing liquid from foods such as sweet

corn to measure juiciness; (v) a triple-beam bending test cell; (vi) a drill chuck that holds 3⁄16- and 5⁄16-in.-diam punches for puncture tests; (vii) a tension test cell; (viii) a thin-slice tension test cell.

3. The digital texture gauge measures peak force and displays it on a direct digital display.

4. A series of five model FTA force transducer load cells with force capacities of 0–30, 0–100, 0–300, 0–1000, and 0–3000 lb that can measure force in both compression and tension.

5. The model TRI integrating Texturecorder continuously measures the total area under the curve and shows the result on a digital display. This allows work functions to be obtained without having to measure the area under the curve by a planimeter.

Table 9 lists the test conditions that have been developed to measure textural properties of a number of foods in the FTC Texture Test System.

The Instron Universal Testing Machine

This instrument was designed for studying the stress–strain properties of materials. It can be set up to perform conventional tests in tension, compression, or bending, as well as a variety of more sophisticated tests such as hysteresis, stress relaxation, stress recovery, strain rate sensitivity, energy of deformation, and rupture. It is used for testing metals, wood, rubber, plastics, packaging materials, fibers, fabrics, adhesives, and many other materials. The machine is built on the modular principle: A large number of accessories for performing specific tests are available and can be attached to the machine as needed. A full account of the construction and operating characteristics of the machine has been published (Hindman and Burr, 1949; Burr, 1949).

A floor model Instron purchased for the author's laboratory in 1962 was one of the first units purchased specifically for research in physical properties of foods. The acquisition of this instrument was noted the following year (Anonymous, 1963). The basic paper describing how the machine was adapted to research on foods was published several years later (Bourne *et al.*, 1966), but several reports citing data obtained from this Instron were published prior to the 1966 paper (La Belle *et al.*, 1964; Bourne, 1965a,b, 1966b).

For the purposes of this chapter, model 1132, which was developed for the needs of the food industry, will be described. It is a bench model that is smaller than the floor model in the author's laboratory. A number of other models are available. Model 1132 is 100 cm wide, 58 cm deep, 135 cm high, weighs 140 kg, and can be supplied in metric or English units of measurement. The author recommends the metric model. A rectangular space of approximately 28 × 27 cm is available for working parts, and the moving crosshead has a stroke length of 70 cm.

Briefly, the machine consists of two parts: (1) the drive mechanism, which drives a horizontal crosshead in a vertical direction by means of twin lead screws at selected speeds in the range 0.02–50 cm min^{-1}; and (2) the force-sensing and recording system, which consists of load cells whose output is fed to a strip chart recorder. The A, B, and C load cells have a maximum capacity of 0.5, 5, and 50 kg, respectively, and are provided as compression cells or tension cells only. The D,F, and G load cells have a maximum capacity of 500, 5000, and 10,000 kg, respectively, and are reversible, that is, they can be used in both tension and compression. A sensitivity selector switch and several different load cells make it possible to obtain full-scale deflection of the recorder pen over the force range 0.2–500 kg (approximately 2 N to 5 kN). The time axis of the chart is either a direct measure of, or a simple multiple of the movement of the crosshead, depending upon the change gears used in the chart and crosshead drive chain. This attribute of the machine arises from the fact that both the recorder chart and the moving crosshead are synchronously driven from the same power supply. When the same gears are used in the chart power train as in the crosshead power train, 1 cm along the time axis of the chart represents exactly 1 cm of crosshead travel. When a set of gears is placed in the chart train different from those in the crosshead drive train, the distance along the time axis of the chart becomes some simple multiple of the distance that the crosshead moves. See Fig. 24.

The crosshead is driven by a synchronous motor and gear train that provides a crosshead speed accuracy of 0.1% of set speed and may be stopped, started, and reversed almost instantaneously. The stiffness of the machine, the backlash-free drive system, and the very low deflection of the weighing system make the Instron a "hard machine" that exhibits essentially no deflection under an applied load. This factor, coupled with the fact that the load system has very low mechanical inertia, means that the machine produces data that are essentially free from interactions of the test machine and the specimen.

In standard operation the load cell is inverted and mounted on the moving crosshead, the moving parts of the test cell are attached to the load cell, and the stationary parts of the load cell are attached to the machine base. Having the load cell in the inverted position eliminates any possibility of crumbs or juice falling down into the load cell and also eliminates having to compensate for sample weight for each test. Some operators prefer to mount the load cell on the bed of the machine with the test cell resting on top of the load cell. This has the advantage of simplifying the load cell calibration in compression with weights.

The interchangeable load cells measure the load directly and linearly, with high response speed and with no inertia. The load weighing accuracy is within ±0.5% of the full-scale force. The load cells should be calibrated daily or whenever freshly installed in the machine. The load cells may be calibrated by weights or by an electronic calibration system accessory module supplied by the manufacturer.

TABLE 9

GUIDELINES FOR TESTING VARIOUS FOODS IN THE FOOD TECHNOLOGY CORPORATION TEXTURE TEST SYSTEM[a,b]

Commodity	Type of test cell					Test principle	Parameters measured
	Shear–compression	Universal	Single blade	Meat shear	Other		
Fruits							
Apples	X	X				Compression–shear	Maximum force
					Penetration	Puncture	Yield point
Applesauce		X				Compression - orifice or back extrusion	Curve peak(s) Frequency and height
Apricots	X	X				Compression–shear	Maximum force
					Penetration	Puncture	Yield point
Cherries	X	X				Compression–shear	Maximum force
					Penetration	Puncture	Yield point
Fruit cocktail	X	X				Compression–shear	Maximum force–curve peaks
Citrus fruit	X	X				Compression–shear	Maximum Force
Mangoes	X	X				Compression–shear	Maximum force
					Penetration	Puncture	Yield point
Olives	X					Compression–shear	Maximum force
Peaches	X	X				Compression–shear	Maximum force
					Penetration	Puncture	Yield point
Pears	X	X				Compression–shear	Maximum force
					Penetration	Puncture	Yield point
Raisins	X					Compression–shear	Maximum force
Strawberries	X	X				Compression–shear	Maximum force
Meat and poultry							
Beef	X					Compression–shear	Maximum force
			X	X		Cutting	Maximum force

Frankfurters	X					Compression–shear	Maximum force
			X	X		Cutting	Maximum force
Lamb	X					Compression–shear	Maximum force
			X	X		Cutting	Maximum force
Pork	X					Compression–shear	Maximum force
			X	X		Cutting	Maximum force
Rabbit	X					Compression–shear	Maximum force
			X	X		Cutting	Maximum force
Luncheon meat	X					Compression–shear	Maximum force
					Tension	Tensile	Maximum force–curve slope
					Thin-slice tensile	Tensile	Maximum force–curve slope
Chicken	X					Compression–shear	Maximum force
			X	X		Cutting	Maximum force
Turkey	X					Compression–shear	Maximum force
			X	X		Cutting	Maximum force
Pheasant	X					Compression–shear	Maximum force
			X	X		Cutting	Maximum force
Poultry bones					Bending	Breaking force	Maximum force
Eggs					Compression	Breaking force	Maximum force
Miscellaneous							
Gels and semi-solids such as fats, jelly, paste, and pharmaceuticals	X					Compression–shear	Maximum force–curve slope
			X			Cutting	Maximum force–curve slope
					Penetration	Yield point	Maximum force
		X				Compression-orifice or back extrusion	Maximum force or curve slope and area
Cottage cheese	X	X				Compression–shear	Maximum force
Mushrooms	X	X				Compression–shear	Maximum force
			X	X		Cutting	Maximum force
Pasta	X	X				Compression–shear	Maximum force
Rice	X				Thin bladed	Compression–shear	Maximum force
Peanuts	X	X				Compression–shear	Maximum force

(continued)

TABLE 9 *Continued*

Commodity	Type of test cell: Shear–compression	Universal	Single blade	Meat shear	Other	Test principle	Parameters measured
Seeds	X	X				Compression–shear	Maximum force
Dough	X					Compression–shear tensile	Texture profile
Stems of flowers and plant material	X					Compression–shear	Maximum force
			X	X		Cutting	Maximum force
					Bending	Breaking point	Maximum force
Canned tuna fish	X					Compression–shear	Maximum force
					Succulometer	Compression	Liquid expressed
Dry or moist pet food	X		X	X		Compression–shear	Maximum force
Canned pet food	X	X			Penetration	Puncture	Yield point
Bread and bread products	X					Compression–shear	Maximum force
			X			Cutting	Maximum force
					Thin-slice tensile	Tensile	Maximum force–curve slope
Cakes and other similar bakery products	X					Compression–shear	Maximum force
			X			Cutting	Maximum force
Vegetables							
Asparagus			X	X		Cutting	Maximum force
Beans, dry-baked	X	X				Compression–shear	Maximum force
Beans, lima	X					Compression–shear	Maximum force
Beans, green	X					Compression–shear	Maximum force
			X			Cutting	Maximum force

Broccoli	X					Compression–shear	Maximum force
			X			Cutting	Maximum force
Carrots	X					Compression–shear	Maximum force
Celery	X					Compression–shear	Maximum force
			X			Cutting	Maximum force
Greens	X					Compression–shear	Maximum force
Corn, sweet	X					Compression–shear	Maximum force
					Succulometer	Compression	Liquid expressed
Eggplant	X					Compression–shear	Maximum force
Onions	X					Compression–shear	Maximum force
Peas	X					Compression–shear	Maximum force
Peppers	X					Compression–shear	Maximum force
Potatoes, white	X					Compression–shear	Maximum force
Potatoes, sweet	X					Compression–shear	Maximum force–curve ratio or slope
			X	X		Cutting	Maximum force
Tomatoes	X					Compression–shear	Maximum force
Tomatoes, paste		X				Compression–orifice or back extrusion	Maximum force–flow rate

[a]Compiled by Dr. B. A. Twigg.

[b]Note: Sample size and method of presenting sample (diced, whole, stripped, etc.) will vary within and among commodities, depending on purpose of test, test cell used, and the nature of the commodity. For example, size could be one discreet unit, full cell filled by volume or weighted sample that best accomodates the commodity and test cell. Various sample sizes and methods of presenting sample should be investigated, unless specific working conditions are recommended by the literature or by the manufacturer.

FIG. 24. The model 1132 Instron for work with foods. (Courtesy of Instron Corp.)

The electronic calibration system can be used with the load cell in any position. When weights are used for calibration, the calibration procedure depends on the position of the load cell:

A. Compression or tension/compression load cell positioned on bed of machine. Attach a circular plate to the load cell and place weights on this plate using the procedure outlined in the Instron manual.

B. Compression load cell attached to underside of crosshead (inverted position normally used for food work). Leave the load cell in the inverted position and proceed with the sequence of steps described below that calibrate the chart in reverse. The manufacturer recommends that the procedure described below be used for tension and tension/compression load cells only, but the author has found that this procedure is effective for compression cells provided the full-scale load switch is in the "1" position.

(i) Attach a hook or hanger to the load cell. The Instron Corporation supplies hangers for this purpose or a simple hanger can be made in the workshop.

(ii) Set the full-scale load switch to the "1" position. Use the ZERO and BALANCE control knobs to bring the pen exactly to the right-hand border of the chart (i.e., the normal full-scale position 10).

(iii) Hang the required weight, depending on the load cell capacity from the load cell. Use the CALIBRATION control knob to bring the pen exactly to the normal zero position at the left-hand border of the chart (position 0).

(iv) Remove the weight, check to see that the pen returns to the 10 position. Repeat steps (iii) and (iv) iteratively until both positions of the pen are correct.

(v) Reset the pen to the left-hand border of the chart (position 0) by adjusting the ZERO control knob. The load cell has now been calibrated. Attach the test cell to the load cell, use the BALANCE control knob to bring the pen back to zero, and begin testing.

C. Tension or tension/compression load cell attached to the underside of the crosshead. Follow the calibration procedures outlined in the manufacturer's operation manual.

Sometimes it is necessary to zero the pen at some point other than zero. For example, when performing Texture Profile Analysis (TPA) on foods that exhibit adhesiveness the pen should be zeroed at a point high enough on the chart to measure the adhesive force that will be below the zero line. This does not affect the calibration procedures described above; it simply means that in step B(v) the pen is set at a suitable level above the normal zero position.

Another way to calibrate the compression load cell in the inverted position is to fasten a ball-bearing pulley to the top of the Instron frame with a cord passing

FIG. 25. The wire frame, cord, and pulley arrangement that allows compression cells to be calibrated in the inverted position. (Courtesy of P. W. Voisey.)

over the pulley. A hook to hold the calibration weight is attached to one end of the cord, and a wire frame that spans the crosshead and transmits the force to the load cell is attached to the other end, as shown in Fig. 25 (Voisey, 1977c).

SI Calibration of the Load Cell

Since it is desirable to measure and record force directly in the SI unit of newtons (see p. 101), it is convenient to calibrate the load cell directly in newtons force instead of in grams, kilograms, or pounds force. One newton is equivalent to 101.9716 g force; hence, 100 g force is equal to 0.9807 N and 1 kg force is equal to 9.807 N. Therefore, any metric weight can be used for calibration of the load cell directly in newtons by calibrating to 98% of full scale instead of to 100%. A 98% calibration will give an error of 0.07%, which is less than the error in the recorder.

For example, if the "B" load cell is in the inverted position it is calibrated directly in newtons force by following steps (i), (ii), and (iv) outlined above, but modifying step (iii) to read: "Hang the required weight from the load cell. Use the CALIBRATION control knob to bring the pen exactly to the 0.2 position near the left-hand border of the chart (i.e., 98% of full scale)." Then continue with steps (iv) and (v). Full-scale force is now 1.0 N. Table 10 shows the SI forces for conventional metric load cells that have been calibrated to 98% of full scale. This procedure enables the researcher to record all data directly in newtons without having to record data in units of mass and then convert to units of force.

A metal adapter attaches to the inverted load cell by means of a threaded collar. This allows various test cell components to be attached to the load cell. The system is designed to accept a variety of test fixtures, including puncture, Warner–Bratzler Shear, all the test cells of the FTC Texture Test System, and cyclic deformation, which allows one to perform Texture Profile Analysis. An adapter is provided to accept the test cells of the Ottawa Texture Measuring System. The standard fixtures are easily mounted and readily interchangeable.

Almost any instrument that uses a linear motion to measure food texture can be duplicated in the Instron by fitting the appropriate test cell components into it. The complete force–distance curve that is obtained from the Instron frequently gives more information from the test than does the original device. Instruments that use a rotary or blending motion cannot be duplicated in the Instron.

One precaution that should always be taken when setting up a test cell in the Instron is to ensure that the crosshead travel stops are set in positions that positively prevent accidental metal-to-metal contact of the test cell parts. Most food tests involve compression in which metal parts approach each other, often at high speeds. Unless the crosshead travel stops are properly set, the metal parts may collide and, since there is no "give" in the machine, the load cell will be broken. Repair of a damaged load cell is expensive and time consuming.

TABLE 10

SI Calibration of Instron Metric Load Cells, Full-Scale Force[a]

Multiplier switch position	A load cell		B load cell		C load cell		D load cell		F load cell	
	98% (N)	100% (g)	98% (N)	100% (g)	98% (N)	100% (kg)	98% (N)	100% (kg)	98% (kN)	100% (kg)
1	0.1	10	1.0	100	10	1	100	10	1	100
2	0.2	20	2.0	200	20	2	200	20	2	200
5	0.5	50	5.0	500	50	5	500	50	5	500
10	1.0	100	10.0	1000	100	10	1000	100	10	1000
20	2.0	200	20	2000	200	20	2000	200	20	2000
50	5.0	500	50	5000	500	50	5000	500	50	5000

[a]Note: This is *not* a conversion table. When the load cell is calibrated to 100% of full scale the metric system of grams is in effect. When it is calibrated to 98% of full scale the SI units of newtons force applies.

Because of the wide range of test conditions and high sensitivity this machine is being widely used for research on textural properties of foods. It is a laboratory instrument. The construction of the machine and the electronic components render it unsuitable for operating in most food processing environments.

Depending on the information that is needed one can measure forces, including maximum force and multiple peaks, distances, area under the curve which represents work, and slopes of lines. The meaning of these various features needs to be interpreted for the particular test and the commodity on which it is used. Table 11 lists a number of types of tests and the experimental conditions under which the tests were performed in the Instron in the author's laboratory. This table shows the wide range of test principles, test cells, speed of travel, and range of forces that can be used in the Instron.

Since a number of different test principles may be used in the Instron (puncture, extrusion, shear, compression, bending, etc.), any reports of results obtained with the Instron should carefully specify the type of test and the conditions that were used. It is not satisfactory to report "the Instron test correlated highly (or poorly) with sensory scores." The test principle and all the operating conditions must also be stated. Another type of test performed in the same Instron may have given a better (or worse) correlation. The Instron will give poor results if an inappropriate test principle or the wrong conditions (e.g., speed of travel) are used.

Ottawa Texture Measuring System (OTMS)

This general purpose testing machine (see Fig. 26) was developed by the Engineering Research Service of Agriculture Canada under the direction of P. W. Voisey (Voisey, 1971b; Voisey *et al.*, 1972).

It consists of a single screw operated press that is driven by an electric motor via a gear box. A synchronous motor is used when a single constant speed is required for a specific commodity (e.g., the Ottawa Pea Tenderometer; see p. 141). Provision for four speeds can be arranged by using a pair of stepped pulleys in the drive train. A variable-speed motor with an infinite selection of speeds over the range 2–29 cm min^{-1} is used for research applications. This range may be changed by changing the gear ratio between the motor and the gear box input. The screw is driven via a brass pin, which breaks when overloaded, forestalling damage to the press.

The screw drives a carriage up and down two vertical guide rods. Adjustable stops on the guide rod can be set to stop, start, and reverse the motion of the carriage.

Force is detected by strain gauge force transducers attached to the carriage. A set of force transducers cover any force range up to the nominal 1500-kg force capacity of the instrument. A chuck attached to the bottom of the load cell is used

TABLE 11

CONDITIONS FOR TESTING FOODS IN THE INSTRON[a]

Commodity	Test principle	Test cell	Crosshead speed ($cm\ min^{-1}$)	Chart speed ($cm\ min^{-1}$)	Full-scale force (N)	Parameter measured
Agar gel	Deformation	Flat plate, 14.5 cm diam	0.1	20	2	ΔD between 1 and 2 N
Agar gel	Puncture	Rectangular punches	30	20	50	Yield point force
Apple, whole, skin off	Puncture	7/16-in.-diam Magness–Taylor tip	20	20	100–200	Yield point force and force at 5/16-in. penetration
Apple, slice	Back extrusion	10.15-cm i.d., 4-mm annulus	20	20	500–1000	Maximum plateau force
Apple, slice	Compression–extrusion–shear	Standard Kramer Shear cell	20	20	1000–2000	Maximum force
Apple, slice	Puncture	1/8-in. diam punch	30	10	10	Maximum force
Apple, 1.2 cm cube	TPA[b]	Flat plate, 7 cm diam	5	50	50–200	All TPA parameters
Apple, whole	Relaxation, recovery	Flat plate, 14.5 cm diam	1	5	500	ΔF for 2 min
Apricot, whole	Deformation	Flat plate, 14.5 cm diam	1	30	1	ΔD between 0.05 and 0.35 N
Apricot, skin off	Puncture	5/16-in.-diam Magness–Taylor tip	30	10	10–20	Yield point force and force at 5/16-in. penetration
Banana, skin off	Deformation	3/8-in.-diam punch	0.2	50	5	ΔD between 0.5 and 1.5 N
Beans, pea, red kidney soy, cooked	Puncture	1/8-in.-diam punch	30	10	20–50	Maximum force
Beans, green, raw	Puncture	1/8-in.-diam punch	20	10	50	Maximum force
Beans, green, raw	Back extrusion	7.35 cm i.d., 4-mm annulus	20	10	1000–2000	Maximum plateau force

Beans, green, canned	Back extrusion	10.15-cm i.d., 4-mm annulus	30	10	1000	Maximum plateau force
Beef, cooked, 1-cm cube	TPA	Flat plate, 7 cm diam	5	50	100–200	Texture profile parameters
Beets, red, canned	Back extrusion	10.15-cm i.d., 4-mm annulus	20	20	2000	Maximum plateau force
Beets, red, whole, raw	Deformation	Flat plate, 14.5 cm diam	0.5	50	2	ΔD between 0.1 and 1.1 N
Beets, blanched, 1-cm cube	TPA	Flat plate, 7 cm diam	5	50	10	Texture profile parameters
Bologna	Deformation	Flat plate, 14.5 cm diam	1	10	20	ΔD between 0.5 and 10.5 N
Bread, roll	Deformation	Flat plate, 14.5 cm diam	1	10	20	ΔD between 0.5 and 10.5 N
Bread, ½-in. cube	TPA	Flat plate, 7 cm diam	5	50	2	Texture profile parameters
Bun, hamburger	Deformation	Flat plate, 14.5 cm diam	1	5	20	ΔD between 0.5 and 10.5 N
Candy, rock	Deformation	Flat plate, 7 cm diam	0.1	50	20	ΔD between 0.5 and 10.5 N
Carrot, raw, cylinders 2 mm thick × 7.3 mm diam	Deformation	Flat plate, 7 cm diam	0.2	50	2	ΔD between 0.05 and 1.05 N
Carrot, raw	Puncture	3/32-in.-diam punch	20	10	20	Yield point
Carrot, raw, 1 cm diam × 2 cm high	TPA	Flat plate, 7 cm diam	1	20	100–500	Texture profile parameters
Carrot, canned	Back extrusion	10.15-cm-i.d., 4-mm annulus	30	10	1000	Maximum force
Cheese, cream	TPA	Flat plate, 7 cm diam	5	50	500	Texture profile parameters
Cheese, cheddar	Puncture	Rectangular punches	10	10	20	Maximum force

(continued)

TABLE 11 *Continued*

Commodity	Test principle	Test cell	Crosshead speed (cm min^{-1})	Chart speed (cm min^{-1})	Full-scale force (N)	Parameter measured
Cherries (sweet and Montmorency)	Puncture	Single Dunkley pitter	20	5	100	Maximum force
Cherries (sweet and Montmorency)	Multiple puncture	Array of 30 Dunkley pitters	10	20	1000–2000	Maximum force
Cherries (sweet and Montmorency), pitted	Puncture	⅛-in.-diam punch	20	5	10	Maximum force
Cherry pie filling	Back extrusion	10.15-cm i.d., 4-mm annulus	10	20	2000	Maximum force
Chocolate bars	Snap	Triple-beam assembly, 4.3 cm clearance between supporting beams	5	50	200	Maximum force
Cookies, ginger snap	Snap	Triple-beam assembly, 4.3 cm clearance between supporting beams	5	50	20	Maximum force
Corn, sweet, cut kernels	Back extrusion	10.15-cm i.d., 4-mm annulus	20	20	5000–20,000	Maximum force
Corn, sweet, on cob	Puncture	Diameters of 0.052, 0.076, 0.101, 0.128 in.	20	20	10–20	Maximum force
Cucumber, whole	Deformation	14.5-cm-diam plate	0.5	50	1	ΔD between 0.05 and 0.55 N
Cucumber, whole	Puncture	⅛-in.-diam punch	20	20	20	Yield point
Custard, egg	Puncture	1.13-cm-diam punch	0.5	50	0.2	Yield point
Cranberries, raw	Deformation	2-cm-diam plate	0.2	50	1	ΔD between 0.02 and 0.27 N

Egg, whole	Deformation	Flat plate, 7 cm diam	0.05	50	10	ΔD between 0 and 10 N
Frankfurter, beef	Deformation	Flat plate, 14.5 cm diam	1	10	20	ΔD between 0.5 and 10.5 N
Frankfurter, beef	TPA	Flat plate, 7 cm diam	5	50	1000	Texture profile parameters
Gari	TPA	Flat plate, 7 cm diam	5	50	20–100	Texture profile parameters
Grapes, single berry	Puncture	1/16-in.-diam punch	2	40	5–20	Maximum force
Grapes, single berry	Deformation	2-cm-diam plate	2	50	1	ΔD between 0.05 and 0.65 N
Lettuce, whole head	Deformation	Flat plate, 14.5 cm diam	1	10	20	ΔD between 0.5 and 10.5 N
Marshmallow	Deformation	Flat plate, 7 cm diam	1	10	2	ΔD between 0.05 and 1.05 N
Onions, boiled	Puncture	1/8-in.-diam punch	30	30	5–10	Yield point
Peas, green, raw	Multiple puncture	Matuometer, 143 × 1/8-in.-diam punches	20	10	2000	Maximum force
Peas, green, raw	Puncture	Single 1/8-in.-diam punch	30	10	20–50	Maximum force
Peas, green, raw	Back extrusion	10.15-cm i.d., 4-mm annulus	20	20	2000–10,000	Maximum force
Peas, green, canned	Back extrusion	10.15-cm-i.d., 4-mm annulus	20	20	2000	Maximum force
Peas, green, raw	Back extrusion	OTMS	20	20	5000	Maximum force
Peaches, raw, sliced	Puncture	1/8-in.-diam punch	20	20	5–10	Yield point
Peaches, sliced, canned or frozen	Back extrusion	10.15-cm i.d., 4-mm annulus	20	20	2000–5000	Maximum force
Peaches, whole, raw	Puncture	7/16-in.-diam Magness–Taylor tip	20	20	50	Maximum force
Peaches, fresh, 2 cm diam × 1 cm high	TPA	Flat plate, 7 in. diam	5	50	20–100	Texture profile parameters
Pears, whole, raw	Deformation	Flat plate, 14.5 cm diam	2	50	2	ΔD between 0.25 and 1.75 N

(continued)

TABLE 11 *Continued*

Commodity	Test principle	Test cell	Crosshead speed (cm min^{-1})	Chart speed (cm min^{-1})	Full-scale force (N)	Parameter measured
Pears, fresh, 2 cm diam × 1 cm high	TPA	Flat plate, 7 cm diam	5	50	50–200	Texture profile parameters
Pickle, cucumber slice	Puncture	0.086-in.-diam punch	20	10	20	Maximum force
Pretzel, large, 1-cm-high piece	TPA	Flat plate, 7 cm diam	5	50	5000	Texture profile parameters
Potato, whole, raw	Deformation	0.77-cm-diam punch with potato nestled in a bed of sand	0.2	50	1	ΔD between 0.06 and 0.26 N
Potato chip	Deformation	5/16-in.-diam probe, chip rests on a 3-cm-ring	1	50	2	Slope of initial line
Protein foam	Puncture	1.25-cm-diam punch	10	50	0.1–0.5	Inflection point
Plums, fresh, pitted	Back extrusion	10.15-cm i.d., 4-mm annulus	20	20	20,000	Maximum force
Plums, fresh, pitted	Puncture	1/8-in.-diam punch	20	20	20–50	Yield point
Soymeat analog, 1-cm cube	TPA	Flat plate, 7 cm diam	5	50	100	Texture profile parameters
Strawberries, whole, raw	Punch	Flat face, Dunkley pitter	50	5	20	Maximum force
Sauerkraut, 25-g sample	Compression	Flat plate, compress to 1 mm clearance	5	40	10,000–20,000	Maximum force
Tofu (soy curd)	Punch	1/8-in.-diam punch	10	10	5	Yield point
Tomatoes, raw, whole	Deformation	Flat plate, 7 cm diam	2	50	5	ΔD between 0.1 and 1.1 N
Turkey, roll, sliced	Tensile	Clamp at each end	1	5	100–200	Maximum force

[a] Compiled by M. C. Bourne and S. Comstock.
[b] TPA is Texture Profile Analysis.

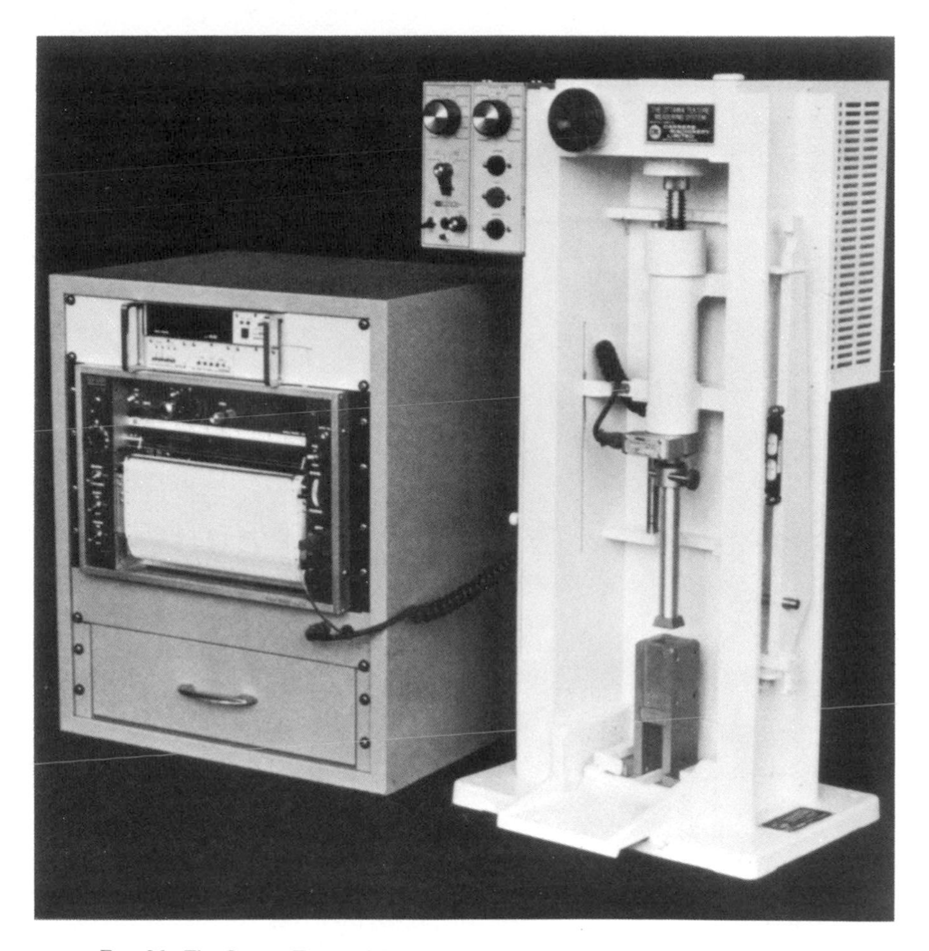

FIG. 26. The Ottawa Texture Measuring System. (Courtesy of Canners Machinery.)

to hold the moving parts of the test cells. The load cell is connected through an electronic signal conditioning unit to a potentiometric-type strip chart recorder.

A number of test cells have been developed for this instrument (Voisey and deMan, 1976). In addition, the test cells from most other instruments can be attached to the OTMS by means of an adapter. Table 12 lists test conditions for a number of foods.

The OTMS is manufactured by Canners Machinery Limited under license from Agriculture Canada. Special test cells and custom test cells and accessories for the OTMS are manufactured by Queensboro Instruments under license from Agriculture Canada. The Engineering Research Service of Agriculture Canada provide a service to select suitable measuring techniques for various products for

TABLE 12

LIST OF COMMODITIES TESTED ON THE OTTAWA TEXTURE MEASURING SYSTEM, INCLUDING CONDITIONS FOR TESTING[a]

Commodity	Test principle	Test cell	Crosshead speed (mm min^{-1})	Full-scale force (N)	Parameter measured
Apple, core	Compression	Flat plates, 114 mm bottom, 25.4 mm top	10	100	Modulus of elasticity, yield stress
Apple, whole, skin off	Puncture	6.0-mm punch	200	100	Maximum force
Apple, whole, skin off	Puncture	11.1-mm Magness–Taylor tip	180	100	Maximum force
Apple, canned slices	Extrustion	30-cm^2 OTMS cell with 9-wire (1.6-mm-diam) grid	162.5	500	Maximum force
Apple sauce	Extrusion	Back extrusion cell, 0.25-mm annulus	180	400	Graininess, plateau force and amplitude of force fluctuations during extrusions
Bacon	Puncture	2-mm-diam punch	20	50	Maximum force, peak angle, and number of peaks
Beans, baked	Extrusions	30-cm^2 OTMS cell with 9-wire grid	100	100	Maximum force
Bean sprouts	Shear	Warner–Bratzler shear cell	50	1500	Maximum force
Beef, ground and cooked	Extrusion	10-cm^2 OTMS cell with 8-wire (1.0-mm-diam) grid	50	1000	ITP parameters[b]
Beef, smoked	Extrusion	15-cm^2 OTMS cell with 7-wire (2.4-mm-diam) grid	50	1500	ITP parameters
Beef, raw	Shear	Warner–Bratzler meat shear cell	198	100	Maximum force
Beef, cooked	Shear	Warner–Bratzler meat shear cell	198	100	Maximum force
Candy bar, granola type	Shear	Modified Warner–Bratzler meat shear cell	20	200	Maximum force
Carrots, fresh	Deformation	114-mm-diam flat plates	180	50	ΔD between 4 and 30 N

Carrots, canned, frozen and thawed	Extrusion	30-cm^2 OTMS cell with 9-wire grid	100	1000	Maximum force and plateau force
Cherries, stem	Tension	OTMS stem-pulling cell	100	10	Maximum force
Cherries, whole	Compression	11-mm-diam Magness–Taylor probe	15	40	Maximum force and firmness (slope)
Cheese	Bite	OTMS bite tester, 30° wedges	50	200	Maximum force
Cod, filets cooked	Extrusion	20-cm^2 OTMS cell with 7-wire grid	100	1150	Maximum force
Cowpeas	Wedge	30° wedge blade	5–1000	10	Maximum force
French fries	Puncture	3.12 mm punch	100	20	Maximum force
Grains, wheat, oat, and barley	Puncture	0.4 mm punch	8	50	Maximum force
Ham, cured	Extrusion	15-cm^2 OTMS cell with 7-wire grid	100	400	Maximum force, energy
Herring, canned	Shear compression	Modified Kramer Shear cell	300	600	Maximum force
Icing, cake	Extrusion	OTMS tube extrusion cell	50	200	ITP parameters
Onions	Deformation	114-mm flat plates	150	40	ΔD between 2 and 28 N
Meat paste, raw	Extrusion	50-cm^2 OTMS cell with perforated plate insert	25	1000	Plateau force
Peas, fresh, cooked, canned	Extrusion	30-cm^2 OTMS cell with 9-wire grid	200	2500	Maximum force
Peaches, fresh, 20-mm core	Puncture, compression	2.4-mm-diam-punch 114-mm-diam flat plates	50 10	50 100	Bioyield force, modulus of elasticity, yield stress, energy
Peaches, canned	Shear compression	Standard Kramer Shear cell	100	800	Maximum force, energy
Potatoes, boiled, baked	Extrusion	30-cm^2 OTMS cell with perforated plate grid, 11.5-mm-diam holes	200	100	Maximum force
Potatoes, raw	Puncture	3.12-mm punch	200	20	Maximum force
Scallops, cooked	Extrusion	20-cm^2 OTMS cell with 7-wire grid	100	700	Maximum force
Soybeans, cooked	Extrusion	20-cm^2 OTMS cell with 7-wire grid	100	1000	Maximum force

(continued)

TABLE 12 *Continued*

Commodity	Test principle	Test cell	Crosshead speed (mm min^{-1})	Full-scale force (N)	Parameter measured
Spaghetti, cooked	Shear	OTMS multiblade shear cell	50–1550	200	ITP parameters
Spaghetti, cooked	Tensile (stickiness)	OTMS stickiness test cell	10	40	ITP parameters
Spaghetti, raw	Stress (bending)	OTMS bending test cell	25–1550	60	Maximum force
Strawberries, fresh	Puncture	2.4-mm-diam punch	50	20	Bioyield force
Tomato, fresh, frozen and thawed	Puncture	4.75-mm-diam punch	50	50	Bioyield force
Tomato, fresh, skin off	Extrusion	30-cm^2 OTMS cell with 9-wire grid	180	500	Maximum force
Tomato juice	Back extrusion	30-mm-i.d., 1-mm annulus	50	20	Plateau force, force fluctuations during extrusion (consistency, graininess)
TVP[c]	Extrusion	10-cm^2 OTMS cell with 8-wire grid	50	1000	ITP parameters
Tobacco, stalks from seedlings	Bending	OTMS bending cell	50	200	Maximum force
Wieners, core	Compression	114-mm-diam flat plates	50	50	Maximum force (rupture)
Wieners, core	Compression	114-mm-diam flat plates	50	100	Relaxation force/time
Wieners	Shear	Warner–Bratzler meat shear cell	50	100	Maximum force

[a]Compiled by P. W. Voisey and M. Kloek.
[b]Instrumental Texture Profile analysis.
[c]Textured Vegetable Protein.

organizations that wish to use the OTMS for quality control or research purposes. Inquiries should be addressed to Engineering and Statistical Research Institute, Agriculture Canada, Ottawa, Ontario K1A OC6, Canada.

The Tensipresser

This instrument, designed and manufactured in Japan, is intended for testing of foods, chemicals, and packaging materials in compression or tension. It features components that are particularly well suited for performing texture profile analysis under a wide range of test conditions.

The compression of the sample is accomplished by a single vertical screw driven by a step pulse motor that gives precise control of distance to within a few microns. Speeds of 1–7 mm sec^{-1} are available, and the speed is constant at each speed setting. A speed changer can reduce all seven speeds by a factor of 10. The speed can be converted to a sinusoidal mode by using a special adapter. The position of the compression head can be controlled to within very close limits between 0.1 and 240 mm clearance above the horizontal sample table. This allows precise control of the start, stop, and recycle set points. The maximum available travel of the compression head is 240 mm. The sample table is approximately 100 mm diam. A display indicates the position of the face of the compression plunger to 0.01 mm. The Tensipresser can be programmed to stop at any point in the compression cycle for 0–9.9 sec to allow relaxation to occur before continuing the test. It can be programmed to cycle from 1 to 999 times.

Force is measured by electronic force transducers with force ranges of 0–2, 0–10, and 0–20 kg. An overload protection mechanism stops the machine when the load cells reach 150% of their rated force capacity.

The instrument is provided with a strip chart recorder that uses a 25-cm-wide chart and has 0.25-sec pen response time, an oscilloscope monitor, a differentiater for measuring slopes and changes in slope, a digital integrater for measuring area under the force–time curve, and a data memorizer and processing system. The output from the force transducer is displayed on the oscilloscope and stored in the data memorizer. When a permanent record of the force–time curve is needed, the data is transmitted to the strip chart recorder at a rate slow enough to preclude pen response error. The force–time curve, or selected parts of it, can be magnified up to 19-fold to provide greater detail on points of particular interest.

The Tensipresser model TTP-50BX is 23 × 34 × 64 cm high and weighs 30 kg; the processing system type B is 23 × 23.5 × 64 cm high and weighs 16 kg; the recorder is 44.4 × 48 × 18 cm high and weighs 13 kg; the monitor scope is 22.3 × 30 × 28 cm high and weighs 8.5 kg. It has been used in Japan since 1975 for texture profile analysis of a number of foods. These studies, originally published in Japanese, are now becoming available in English (Tsuji, 1981, 1982).

Stevens Compression Response Analyzer

The most recent of the multiple measuring instruments is the Compression Response Analyzer, manufactured by C. Stevens and Sons (who also manufactures the Stevens LFRA Texture Analyzer). It provides travel speeds from 5 to 999 mm min^{-1}; four load cells with capacities of 5, 10, 20, and 50 kg; and a number of probes for puncture, compression, and cutting-shear tests. A LED display can be set to show maximum force, force at a preset compression distance, or distance the contact surface moves to achieve a predetermined force. There is no recorder for routine testing because the digital display provides sufficient information. For research purposes a recorder can be attached to the instrument to give a force–time plot.

Other Universal Testing Machines

A number of other manufacturers supply testing machines that provide a drive system and a force measuring and recording system. However, they do not provide test cells for food applications at the present time. With suitable adaptation, many of these testing machines would probably be just as suitable for working with foods as the Instron. The suppliers of some of these universal testing machines are listed in the Appendix. The innovative texture technologist may wish to explore the possibilities of adapting one of them to his needs. Some of them are quite economical.

CHAPTER 5

Viscosity and Consistency

Introduction

The previous two chapters dealt with texture of solid foods. This chapter deals with fluid foods. Unfortunately, the distinction between solid and fluid is not sharp and clear; consequently there is some overlap between the discussion in this chapter and the previous chapter.

The tendency of a fluid to flow easily or with difficulty has been a subject of great practical and intellectual importance to mankind for centuries. The famous English physicist Sir Isaac Newton (1642–1727) was one of the earliest researchers to study the flow of fluids. In his *Principia,* the section entitled "On the Circular Motion of Liquids," he stated the hypothesis that "the resistance which arises from the lack of slipperiness of the parts of the liquid, other things being equal, is proportional to the velocity with which the parts of the liquid are separated from one another." This principle, that the flow of fluid is directly proportional to the force that is applied, is used to describe the class of liquids known as "Newtonian fluids." Water is the best-known Newtonian fluid.

Other scientists have studied more complex liquids; for example, Schlubler in a 1828 paper on "The Fatty Oils of Germany" included within the physical constants a "fluidity ratio" using an instrument that is similar to some of the simple instruments that are currently used. Poiseuille (1797–1869) performed an elegant study of the flow of fluids in capillary tubes and may be considered as one of the founders of modern viscometry. Sir George Gabriel Stokes (1819–1903), who was president of the Royal Society from 1885 to 1892, studied the flow of liquids through an orifice and can be considered the founder of the efflux type of viscometer.

Some important definitions in viscometry are set out in the following:

Laminar flow is streamline flow in a fluid. *Turbulent flow* is fluid flow in which the velocity varies erratically in magnitude and direction.

The difference between laminar flow and turbulent flow is illustrated in Fig. 1. Suppose a fluid is being pumped through a pipe at a constant rate and a thin thread of colored solution is injected into the flowing stream. If laminar flow is occurring, the thread of colored solution will move straight down the tube. In the case of turbulent flow there are many eddies and currents, which are shown by the line of colored solution breaking up and forming eddies and vortices as it moves down the pipe. Laminar flow occurs at slow rates of flow and turbulent flow occurs at high rates of flow.

The Reynolds number (Reynolds, 1883) is a dimensionless number defined by an equation that can take several forms, one of which is the following:

$$\mathrm{Re} = 2\rho Q/\pi r\eta,$$

where Re is the Reynolds number (a dimensionless number); ρ, the density of liquid; Q, the rate of flow; r, the radius of pipe; and η, the viscosity. The point at which the onset of turbulence occurs is known as the critical Reynolds number Rc. The critical value of Reynolds number denotes the rate of flow at which the flow changes from laminar to turbulent flow. For pipe flow this occurs at approximately Rc = 2200 and is shown schematically in Fig. 1. Newtonian flow only occurs in the laminar region. Even a Newtonian fluid will lose its Newtonian behavior when turbulent flow begins. This critical Reynolds number determines the lowest velocity at which turbulent flow can take place, but it does not determine the highest velocity for the appearance of laminar flow. It is possible to obtain a laminar flow above the critical Reynolds number, particularly if the fluid is free of colloidal or suspended material and the pipe is very smooth, giving a metastable region that is somewhat analogous to supercooling and superheating effects that can be found when heating or cooling pure liquids. The point to remember is that a Newtonian fluid *appears* to be non-Newtonian when the shear rate is too high.

Dynamic viscosity. This term is frequently called "viscosity," or "absolute viscosity." It is the internal friction of a liquid or its tendency to resist flow. It is usually denoted by η and is defined by the equation

$$\eta = \sigma/\dot{\gamma},$$

where η is the viscosity; σ, the shear stress; and $\dot{\gamma}$, the shear rate.

The conventional unit of viscosity is the poise (P) (after Poiseuille, 1846). One poise is defined as that viscosity in which a velocity gradient of 1 cm sec^{-1} is obtained when a force of 1 dyne is applied to two surfaces 1 cm apart that encompass the liquid that is flowing. It has the dimensions $ML^{-1}T^{-1}$. Since the

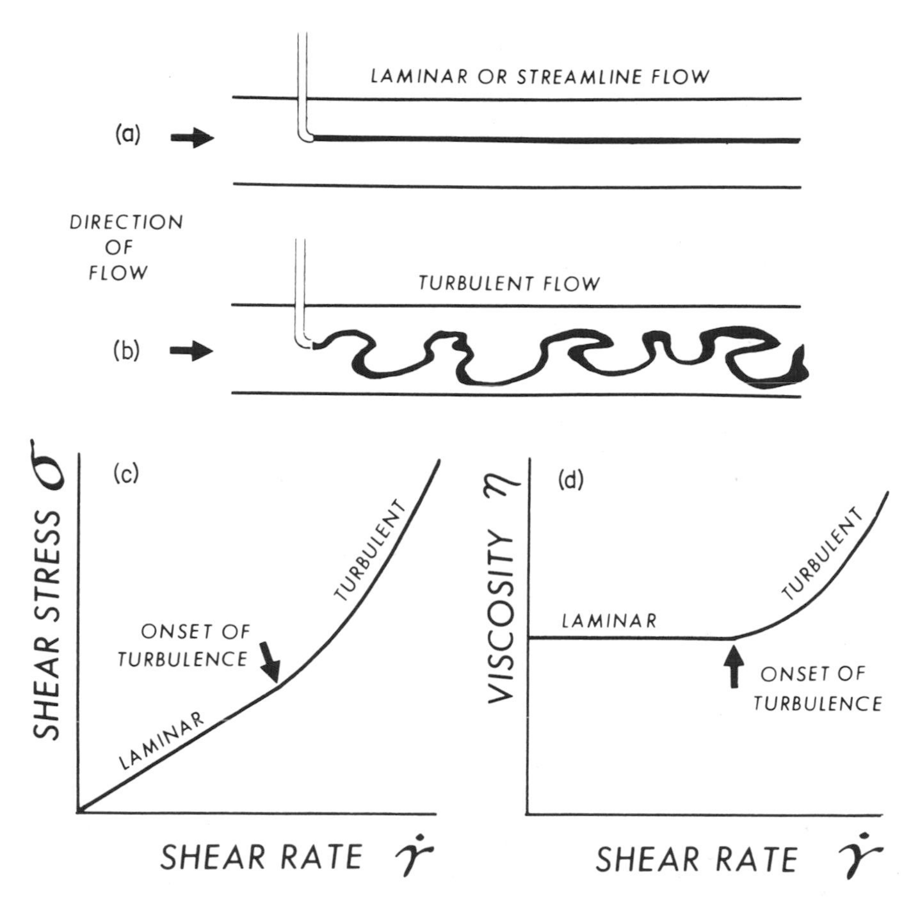

FIG. 1. The difference between (a) laminar and (b) turbulent flow; (c) shear stress versus shear rate for a Newtonian fluid in the laminar and turbulent flow range; (d) viscosity versus shear rate for a Newtonian fluid in the laminar and turbulent flow range.

poise is a fairly large unit of measurement, a more common unit for fluids is the centipoise (cP). One centipose equals 0.01 P. Table 1 shows some typical viscosities of fluids. It is worth noting that water at 20°C has a viscosity of 1.0 cP.

According to the International Organization for Standardization (ISO) the SI unit for dynamic viscosity is the pascal second (Pa-sec). The conversion factor is 1 Pa-sec is 10.000P, or 1 cP is one millipascal second (mPa-sec). However in a 1968 report, the Royal Society (London) recommended that the poise continue to be allowed to be used in conjunction with the SI unit for viscosity. In this chapter we shall continue to use the poise, but the reader is reminded that dynamic viscosity can also be expressed in pascal seconds.

TABLE 1
SOME TYPICAL VISCOSITIES

Substance	Viscosity (cP or mPa-sec)
Water (0°C)	1.7921
Water (20°C)	1.000
Water (100°C)	0.2838
20% Sucrose solution (20°C)	1.967
40% Sucrose solution (20°C)	6.223
60% Sucrose solution (20°C)	56.7
80% Sucrose solution (20°C)	40,000
Diethyl ether (20°C)	0.23
Glycerol (20°C)	1759

Fluidity. This is the reciprocal of dynamic viscosity. It is occasionally used in place of viscosity. It is denoted by ϕ, and is defined by the equation

$$\phi = \dot{\gamma}/\sigma.$$

Kinematic viscosity. This defined as the absolute viscosity divided by the density of the fluid. It is usually denoted by ν:

$$\nu = \eta/\rho = \sigma/\rho\dot{\gamma},$$

where ν is the kinematic viscosity in stokes; η, the absolute viscosity in poise; and ρ, the density in grams per cubic centimeter.

The conventional unit of kinematic viscosity is the stoke (after Stokes, 1819–1903). It has the dimensions M^2T^{-1}. One centistoke equals 0.01 stoke. The SI unit for kinematic viscosity is the meter-square-second but the Royal Society (London) recommends that the stoke continue to be allowed to be used in conjunction with the SI unit.

Kinematic viscosity is measured in efflux viscometers because the rate of flow in this type of viscometer is proportional to density as well as viscosity. Kinematic viscosity is widely used in the petroleum industry where the specific gravity of liquid hydrocarbons does not vary widely. Kinematic viscosity is not used in the food industry to the same extent because a wide range of densitites can be encountered, which compresses the kinematic viscosity into a smaller range than the absolute viscosity. This is exemplified in Table 2, which shows the absolute viscosity and kinematic viscosity of sucrose solutions. The absolute viscosity changes from 1.0 cP for water to 480.6 cP for 70% syrup, while over the same range the kinematic viscosity changes from 1.0 to 357.4 centistokes (cS).

Relative viscosity. This is sometimes called the "viscosity ratio," which is

TABLE 2

VISCOSITY AND DENSITY OF AQUEOUS SUCROSE SOLUTIONS AT 20°C

% Sucrose	Specific gravity	Absolute viscosity η (cP)	Kinematic viscosity ν (cS)
0	1.00	1.00	1.00
20	1.083	1.97	1.82
40	1.179	6.22	5.28
60	1.289	56.7	44.0
70	1.350	480.6	357.4
74	1.375	1628	1188

the ratio of the viscosity of a solution to the viscosity of the pure solvent and is defined by the equation

$$\eta_{rel} = \eta/\eta_s,$$

where η_{rel} is the relative viscosity; η, the viscosity of solution; and η_s, the viscosity of solvent.

Apparent viscosity. This is the viscosity of a non-Newtonian fluid expressed as though it were a Newtonian fluid. It is a coefficient calculated from empirical data as if the fluid obeyed Newton's law. This concept will be discussed in more detail on p. 209. The symbol η_a is used to denote apparent viscosity.

Shear stress. This is the stress component applied tangential to the plane on which the force acts. It is expressed in units of force per unit area. It is a force vector that possesses both magnitude and direction.

The nomenclature committee of the Society of Rheology recommends that σ be used to denote shear stress in simple steady shear flow and that τ be used to denote relaxation time or retardation time [*Rheol. Bull.* **43**(2), 6 (1974)]. In accordance with this convention, σ will be used to denote shear stress in this chapter. However, the reader is cautioned that many rheologists continue to use τ to denote shear stress (for example, see Figs. 9 and 14, pps. 213 and 218, respectively).

Shear rate. This is the velocity gradient established in a fluid as a result of an applied shear stress. It is expressed in units of reciprocal seconds (sec^{-1}).

The nomenclature committee of the Society of Rheology (see above) recommends that $\dot{\gamma}$ be used to denote shear rate and that γ be used to denote shear strain. The use of $\dot{\gamma}$ to denote shear rate is conventional among rheologists and will be used in this chapter.

Krumel and Sahar (1975) give some useful guidelines that enable one to think in practical terms of what various shear rates mean when related to well-known phenomena. A shear rate of 0.1 sec^{-1} approximates rate of film sag, or flow of film over a vertical plate; 0.1–10 sec^{-1} approximates the rate of flow of a normal Brookfield reading; 50 sec^{-1} approximates the shear rate in the mouth; 10–100 sec^{-1} approximates the shear in tumbling or pouring; 100–1000 sec^{-1} approximates the shear rate in most home mixers; >1000 sec^{-1} approximates the shear rate in a blender.

Factors Affecting Viscosity

Temperature

There is usually an inverse relationship between viscosity and temperature. Typical data are shown in Fig. 2, which plots the viscosity of water and some sucrose solutions as a function of temperature. Note also from Table 1 that the viscosity of water at 0°C is 1.79 cP falling steadily to 0.28 cP at 100°C.

Concentration of Solute

There is usually a direct nonlinear relationship between the concentration of a solute and viscosity at constant temperature. Figure 3 shows the viscosity– concentration behavior of salt solution and sucrose solutions at constant temperatures. It is typical of the concentration effect on viscosity. Table 1 also shows this phenomenon: Water at 20°C has a viscosity of 1 cP, while 80% sucrose solution has a viscosity of approximately 40,000 cP.

Molecular Weight of Solute

There is usually a direct nonlinear relationship between the molecular weight of the solute and the viscosity of the solution at equal concentrations. Figure 4 shows the viscosity of corn syrups as a function of molecular weight. Corn syrup is made by hydrolyzing by degrees high molecular weight starch into dextrose, a simple hexose monosaccharide. The abbreviation D.E. refers to "dextrose equivalent" and means the equivalent reducing activity of pure dextrose. A "36-D.E." syrup means that 100 g of corn syrup solids has the same chemical reducing capacity as 36 g of pure dextrose. A low D.E. means a long chain length and high molecular weight oligosaccharide. Figure 4 shows that 5-D.E. corn syrup (consisting principally of long-chain oligosaccharides) has a much higher viscosity at the same solids concentration than lower average molecular weight corn syrups of equal concentration.

Pressure

The viscosity of most liquids is essentially constant over a pressure range of 0–100 atm. Hence the pressure effect can usually be ignored for foods.

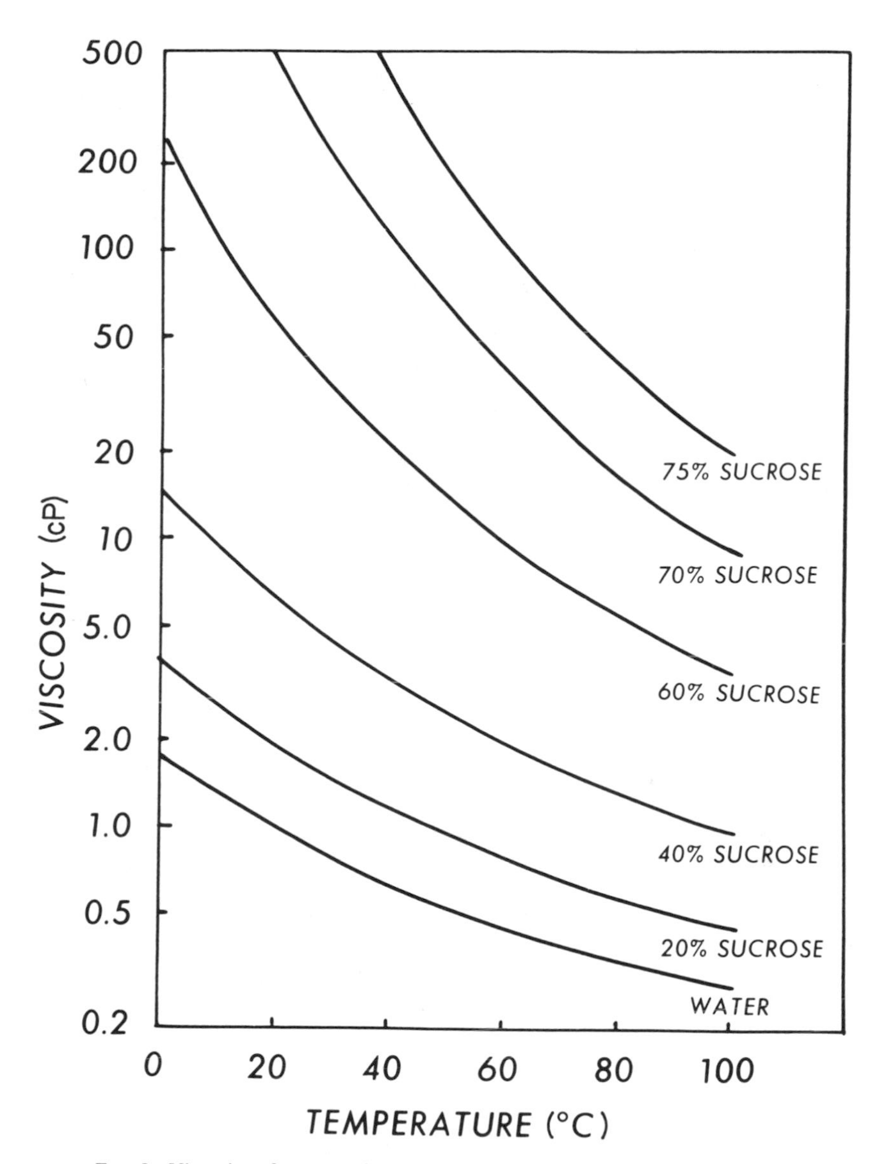

FIG. 2. Viscosity of water and sucrose solutions as a function of temperature.

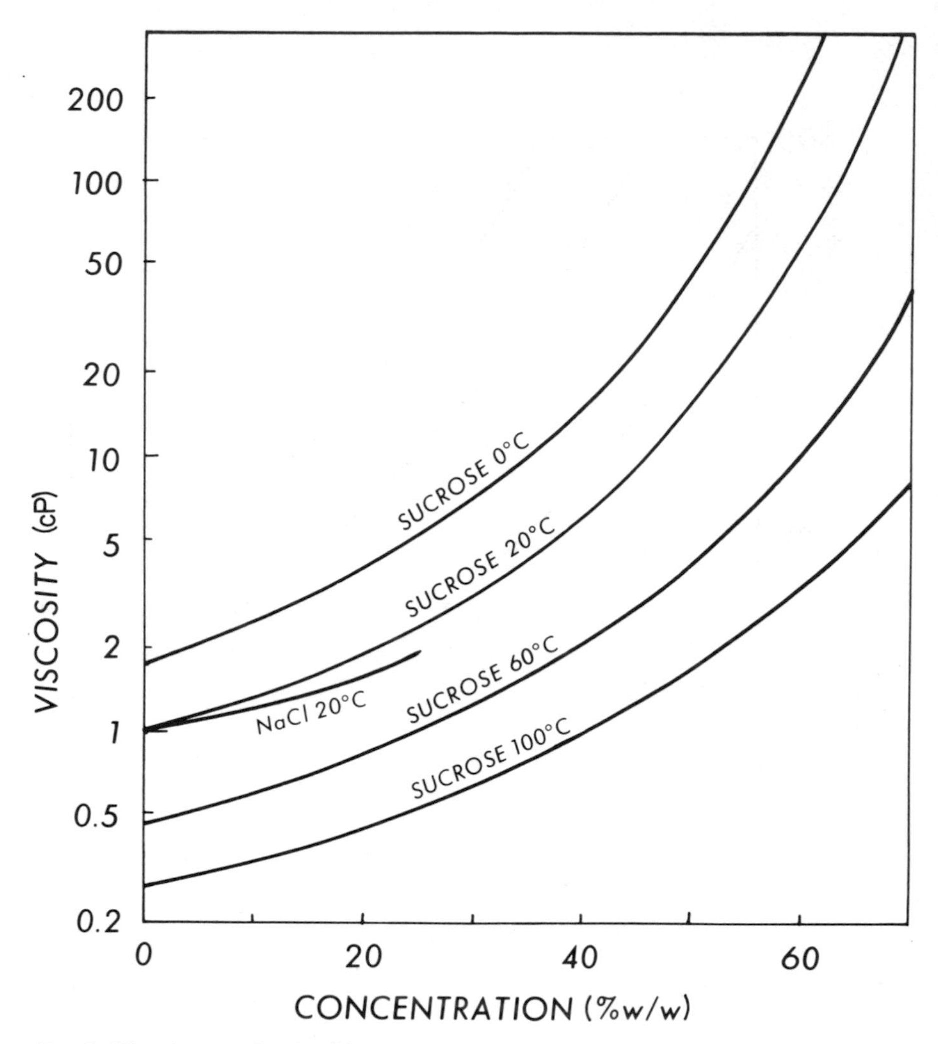

FIG. 3. Viscosity as a function of concentration for sucrose solutions at four temperatures and sodium chloride.

Suspended Matter

This usually increases the viscosity slightly when in low concentrations, but high concentrations of suspended matter can cause substantial increases because of entanglement between the particles. High concentrations of suspended matter usually renders the product non-Newtonian and can lead to plastic flow or dilatant flow (see p. 208).

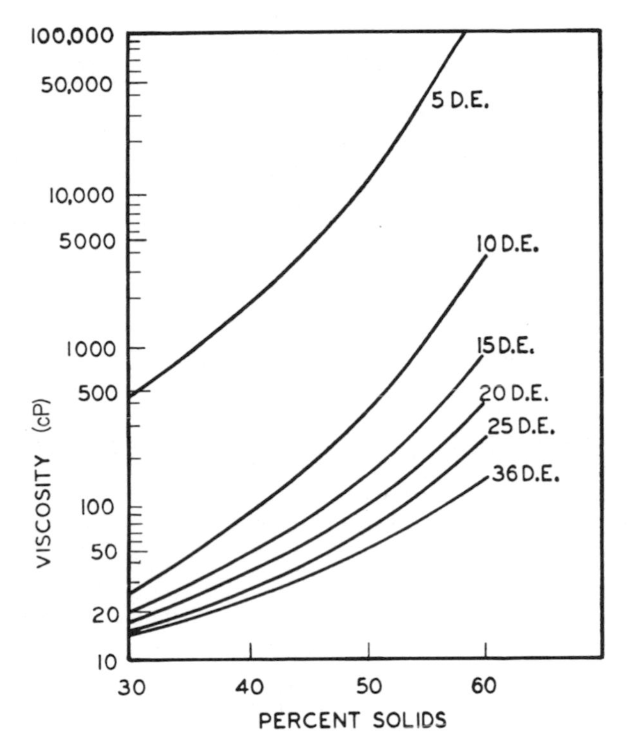

FIG. 4. Viscosity–concentration–molecular weight relationships for hydrolyzed cornstarch syrups. (From Murray and Luft, 1973; courtesy of Grain Processing Corp. Reprinted from Food Technol. **27** (3), 33, 1973. Copyright by Institute of Food Technologists.)

Types of Viscous Behavior

Newtonian

This is true viscous flow. The shear rate is directly proportional to the shear stress and the viscosity is independent of the shear rate within the laminar flow range. The viscosity is given by the slope of the shear stress–shear rate curve (see Fig. 5). Typical Newtonian fluids are water, and watery beverages such as tea, coffee, beer, and carbonated beverages, sugar syrups, most honeys, edible oils, filtered juices, and milk. A Newtonian fluid possesses the simplest type of flow properties. The characteristics of this type of flow are adequately described by the equation given above ($\eta = \sigma/\dot{\gamma}$). A fluid with high viscosity is called "viscous" while a fluid with low viscosity is called "mobile."

Unfortunately, most of the fluid foods encountered in the food industry are not Newtonian in nature; in fact, they deviate very substantially from Newtonian

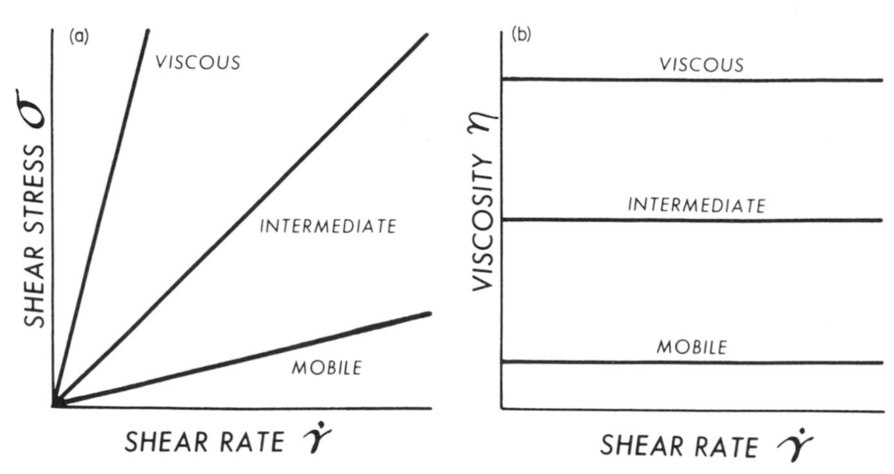

FIG. 5. Newtonian flow: (a) shear stress versus shear rate (note that the straight lines begin at the origin); (b) viscosity versus shear rate (the viscosity remains constant with changing shear rate).

flow. And yet there often seems to be a mental fixation on Newtonian-type flow. Although many instruments that satisfactorily measure Newtonian flow are far from satisfactory for measuring the flow properties of non-Newtonian fluids, one often sees food scientists using equipment designed for Newtonian fluids to measure viscous properties of non-Newtonian fluids. Much confusion is found in the literature because the viscous properties of non-Newtonian fluids has been measured by instruments that are applicable to Newtonian fluids and the data is erroneously interpreted using the concepts of Newtonian fluids.

Non-Newtonian Fluids

Most fluid and semifluid foods fall into one of several classes of non-Newtonian fluids.

Plastic (or Bingham). A minimum shear stress known as the "yield stress" must be exceeded before flow begins. This type of flow is often found in foods. Typical examples of this type of flow are tomato catsup, mayonnaise, whipped cream, whipped egg white, and margarine. This type of flow is named after Bingham (1922), who studied the flow properties of printing inks and discovered the important principle that no flow occurs at low stress. He identified the point at which flow begins as the "yield stress." The term "plastic" refers to materials that exhibit this yield stress; it does not refer to synthetic plastics.

Figure 6 shows the characteristics of plastic flow for three fluid foods. Fluid A has a low yield stress; the rate of flow (shear rate) is directly proportional to the shear rate after the yield stress has been exceeded. Fluids B and C have a higher

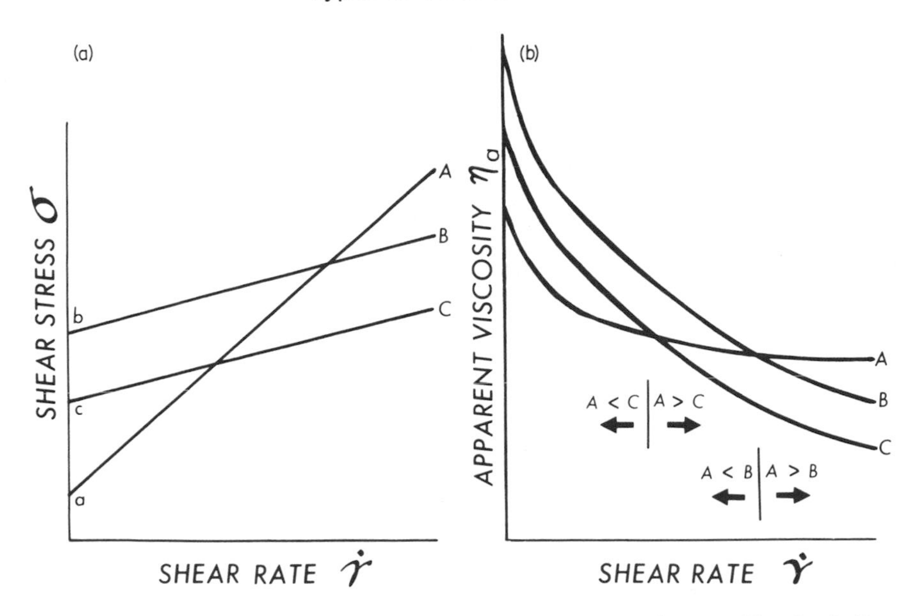

FIG. 6. Plastic flow for three foods A, B, and C. (a) Shear stress versus shear rate. Note that the lines do not begin at the origin. There is always an intercept (''yield stress'') on the vertical axis. (b) Apparent viscosity versus shear rate for same three foods. The apparent viscosity decreases with increasing shear rate. Note that the apparent viscosity of fluid A may be greater or less than that of fluids B and C, depending on the shear rate at which the measurement is taken.

yield stress than A. The rate of flow of fluids B and C is also directly proportional to the shear rate after the yield stress has been exceeded. Table 3 lists published values for yield stress of some plastic foods.

Apparent viscosity was defined as the viscosity of a non-Newtonian fluid. Since, in a Newtonian fluid, the flow rate is directly proportional to the shear stress and the curve begins at the origin, a single-point measurement suffices to establish viscosity. One simply measures the shear stress at a standard shear rate, or the shear rate at a standard shear stress, and by drawing a line from there to the origin obtains the true Newtonian viscosity. This is known as a ''one-point test'' and is quite satisfactory for specifying the viscosity of Newtonian fluids.

When this test is used (as is commonly done) on a plastic fluid, the apparent viscosity will change, depending upon the shear rate. Figure 7 shows how apparent viscosity is measured. Suppose the viscosity of a Newtonian fluid is measured at shear rate *a* and shear rate *b*. The shear stress measured at shear rate *a* (*Na*) is marked on the graph and a line drawn from that point to the origin. Similarly the shear stress is measured at shear rate *b* (*Nb*) and a line drawn from this point back to the origin. The slope of the line at both shear rates is the same; this is characteristic of a Newtonian fluid.

TABLE 3

VALUES FOR PLASTIC YIELD STRESS OF SOME FOODS[a]

Type of food and condition	Yield stress (dyn cm^{-2})
Chocolate, melted	12
Cream, whipped	400
Guar gum, 0.5% solids, in water	20
Guar gum, 1.0% solids, in water	135
Orange juice, concentrated 60° Brix	7
Pear puree, 18.3% solids	35
Pear puree, 45.7% solids	339
Protein from yeast, 10% solids	0
Protein from yeast, 25% solids	42
Protein from soy isolate, 20% solids	1271
Protein, whey, 20%	21
Sucrose, 75% in water	0
Tomato puree, 11% solids	20
Xanthan gum, 0.5% solids, in water	20
Xanthan gum, 1.2% solids, in water	45

[a]From Rha (1980).

In contrast, when a one-point measurement is made at shear rate *a* on a Bingham plastic the apparent viscosity is the slope of the line *OPa;* at shear rate *b* the apparent viscosity is *OPb*. The apparent viscosity changes as the shear rate changes. This explains why the term ''apparent viscosity'' is used because it implies a Newtonian-type measurement on a non-Newtonian fluid. Figure 7 demonstrates the difficulties that can arise from using Newtonian concepts for non-Newtonian fluids. A plot of apparent viscosity versus shear rate for three Bingham fluids is shown in Fig. 6b. This should be compared with Fig. 5b. One problem that arises with the use of the concept of apparent viscosity is that fluid A can appear to be more viscous or less viscous than fluids B and C, depending on the shear rate at which the test was performed (see Fig. 6b).

Plastic flow is not always as simple as shown in Fig. 6. Houwink (1958) pointed out that the shear stress–shear rate curve for plastic fluids is usually curved at low shear rates and he postulated three yield values, which are shown in Fig. 8. The extrapolation of the straight-line position of the experimental curve to zero shear rate gives true plastic or Bingham flow. The downward curvature of the experimental curve at low shear rates is often found in practice. The shear stress at which curvature begins in the shear stress–shear rate plot is defined as the ''upper Houwink yield value''; the intercept on the vertical axis from the extrapolation of the straight-line part of the curve is known as the ''extrapolated yield value'' or Bingham value; and the actual intersection of the shear stress–

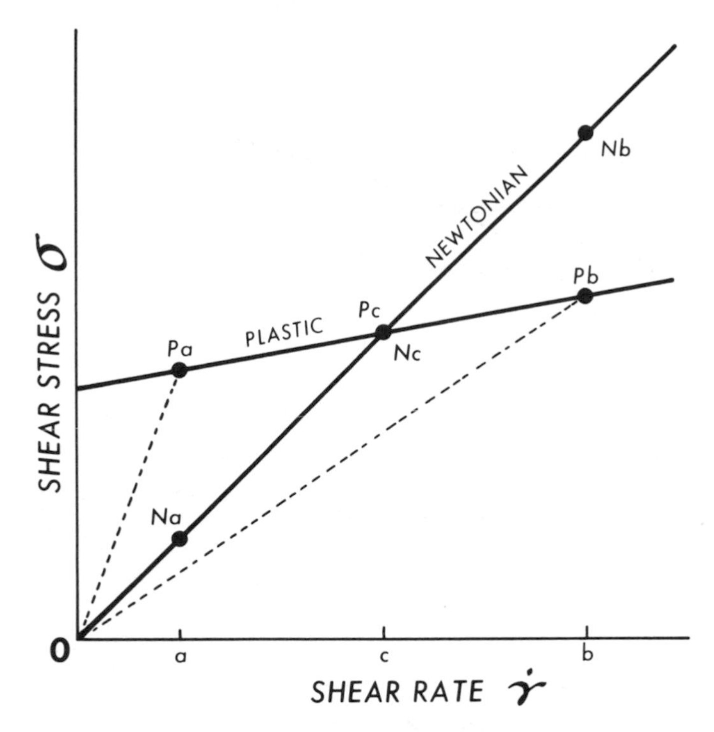

FIG. 7. Shear stress–shear rate plots for a Newtonian fluid and a plastic fluid. Note that the viscosity of the Newtonian fluid *N* is the same when measured at shear rates *a*, *b*, and *c*, whereas the apparent viscosity of the plastic fluid *P* is different at each shear rate.

shear rate plot on the vertical axis is known as the "lower Houwink yield value." The deviation from linearity of plastic flow at low shear rate is sometimes of importance but for some foods the deviation is so small that it can be safely ignored. For example, Fig. 9 shows the experimental shear stress–shear rate plot of a meat extract that shows true Bingham behavior with no curvature at low shear rates.

Another type of plastic flow is the type in which the shear stress–shear rate plot is nonlinear above the yield stress. The curve may be concave downward (dilatant with a yield stress), or convex downward (pseudoplastic with a yield stress). It is sometimes known as the "mixed type." This type of flow is described by the Herschel–Bulkley equation, which is discussed on p. 217 and is illustrated in Fig. 12.

Pseudoplastic. In this type of flow an increasing shear force gives a more than proportional increase in shear rate, but the curve begins at the origin. The term "pseudoplastic" was originated by Williamson (1929); it does not refer to synthetic plastics. Salad dressings are a good example of this type of flow. Fig.

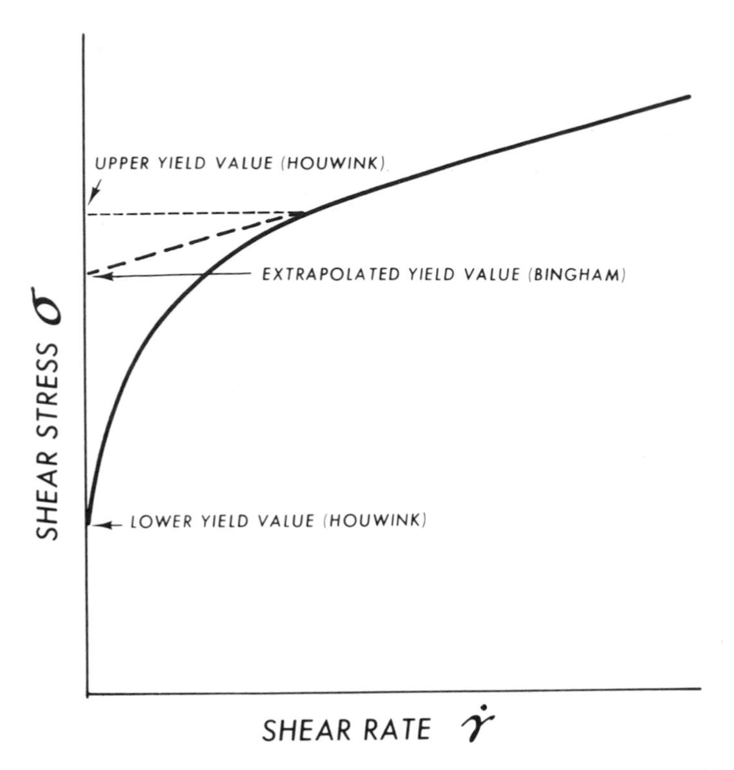

FIG. 8. The upper yield value, extrapolated yield value, and lower yield value that is found in some plastic fluids.

10b shows that the apparent viscosity of a pseudoplastic fluid is dependent upon the shear rate and, as in the discussion of plastic flow, it illustrates the danger of using a single-point measurement and Newtonian concepts for specifying the flow characteristics of a pseudoplastic fluid. Many pseudoplastic fluids exhibit nearly linear shear stress–shear rate behavior at low shear rates. This is called the "Newtonian regime."

Dilatant flow. The shear stress–shear rate plot of this type of a flow begins at the origin but is characterized by equal increments in the shear stress giving less than equal increments in the shear rate (Fig. 11). Examples are high solids, raw starch suspensions, and some chocolate syrups. This type of flow is only found in liquids that contain a high proportion of insoluble rigid particles in suspension. It can be demonstrated quite simply by making a 60% cornstarch suspension. Dilatant flow is fairly rare in the food industry and extremely rare in finished food products.

This type of flow is described as "dilatant" because it is associated with an increase in volume of the fluid as flow occurs, and it only occurs in high

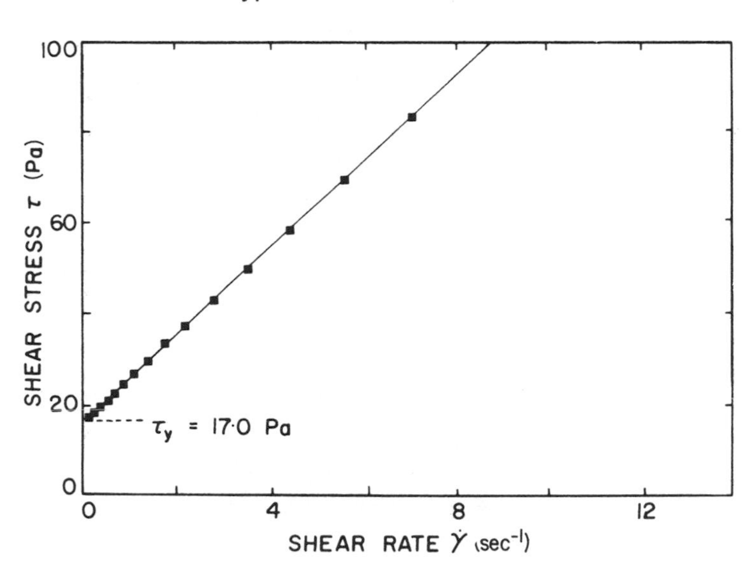

FIG. 9. Shear stress–shear rate plot for a concentrated meat extract (T = 77°C). This is a true Bingham plastic that shows a linear relationship all the way down to zero shear rate. (Courtesy of Dr. A. L. Halmos and Dr. C. Tiu. Reprinted from *J. Texture Stud.*; with permission from Food and Nutrition Press.)

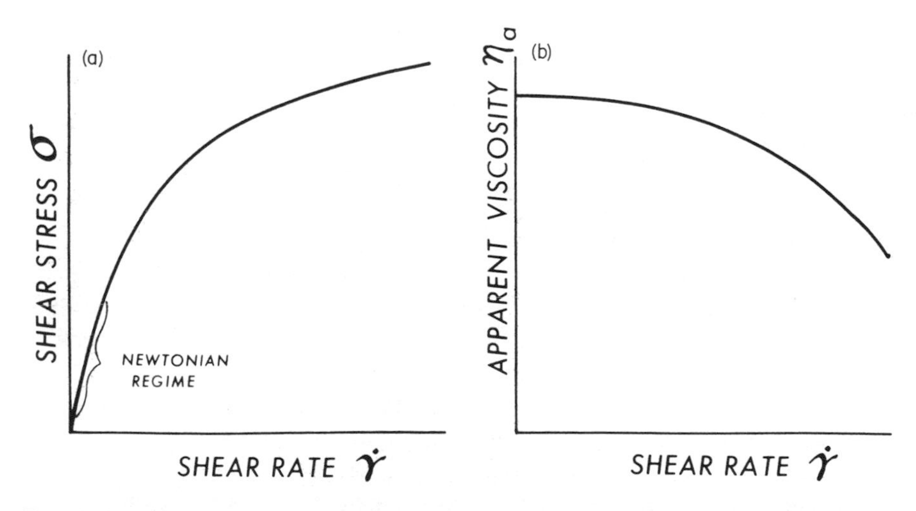

FIG. 10. Pseudoplastic flow: (a) shear stress versus shear rate (note the convex line that begins at the origin); (b) apparent viscosity versus shear rate (note that the apparent viscosity decreases with increasing shear rate).

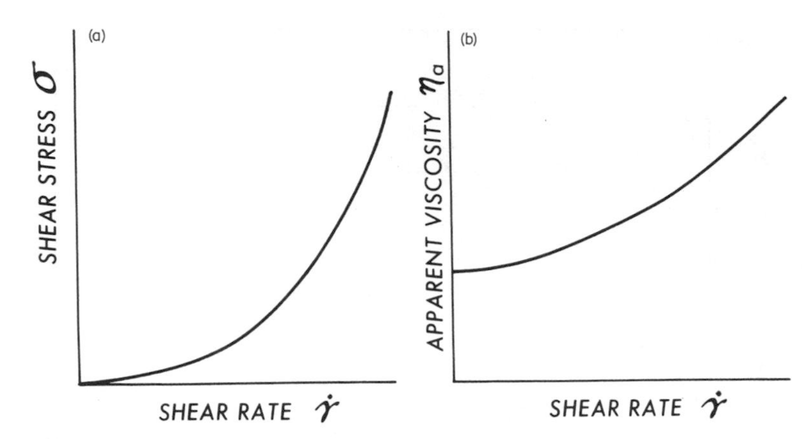

FIG. 11. Dilatant flow: (a) shear stress versus shear rate (note the concave line that begins at the origin); (b) apparent viscosity versus shear rate (note that the apparent viscosity increases with increasing shear rate).

concentration suspensions. Reynolds (1883), who introduced the term "dilatancy," gave quicksand as as an example, stating:

> When the water-to-sand ratio is such that there is just enough water to fill all the voids, and when the volume of voids is at a minimum, any shear applied to force that material to flow disturbs the position of the particles and causes a dilation of the voids. This leads to the situation in which the total volume of the voids is greater than the volume of water present. This results in an apparent partial dryness which increases the resistance of the material to shearing stress. The dryness is the result of the time necessary for the capillary forces to provide the additional water required for complete saturation. When the pressure is removed, the sand becomes wet because the voids contract, and the water which has become excess escapes at the surface.

An equally good example of this type of behavior can be found with a 60% suspension of cornstarch in water.

True dilatancy can probably exist in any suspension so long as the concentration is high enough for the material to exist in closely packed form. The property of dilatancy disappears when the suspension is diluted. For example, a 40% cornstarch suspension in water shows no dilatant properties. The densest packing of spheres is about 74% and one of the least-dense packing is about 37%. Hence it is usual to find that the property of dilatancy only appears in suspensions between about 40 and 70% solids concentration.

Some fluids that do not dilate when sheared may still exhibit a dilatant type of shear stress–shear rate behavior; that is, equal increments in shear stress give less than equal increments in shear rate. The general term "shear thickening" applies to these fluids as well as to dilatant fluids.

The General Equation for Viscosity

All the above types of flow can be described by the equation

$$\sigma = b\dot{\gamma}^s + C,$$

where σ is the shear stress; b, a proportionality factor (for a Newtonian fluid this factor is the viscosity η); C, the yield stress; s, the pseudoplasticity constant, which is an index of the degree of nonlinearity of the shear stress–shear rate curve; and $\dot{\gamma}$, the shear rate. Figure 12a shows all types of flows in a single graph. Newtonian flow is represented by a straight line starting at the origin; dilatant flow starts at the origin and is concave downward, while pseudoplastic flow starts at the origin and is concave upward. Plastic flow does not begin at the origin and is linear while mixed-type flow is curvilinear with a yield stress and may be concave upward or downward.

Some authors publish a shear rate–shear stress curve instead of the conventional shear stress–shear rate curve. Figure 12b plots the same types of flow as Fig. 12a but with the position of the axes interchanged. One should learn to recognize the identity of the various types of flow on both types of plot.

The general equation for viscosity can be used for all of the above types of flow. Table 4 lists the values for the exponent s and the intercept C for the various types of flow, the form of the general equation that can be used, and a simplified version of the general equation that can be used for that particular type of flow. For example, the constant C (yield stress) can be dropped out of the equation for dilatant, Newtonian, and pseudoplastic flow because there is no yield stress.

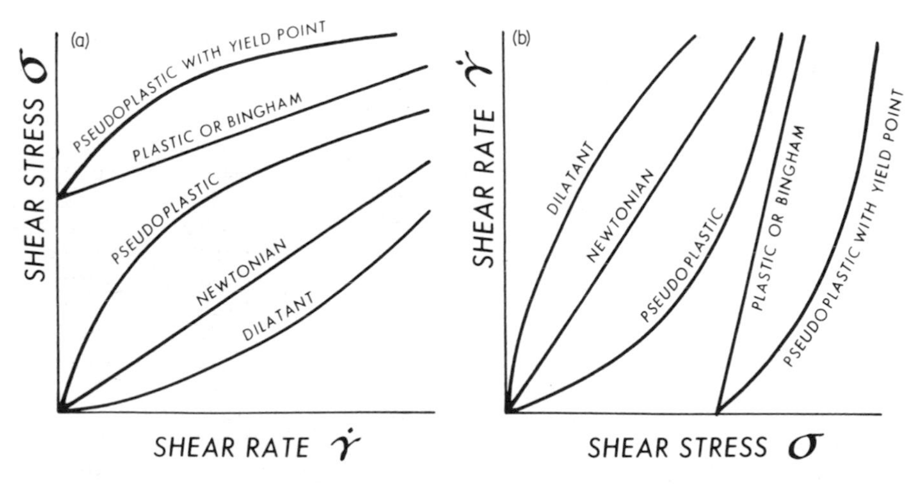

FIG. 12. (a) Shear stress versus shear rate plots for various types of flow; (b) shear rate versus shear stress plots for the same types of flow.

TABLE 4

RELATIONSHIP BETWEEN TYPE OF FLOW AND THE GENERAL VISCOSITY EQUATION[a]

Type of flow	s	C	Equation form
Newtonian	1	0	$\sigma = b\dot{\gamma} = \eta\dot{\gamma}^{b}$
True plastic	1	>0	$\sigma = b\dot{\gamma} + C$
Pseudoplastic	$0 < s < 1$	0	$\sigma = b\dot{\gamma}^s$
Dilatant	$1 < s < \infty$	0	$\sigma = b\dot{\gamma}^s$
Pseudoplastic with a yield value	$0 < s < 1$	>0	$\sigma = b\dot{\gamma}^s + C$
Dilatant with a yield value	$1 < s < \infty$	>0	$\sigma = b\dot{\gamma}^s + C$

[a]The general viscosity equation is $\sigma = b\dot{\gamma}^s + C$.
[b]Term b is the true viscosity η.

TABLE 5

POWER EQUATION CONSTANTS FOR SOME FRUIT PUREES[a]

Product	Solids (%)	Temperature (°C)	Rheological constants	
			n	K
Applesauce	11.0	30	0.34	116
Applesauce	11.0	82	0.34	90
Apricot puree	15.4	4.5	0.37	130
Apricot puree	15.4	60	0.46	38
Apricot puree	19.0	4.5	0.32	220
Apricot puree	19.0	60	0.34	88
Apricot concentrate	26.0	4.5	0.26	860
Apricot concentrate	26.0	60	0.32	400
Banana puree	—	24	0.458	65
Orange juice concentrate	—	0	0.542	18.0
Orange juice concentrate	—	15.0	0.584	11.9
Pear puree	18.3	32	0.486	22.5
Pear puree	18.3	82	0.484	14.5
Pear puree	26.1	32	0.450	62.0
Pear puree	26.1	82	0.455	36.0
Pear puree	31.0	32	0.450	109.0
Pear puree	31.0	82	0.459	56.0
Pear puree	37.2	32	0.456	170.0
Pear puree	37.2	82	0.457	94.0
Pear puree	45.7	32	0.479	355.0
Pear puree	45.7	82	0.481	160.0
Peach puree	11.9	30	0.28	72
Peach puree	11.9	82	0.27	58
Plum puree	14	30	0.34	22
Plum puree	14	82	0.34	20

[a]Data from Holdsworth (1971).

Other Flow Equations

A number of other equations, almost all of which are empirical in nature, have been described in the literature. These equations usually have no theoretical foundation, but because they facilitate the handling of empirical data they have some usefulness. Some of the most common ones are listed below.

The power equation. Although this is often described as the power *law* it is in fact an empirical relationship. This widely used equation takes the form

$$\sigma = K\dot{\gamma}^n,$$

where σ is the shear stress; K, a consistency index; $\dot{\gamma}$, the shear rate; and n, a dimensionless number that indicates the closeness to Newtonian flow. For a Newtonian liquid n is 1; for a dilatant fluid n is greater than 1; and for pseudoplastic fluid n is less than 1. Taking logarithms reduces this equation to the form

$$\log \sigma = \log K + n \log \dot{\gamma}.$$

A plot of the log shear stress versus log shear rate is linear with a slope equal to n for those fluids that obey the power equation. The power equation is frequently used by engineers in designing systems for handling fluid foods. Many systems reduce to a linear relationship over a wide range of shear rates when reduced to a log–log plot. Table 5 lists experimentally determined power equation constants for some fruit purees.

Herschel–Bulkley model. Fluids that obey this model are characterized by the presence of a yield stress and a linear log shear stress–log shear rate plot (Herschel and Bulkley, 1926). The equation for this model is

$$\sigma = \sigma_0 + K\,\dot{\gamma}^n,$$

where σ_0 is the yield stress.

This equation is of the same form as Eq. (5) and (6) in Table 4, the only difference being in some of the symbols. It takes the same form as the power equation but with the addition of the yield stress term σ_0.

The numerical value of the exponent n indicates the closeness to a linear shear stress–linear shear rate plot; the plot is rectilinear when n is 1 and the degree of curvature of the plot on linear axes increases as the value of n moves away from unity.

Casson equation. This equation was developed for printing inks by Casson (1959), but has been found to be effective for some foods, particularly chocolate and some other filled fluids. The equation is

$$\sqrt{\sigma} = \sqrt{\sigma_0} + \eta_a \sqrt{\dot{\gamma}},$$

where σ is the shear stress; σ_0, the yield stress; η_a, the apparent viscosity; and $\dot{\gamma}$,

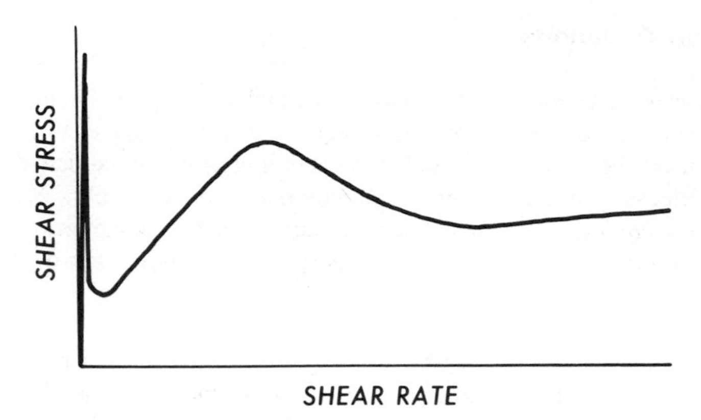

FIG. 13. Experimentally determined shear stress–shear rate plot for an instant pudding. (Courtesy of Dr. A. S. Szczesniak.)

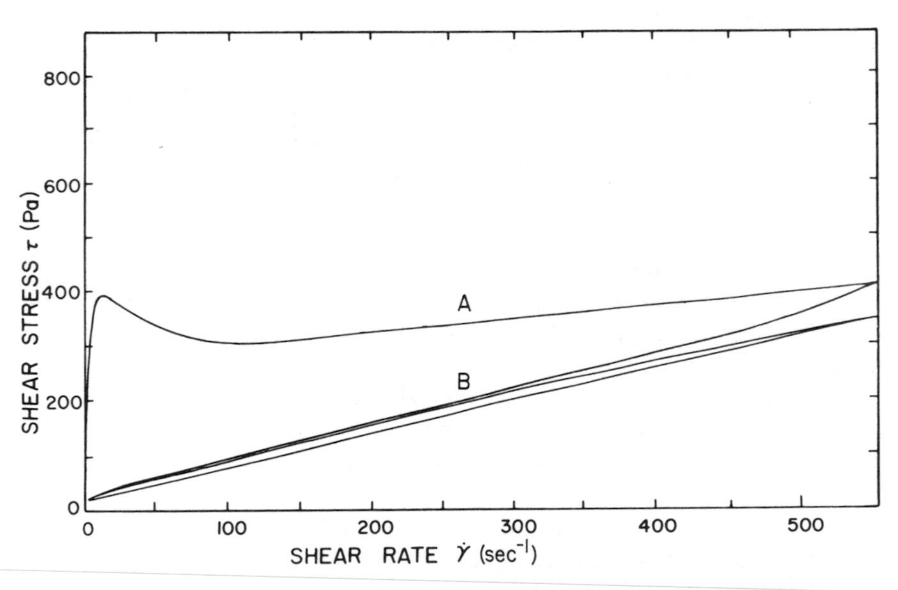

FIG. 14. Repeated shear stress–shear rate curves on the same sample of concentrated yeast extract (T = 25°C). Curve A, first leg "virgin" sample; curves B, "destroyed" sample. (Courtesy of Dr. A. L. Halmos and Dr. C. Tiu. Reprinted from *J. Texture Stud.*; with permission from Food and Nutrition Press.)

the shear rate. This equation gives a linear plot for chocolate. It is used as an international standard for measuring the viscosity of chocolate (Rostagno, 1974). Chevalley (1975) reviewed the validity of the Casson equation for chocolate and factors that affect its flow behavior.

Structural viscosity. The shear stress–shear rate plots for a number of fluid foods do not follow any of the types of viscous behavior explained above nor do they obey any of the above equations, including the general equation for viscosity. Figure 13 shows the shear stress–shear rate plot for an instant pudding. It is obvious that this is unlike any of the flow properties discussed above, and it is difficult to reduce this kind of curve to a suitable equation. The first sharp peak in this curve is probably related to some kind of shear stress needed to start the product flowing while the hump in the center probably represents the breakdown of some soft structure. The flow at high shear rates probably approximates pseudoplastic flow. When this test is repeated on the same sample, the second shear stress–shear rate curve frequently gives a smoother line with the bumps absent or much reduced in size.

At the present time there is no accepted method for analyzing this type of curve and extracting rigorously defined viscosity parameters from it. Halmos and Tiu (1981), who found a similar shape curve when working with concentrated yeast extracts, measured the area between the first and second curves, expressing this as the work required to break down the structural viscosity (Fig. 14). Presumably, the curve obtained on the second test and subsequent tests exhibits plastic flow or something close to pseudoplastic flow.

Time Dependency

Thus far we have assumed that the shear stress at a given shear rate remains constant over a period of time. There are a number of fluids in which the shear stress is a function both of the shear rate and the time to which it is subjected to a shearing force. Newtonian fluids are time independent; hence, this discussion does not apply to Newtonian fluids. The four major types of time dependency are as follows:

1. *Thixotropic.* The apparent viscosity decreases with the time of shearing but the change is reversible; that is, the fluid will revert to its original state ("rebuild itself") on standing. Some starch paste gels are in this class.

2. *Shear thinning.* The apparent viscosity decreases with time and the change is irreversible; that is, it stays in the thinner state when the shear stress is removed. This condition is frequently found in food systems. Some gum solutions and starch pastes fall into this class.

A fluid may exhibit both thixotropic and shear thinning properties, for example, when the apparent viscosity decreases with time of shearing and partially recovers its original viscosity after resting.

Figure 15 shows shear stress–shear rate curves for a thixotropic and a nonthixotropic pseudoplastic fluid. Curve A is nonthixotropic; the curve on the way down retraces the same path as on the way up. Curve B is thixotropic; the curve on the way down lies below the curve on the way up. The area between the up and down curve is called a hysteresis loop. The researcher should be warned that some hysteresis loops are artifacts; two examples are (a) a true Newtonian fluid can give an apparent thixotropic hysteresis loop if viscous heating warms the liquid, and (b) inertial forces can cause a hysteresis loop to appear if the experiment is performed too fast or the rotor has a large mass.

3. *Rheopectic.* The apparent viscosity increases with time of shearing and the change is reversible; that is, after resting, the product returns to its original apparent viscosity. It is rare to find this type of behavior in a food system.

4. *Shear thickening.* The apparent viscosity increases with time and the change is irreversible; that is, it stays thick. When egg white or heavy cream are whipped their viscosity increases until they become stiff. This is an example of shear thickening. However, it is not a good example because the change in

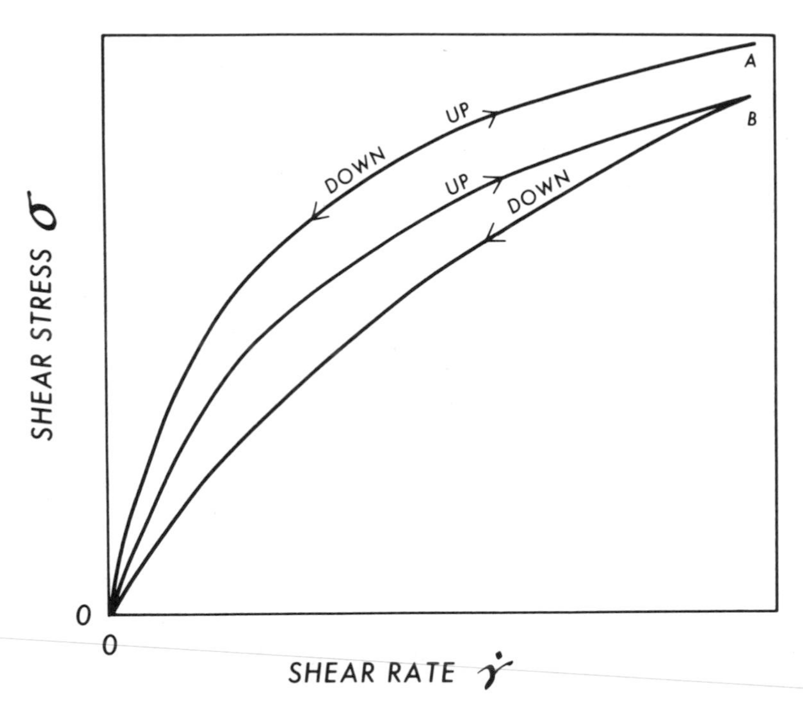

FIG. 15. Shear stress–shear rate curves for a nonthixotropic pseudoplastic fluid (A) and a thixotropic pseudoplastic fluid (B).

viscosity is due to physical changes in the egg protein and the fat globules of the cream. Vernon Carter and Sherman (1980) reported that aqueous solutions of mesquite tree gum exhibited shear thickening when the shear rate exceeded $100 sec^{-1}$.

Figure 16 portrays in graphical form the various types of time-dependent flow. When a fluid is caused to flow at a constant shear rate over a period of time the apparent viscosity is constant for Newtonian fluids, it increases for rheopectic or shear-thickening fluids, and decreases for thixotropic or shear-thinning fluids. On the other hand, when a fluid is caused to flow over a period of time under a constant shear stress, a plot of shear rate versus time is constant for a Newtonian fluid, it increases for a thixotropic fluid (because the product is becoming less viscous), and it decreases for a rheopectic or shear-thickening fluid (because the product is becoming more viscous).

A fluid may exhibit time dependency in addition to other viscous properties. For example, a product may be both plastic *and* thixotropic, or pseudoplastic *and* rheopectic. The combination of non-Newtonian flow plus time dependency brings one into very complex systems, many of which cannot be measured and described well by presently available instrumental methods. Nevertheless, the food technologist is faced with handling these systems and needs to obtain reliable and reproducible measurements, even though there are few guidelines.

Green (1949), who was an associate of Bingham, discusses the unsatisfactory state of analysis for some of these complex fluids. He discusses a practical rheologist, "Bill," who has viscosity measuring equipment in his laboratory and has to produce results describing the flow properties of the commodities being

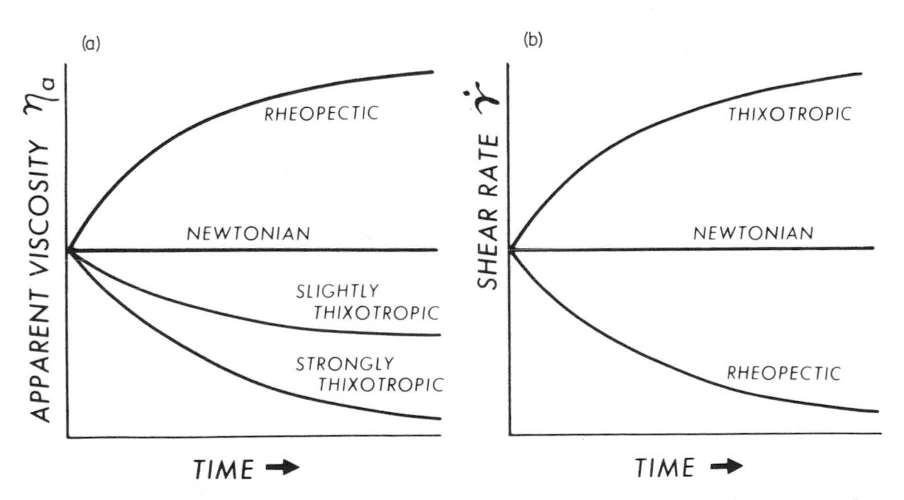

FIG. 16. Time-dependency factors in fluid flow: (a) at constant rate of shear; (b) at constant shear stress.

handled in a manufacturing plant, particularly with regard to the need for quality control purposes. Green writes as follows:

> A dozen theoretical rheologists can give a dozen different explanations as to why Bill's measurements produce the kind of curve they do. Not a single explanation will alter Bill's curves in any visible way. As far as Bill is concerned, the dozen different theoretical explanations might just as well not exist. Bill can, if necessary, get along without them. Bill will find it more desirable, however, to convert his curves into numbers like *U*, *s*, and *M*. Such numbers are easy to enter into reports and are much easier to interpret when making comparisons of different materials. . . . There are many ways of converting consistency curves into numbers. Which method should he choose?

The last sentence in the above quotation is the end of a chapter. Green never attempted to point out the best way for analyzing these complex consistency curves. The best conclusion that can be drawn about handling substances with complex flow properties is to make as complete a shear stress–shear rate study as possible, using adequate instrumentation and taking into account the possibility of time dependency in order to obtain as complete a picture as possible of the

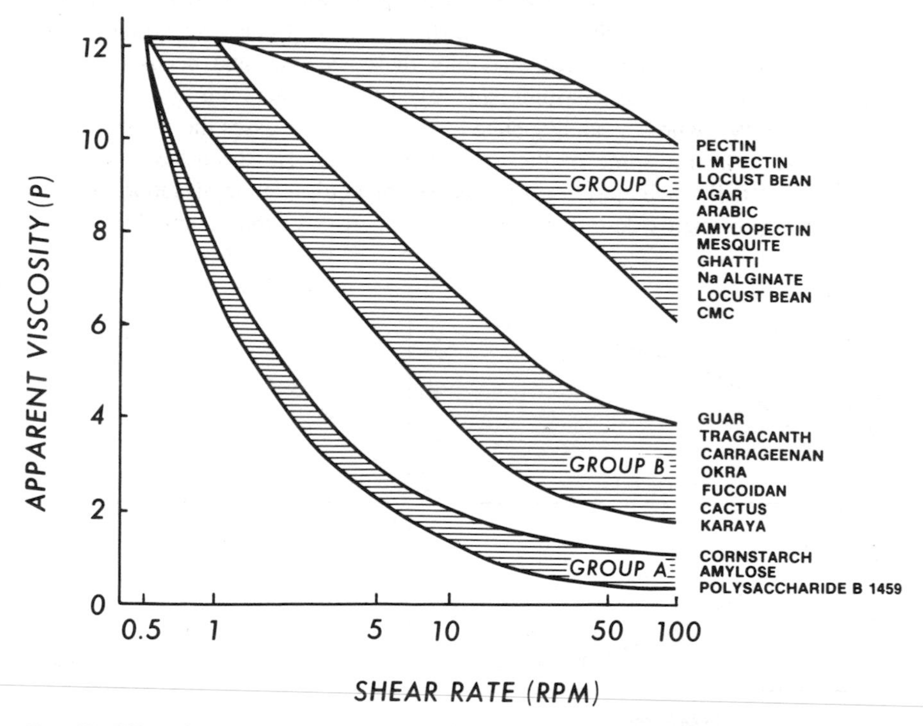

FIG. 17. Effect of shear rate on apparent viscosity of gum solutions. Group C were classed as very slimy by a sensory panel, group B as somewhat slimy, and group A as nonslimy. (Redrawn from Szczesniak and Farkas, 1962.)

rheological properties of the system. A single-point measurement of viscosity, which is satisfactory for Newtonian fluids, will be far from satisfactory for these complex fluids.

Time dependency is an important factor in the quality of some foods. For example, Szczesniak and Farkas (1962) found that aqueous solutions of gums that exhibited no time dependency had a slimy mouthfeel, while gums that exhibited a high degree of shear thinning or thixotropy had no sliminess (see Fig. 17). This finding was confirmed by Stone and Oliver (1966).

Another example is the manner of change of gelatin dessert. When a gelatin gel is put into the mouth it melts into a mobile fluid. This thinning effect (which is temperature controlled rather than mechanically controlled) is an important attribute of the textural quality of gelatin desserts. In contrast, dessert gels made from agar do not melt, because the melting point of agar gels is about 98°C. One has to chew these gels into small lumps for swallowing, and this behavior gives an entirely different type of mouthfeel than a gelatin dessert gel.

The use of the time-dependency terms noted above are presently under discussion and may be changed. For example, Reiner and Scott Blair (1967) propose "anomalous viscosity" for all non-Newtonian viscosities. The International Congress of Rehology has suggested the use of the following terms: "Shear thinning" to replace the word "pseudoplastic," "shear thickening" to replace the term "dilatancy," and "shear degrade" to replace the term "irreversible thinning" or "shear thinning."

The Weissenberg Effect

When a rod is rotated in some viscoelastic fluids, the fluid climbs up the rod against the force of gravity because the rotational force acting in a horizontal plane produces another force at right angles to that plane; this is called a *normal* force. The tendency of a fluid to flow in a direction normal to the direction of shear stress is known as the Weissenberg effect (Weissenberg, 1949). The effect has been observed with some flour doughs, cake batters, melted cheeses, honeys, and aged condensed milk (see Fig. 18).

This characteristic of some viscoelastic liquids is measured in the Weissenberg Rheogoniometer (see p. 237). This is a cone and plate viscometer that is designed to measure forces normal to the plane of shear as well as the shear force. The Ferranti–Shirley cone and plate viscometer can be modified to measure the Weissenberg effect.

Methods for Measuring Viscosity

The first thing to remember in measuring viscosity is that the viscosity of fluids is highly temperature dependent (see p. 204). The Brookfield Engineering

FIG. 18. The Weissenberg effect: the rotation of the glass rod causes the aged sweetened condensed milk to climb up the rod.

Laboratory points out that No. 50 motor oil will change its viscosity about 10% for a 1°C temperature change at 25°C. This company has encountered materials whose viscosity changed 50% per degree centigrade. The viscosity of water at 20°C changes 2.5% per 1°C temperature change. Hence, it is impossible to measure the viscosity of water with an accuracy of 0.1% unless the temperature is controlled to within 0.04°C. Therefore, in all viscosity measurements it is essential that the temperature be closely controlled. The temperature at which viscosity measurements are taken should be stated with all viscosity data because the data are meaningless unless the temperature is known. It is assumed that close temperature control is an essential feature of each system described below.

The various types of viscometers can be classified according to the principle on which they work.

Capillary Type

The time for a standard volume of fluid to pass through a length of capillary tubing is measured. The underlying theory behind capillary viscometers is developed fully a number of authors (see, e.g., Oka, 1960; VanWazer *et al.*, 1963) and will not be given here. This type of flow is described by the Poiseuille equation, which is also known as the Hagen–Poiseuille equation (Poiseuille, 1846; Hagen, 1839):

$$\eta = \pi p r^4 t / 8Vl,$$

where η is the viscosity; p, the driving pressure; r, the radius of capillary; t, the time of flow; V, the volume of flow; and l, the length of capillary.

The driving pressure is usually generated by the force of gravity acting on a column of the liquid, although it can be generated by the application of compressed air or by mechanical means (as in the Instron Capillary Rheometer). The discussion here will be restricted to glass capillary viscometers.

The Ostwald viscometer is one of the simplest of the glass capillary types and is shown schematically in Fig. 19. There are a number of variations in the design of glass capillary viscometers, each with its own specific name and each claiming certain advantages (VanWazer *et al.*, 1963). For example, the Ostwald–Cannon–Fenske Viscometer, which is a widely used modification of the Ostwald viscometer, has both arms bent at an angle that brings the center of the upper bulbs directly over the center of the lower bulb, thus displacing the capillary from the vertical position.

The operation of the Ostwald Viscometer will now be described. Other styles of glass capillary viscometers are operated in a similar manner, the exact details should be provided by the supplier when it is purchased.

In operation, a standard volume of fluid is pipetted into arm A of the Ostwald Viscometer, which should be held in a vertical plane (see Fig. 19). It is not essential that the capillary be exactly vertical but it should be held reproducibly at the same angle. The fluid runs down the wide-bore tube C into bulb D and U-tube E. The apparatus is immersed in a constant temperature water bath until the viscometer and liquid in it reach the standard temperature (about 30 min). Suction is then applied at the top of arm B to draw the fluid through the capillary F

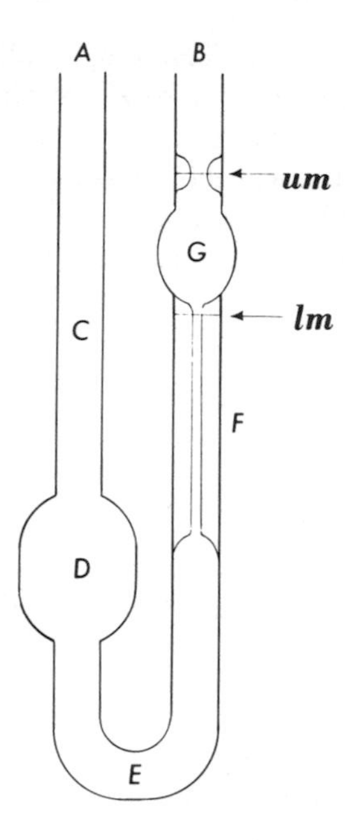

FIG. 19. Schematic representation of the Ostwald Viscometer.

into bulb G until the upper meniscus is above the mark *um*. The suction is removed and the fluid flows from bulb G through the capillary tube F under the force of gravity. A stop watch is started when the meniscus crosses the upper mark *um* and stopped when it crosses the lower mark *lm*. The viscosity is calculated from the elapsed time.

The AVS/N viscometer is a sophisticated glass capillary that uses light barriers to record the time the meniscus passes the set points and displays the elapsed time on a digital indicator to the nearest 0.01 sec. This eliminates errors in operating the stopwatch and allows the operator to attend to other duties once the test has been started.

For a given viscometer of this type, the dimensions of the radius and length of the capillary are constant and the volume is kept constant. The driving pressure P is proportional to the hydrostatic head and the density of the fluid. The head decreases as the liquid falls in reservoir G. This viscometer is designed to minimize the change in head during the measured portion of the efflux time, and the shape of the bulb G is such that most of the efflux time occurs when the head is close to its mean value. Variations in the pressure head have no effect on the viscosity measurement of Newtonian fluids but they do affect measurements on non-Newtonian fluids, the magnitude of the deviations depending on the degree to which the fluid departs from Newtonian behavior. For a Newtonian fluid the driving pressure P can be replaced by $h \times g \times \rho$, where h is the mean head; g, gravity; and ρ, the density of the fluid. Since h is constant for a given viscometer, the Hagen–Poiseuille equation can be simplified to

$$\eta = K\rho t,$$

where K is the instrument conversion factor ($\pi hgr^4/8Vl$) and is a constant for each instrument. This equation can be rearranged to

$$\eta/\rho = Kt \text{ or } \nu = Kt,$$

since kinematic viscosity $\nu = \eta/\rho$.

Hence the kinematic viscosity ν of the fluid is obtained by multiplying the measured efflux time by the instrument conversion factor K. Most laboratory supply houses will provide the K value for each viscometer at a cost of about $20 over the price of the viscometer.

If the instrument conversion factor K is not provided or has been lost, it can be obtained by measuring the efflux time for a fluid of known viscosity:

$$\nu_s = Kt_s,$$

where ν_s is the kinematic viscosity and t_s is the efflux time for a standard fluid of known viscosity.

Rearranging this equation gives

$$K = \nu_s/t_s.$$

The Cannon Instrument Co. (P.O. Box 16, State College, Pennsylvania 16801) supplies a wide range of Newtonian viscosity standards in the form of a series of oils of calibrated viscosity. These are useful for calibrating kinematic viscometers.

Glass capillary viscometers are widely used for measuring low to medium viscosity Newtonian fluids because of their high degree of accuracy, ease of operation, and low cost. Priel *et al.* (1973) developed a system for a Ubbelohde glass capillary viscometer that yielded data with an absolute accuracy within three parts per million. An essential part of their system was a thermostat with a long-time thermal stability of $\pm 2 \times 10^{-4}$°C over a period of 4 weeks.

Because of their low cost it is usual to purchase several glass capillary viscometers if a large number of measurements need to be made. This allows several units to be reaching equilibrium temperature while a measurement is being performed on one unit. It is advisable to purchase a series of viscometers with a range of capillary diameters when a wide range of viscosities are encountered. A capillary diameter should be selected that gives an efflux time between about 200 and 800 sec. Figure 20 shows some typical glass capillary viscometers.

The American Society for Testing and Materials (1918 Race Street, Phila-

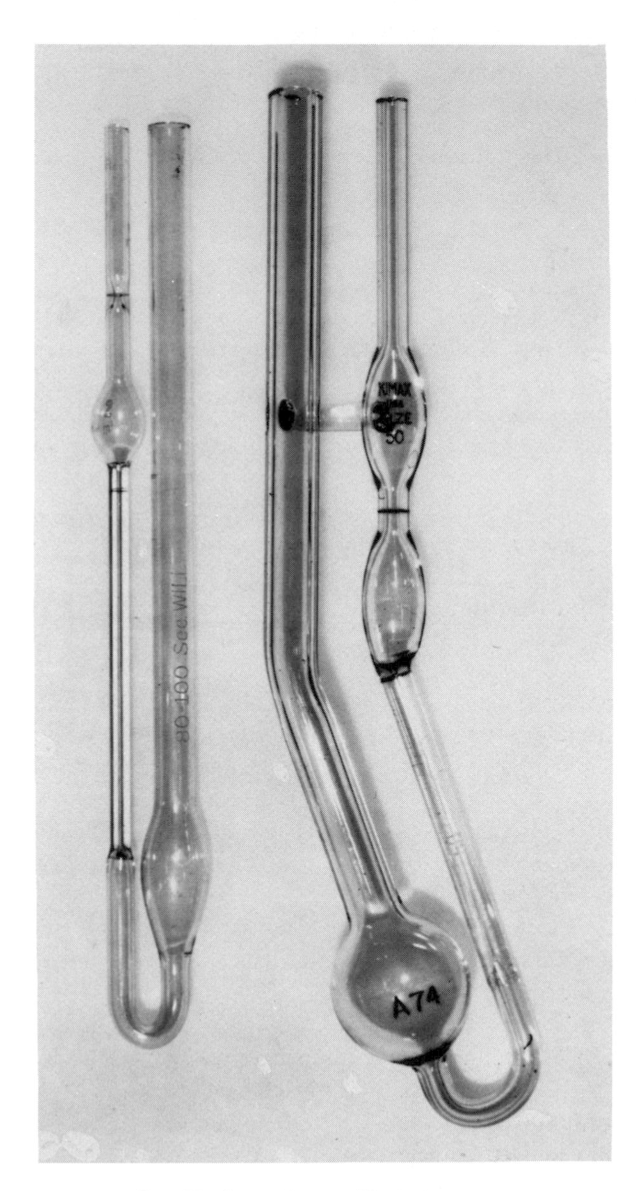

FIG. 20. Some glass capillary viscometers.

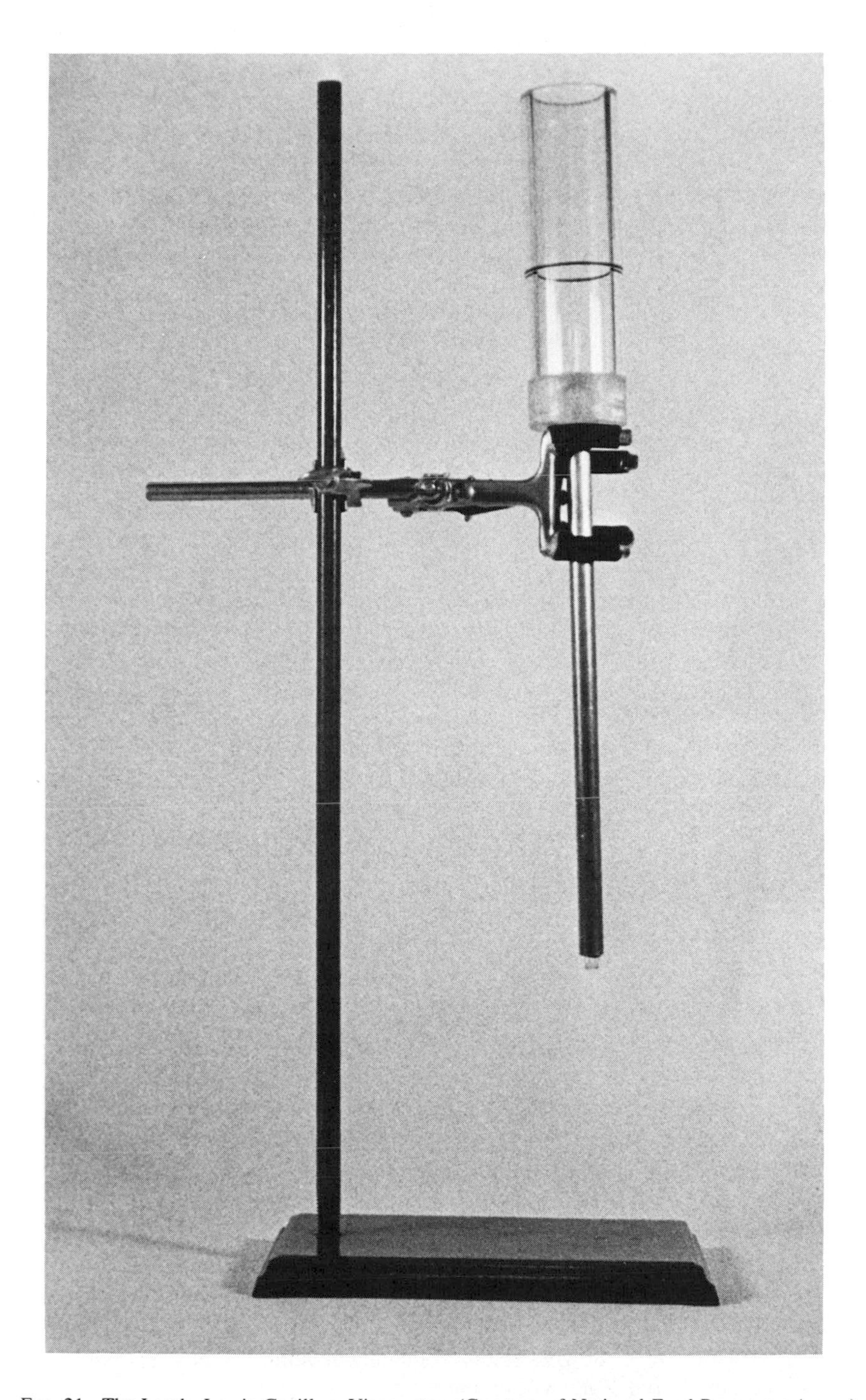

FIG. 21. The Lamb–Lewis Capillary Viscometer. (Courtesy of National Food Processors Assoc.)

delphia, Pennsylvania 19103) has published a standard test method for use of capillary viscometers on Newtonian fluids: (ASTM D445–79, "Kinematic Viscosity of Transparent and Opaque Liquids (and the Calculation of Dynamic Viscosity)." This organization has published another useful document: ASTM-D446–79, "Standard Specifications and Operating Instructions for Glass Capillary Kinematic Viscometers."

A number of corrections need to be made when very accurate results are needed from glass capillary viscometers. These include correction for the kinetic energy lost in the stream as it issues from the bottom of the capillary, and effects due to the change in the meniscus size and shape as it enters or leaves the capillary, possible turbulence in the capillary, and inadequate drainage due to liquid adhering to the walls of the viscometer. These errors, and methods for their correction, are discussed in detail by VanWazer *et al.* 1963). The same authors show how the capillary viscometers may be used for certain non-Newtonian fluids.

The Lamb–Lewis Capillary Viscometer was developed by the National Food Processors Association (formerly known as the National Canners Association) as a low-cost quality-control instrument for use on tomato juice, fruit nectars, and similar fruit or vegetable juices and blends (Lamb and Lewis, 1959). It is used by the fruit and vegetable juice industry as an internal quality standard (Lamb, 1967). It consists of a 1½-in.-i.d. Lucite chamber from the bottom of which protrudes a precision Pyrex glass tube 3 ± 0.01 mm i.d. and approximately 11½ in. long (see Fig. 21). The cup is filled with liquid, which is allowed to flow through the capillary until a steady flow is obtained. A finger is placed over the capillary outlet to stop flow, the chamber is filled level with the top, the finger is removed as a stopwatch is started, and the time for the meniscus to reach the calibration line is recorded to the nearest 0.1 sec.

Orifice Type

This can be considered as a very short capillary type of viscometer. The time for a standard volume of fluid to flow through an orifice is measured. This is a simple, inexpensive rapid method that is widely used in quality control of Newtonian or near-Newtonian liquids where extreme accuracy is not needed.

Possibly the best known of the orifice viscometers in the food industry is the dipping-type Zahn Viscometer. These consist of a stainless-steel 44-ml-capacity cup attached to a handle with a calibrated circular hole in the bottom. In operation, the cup is filled by dipping it into the fluid and withdrawing it. A stopwatch is started as soon as it is withdrawn and stopped when the first break occurs in the issuing stream. The elapsed time gives an empirical value of viscosity. Table 6 gives specifications for the five standard models of Zahn viscometers, and Fig. 22 shows a set of four Zahn viscometers.

TABLE 6
SPECIFICATIONS FOR ZAHN VISCOMETERS

Zahn No.	Orifice diameter (mm)	Approximate viscosity range (cP)
1	2.0	14–40
2	2.7	21–196
3	3.8	88–614
4	4.3	148–888
5	5.3	345–1265

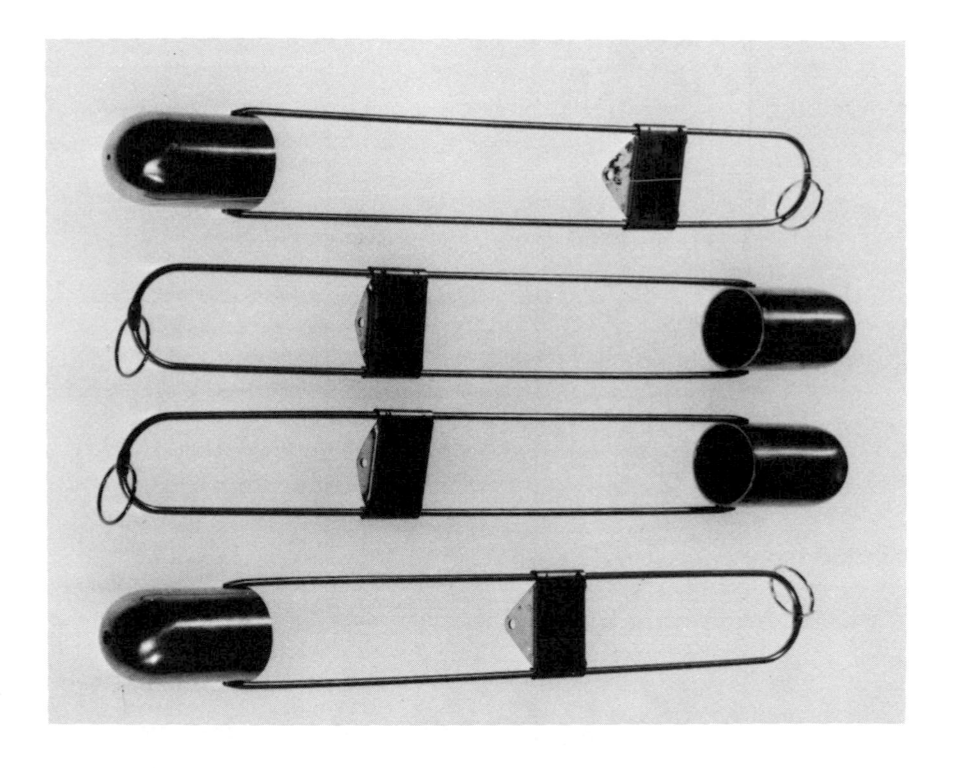

FIG. 22. Some Zahn Viscometers (dipping orifice type).

Coaxial Rotational Viscometers

These are also known as concentric cylinder or couette viscometers, in honor of the developer of the first practical viscometer of this type (Couette, 1890). The principle is shown schematically in Fig. 23. A bob that is circular in cross section is placed concentrically inside a cup containing the test fluid. Either the cup or the bob is rotated and the drag of the fluid on the bob is measured by means of a torsion wire or some other kind of torque sensor. The shear rate–shear stress relationship is the same whether the bob is rotated and the cup held stationary or vice versa. This type of viscometer permits continuous measurements to be made under a given set of conditions and allows time-dependent effects to be studied.

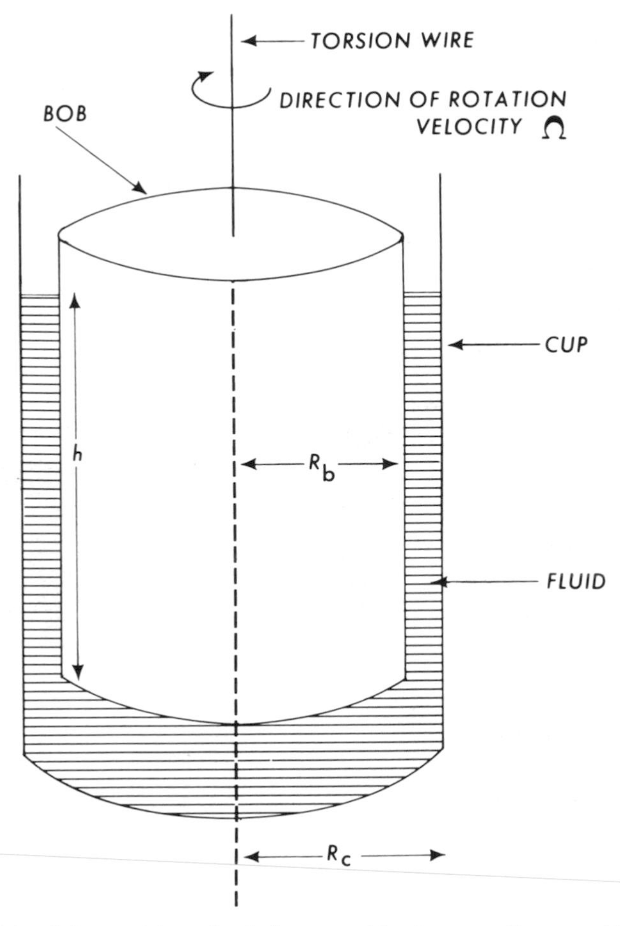

FIG. 23. Principle of the coaxial rotational viscometer (also known as the concentric cylinder or couette type of viscometer).

By changing the rate of shear or magnitude of stress it is possible to obtain viscosity measurements over a range of shearing conditions on the same sample. It can be used for both Newtonian and non-Newtonian foods. This is the most common type of viscometer that is used in the food industry. It might be called the ''workhorse'' viscometer.

The Margules equation (Margules, 1881) apples to the flow of Newtonian fluids in coaxial rotational viscometers:

$$\eta = (M/4\pi h\Omega)\ (1/R_b^2 - 1/R_c^2),$$

where η is the absolute viscosity; M, the torque on the bob or cup; Ω, the angular velocity of rotating member; h, the length of bob in contact with the fluid; R_b, the radius of the bob; and R_c, the radius of the cup. For a given instrument with a given geometry and fill of container this equation reduces to

$$\eta = KM/\Omega,$$

where K is the instrument constant $(1/4\pi h)\ (1/R_b^2 - 1/R_c^2)$.

The Margules equation is not applicable to flow of non-Newtonian fluids in coaxial rotational viscometers. More complex equations have been derived to represent, or approximately represent, the flow of these complex fluids. The reader is referred to standard texts on viscometry for the development of equations applicable to non-Newtonian fluids (see, e.g., VanWazer *et al.*, 1963; Whorlow, 1980).

This type of viscometer may be divided into two classes. In the first class (Stormer type) the shear rate at a constant torque is measured while in the second class (MacMichael type) the torque is measured at a constant speed of rotation.

1. *Stormer type.* The Stormer Viscometer maintains a constant torque by means of a falling weight attached to a thin cord that passes over a pulley and is wrapped around a drum that is connected to the rotor bob (Fig. 24). When the brake is released, the rotor turns at an accelerating rate until the angular velocity reaches an equilibrium speed when the viscous drag of the fluid on the rotor exactly matches the power output of the falling weight. A revolution counter records the number of times the rotor has turned. When equilibrium speed has been reached, a stopwatch is used to record the time for a given number of revolutions of the rotor (usually 100 revolutions). Several different types of rotor and cup are available giving different geometries.

Although the Stormer viscometer is claimed to be used on a number of foods, including canned corn, catsup, condensed milk, edible oils, sugar solutions, gum solutions, mayonnaise, and tomato products, the author has not yet seen a Stormer Viscometer in regular use in a food laboratory. Nevertheless, VanWazer *et al.* (1963) recommend the Stormer Viscometer for vexatious slurries.

The Deer Rheometer is a constant-stress viscometer. It features a frictionless

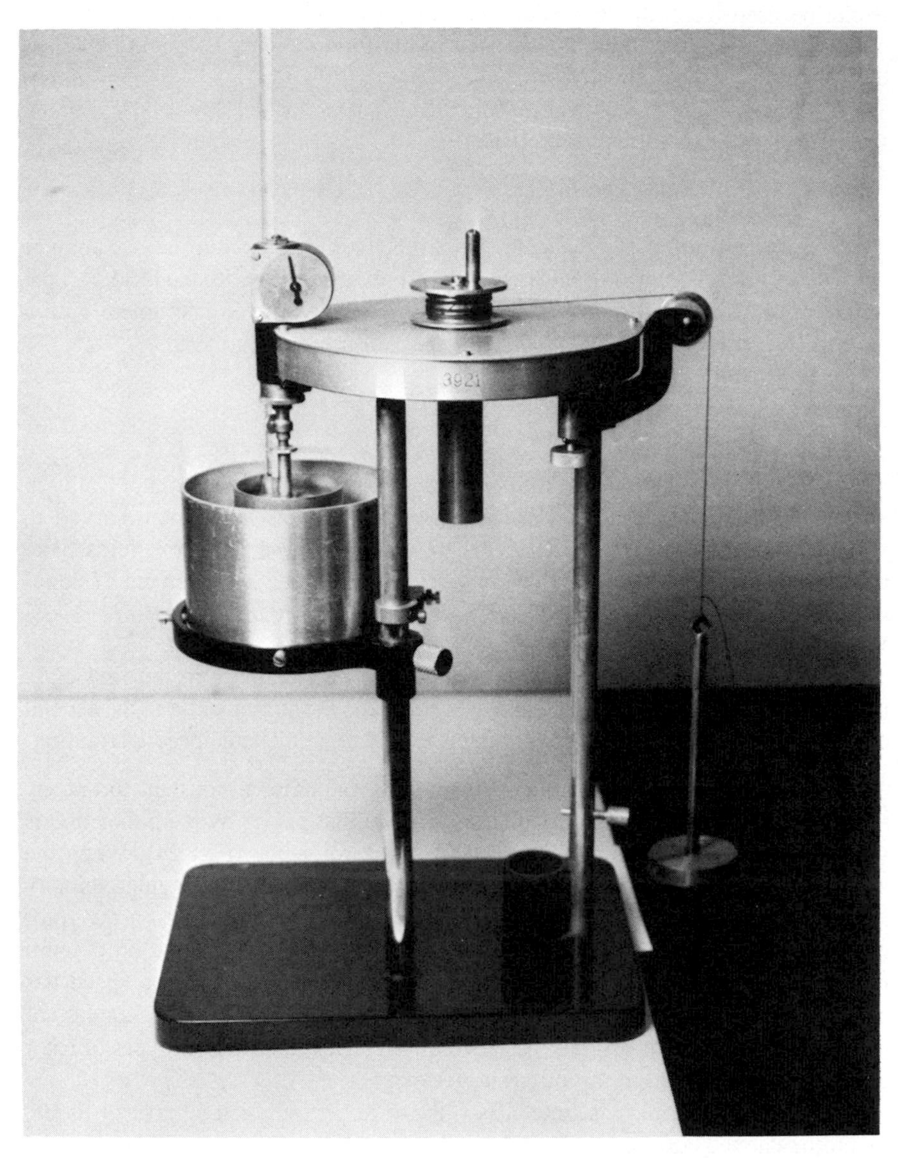

FIG. 24. The Stormer Viscometer.

air-bearing support for the rotating parts and no mechanical connections between the fixed and rotating parts of the instrument. It provides an equivalent shear stress of approximately 0.01 to 1.2×10^4 dyn cm^2 and a shear-rate range from zero to approximately 5000 s^{-1}. Concentric cylinder and cone-and-plate geometries are available. It is particularly suited for measuring yield stress and for recording viscoelastic creep curves.

2. *MacMichael type.* The first successful viscometer that used a rotating cup and measured the torque on the bob at constant shear rates was developed by MacMichael (1915). The MacMichael Viscometer was carried as a stock item for many years by Fisher Scientific but is no longer listed in the Fisher catalogs. A number of commercially available instruments are now available that are based on this principle. A few of the better known ones are briefly described below. A number of geometries of cup and bob have been developed for specific uses or to reduce certain errors that may occur.

The major error that occurs in coaxial viscometers is the "end effect," which arises from the drag of the fluid on the ends of the bob. The derivation of the Margules equation (and other similar equations) assumes an infinitely long bob with no ends. The end effect of the top of the bob is easily eliminated by filling the cup to a level below that necessary to cover the bob. Since the top of the bob is not in contact with the liquid there is no drag. The end effect of the bottom of the bob can be determined experimentally by measuring the torque/angular-speed ratio with the cup filled to several different heights. A rectilinear plot should be obtained when the data is plotted (see Fig. 25). The plot is extrapolated to zero on the torque/angular-speed ratio axis. The negative intercept on the horizontal axis (h_0) gives the end effect in terms of the equivalent length of bob with no ends. In any mathematical exercises the depth of immersion h should be replaced by $h + h_0$ in order to account for the end effect.

Haake Rotovisco. This popular and versatile viscometer is provided in four basic models: (1) Rotovisco RV12 is a low cost model suitable for routine quality control and product development. (2) Rotovisco RV2 is a standard model that is suitable for most industrial applications. It is designed on the modular system, enabling the researcher to add on additional features to expand the capability of the instrument. (3) Rotovisco RV3 is similar to model RV2 but is provided with a programmer. (4) Rotovisco RV100 is the top of the range instrument with a built-in programmer, X–Y–t recorder, and provision for the addition of a very low shear measuring system. The company has available a wide range of geometries and sizes of cups and rotors, including cone and plate geometry.

Contraves. This company supplies several models of coaxial rotational viscometers of which the Rheomat 30 is the best known in the food industry (Cavigelli and Schnyder, 1980). A wide range of cup and bob sizes as well as cone and plate geometry are available. The Rheomat 30 operates over the shear

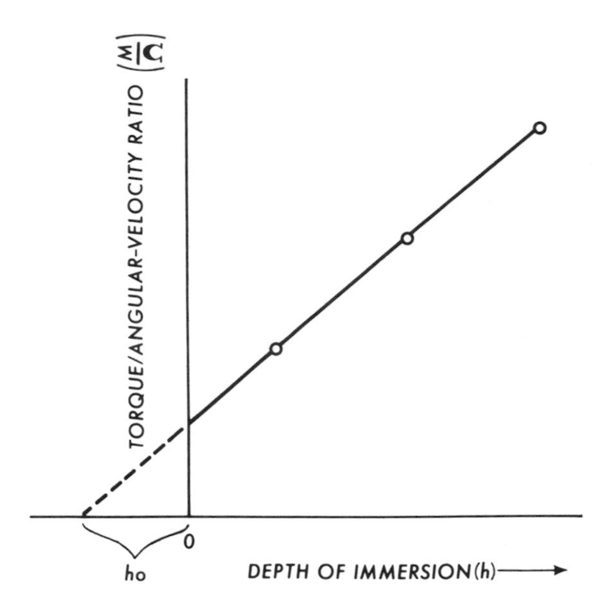

FIG. 25. Measurement of the end effect in a coaxial rotational viscometer. The torque/angular-velocity ratio (M/Ω) is plotted against the depth of immersion of the bob in the fluid h. the intercept of the extrapolated line on the horizontal axis gives the numerical value of the end effect (h_0).

rate range 12×10^{-3} to 4×10^3 sec^{-1} and shear stress range 0.6–200 $\times 10^3$ Pa. For many purposes it is sufficient to read shear stress from a dial at one or more of 30 constant shear rates. An X–Y recorder is offered as an optional accessory when a complete shear stress–shear rate curve is needed.

Rheometrics Fluids Rheometer. This is a sophisticated rotational viscometer that measures both torque and normal force when a sample is under steady strain at constant rate of shear, or dynamic strain when the sample is subjected to oscillating rotation. The instrument is normally used with cone and plate geometry, but the manufacturer can provide fixtures to give parallel plate and concentric cylinder geometries (Starita, 1980). It is particularly well suited for work with viscoelastic materials (Duke and Chapoy, 1976; Whitcombe and Macosko, 1978).

Cone and Plate Viscometers

The fluid is held by its own surface tension between a cone of small angle that just touches a flat surface (Fig. 26). The torque caused by the drag of the fluid on the cone is measured as one of the members is rotated while the other member remains stationary. For a Newtonian fluid the following equation applies:

$$\eta = 3\alpha M/2\pi R_b^3\Omega,$$

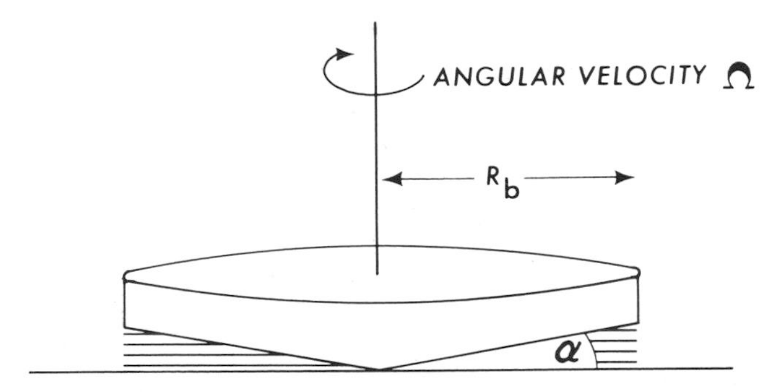

FIG. 26. Schematic of cone and plate viscometer.

where η is the absolute viscosity; α, the angle of cone (usually less than 2°); M, the torque; R_b, the radius of the cone; and Ω, the angular velocity of the rotating member. For a given instrument with a given geometry this can be reduced to

$$\eta = KM/\Omega,$$

where K is the instrument constant $3\alpha/2\pi R_b^3$.

A more detailed analysis of the cone and plate viscometer is given by Slattery (1961).

The special feature of the cone and plate viscometer is that the shear rate is uniform at all points in the fluid, provided that the angle of the cone is small. This makes the cone and plate viscometer of particular use for non-Newtonian fluids because the true rate of shear can be obtained comparatively easily. Other features of this type of viscometer are (1) end effects are negligible, (2) a small amount of fluid is needed (usually less than 2 ml), and (3) the thin layer of fluid in contact with temperature-controlled metal base plate enables measurements to be made at high rates of shear without the need to compensate for the heating effect of the high shear rate.

Ferranti–Shirley Viscometer. This is the best known cone and plate viscometer. It was developed at the Shirley Institute in Manchester, England (McKennell, 1956). Three standard cones (2, 4, and 7 cm diam) are provided but other cone sizes and angles are available, including truncated cones for fluids containing large particles, and wide-angle cones. Shear rates from 0.18 to 18,000 sec^{-1} and shear stresses from 2600 to 563,000 dyn cm^{-2} are available. The instrument is provided with an X–Y recorder and a constant temperature unit.

The Weissenberg Rheogoniometer is basically a cone and plate viscometer. The Haake Rotovisco can be converted into a cone and plate viscometer by means of special attachments. The Rheometrics Fluids Rheometer can also be set up with cone and plate configuration.

There are some more exotic geometries that have been used in coaxial rotational viscometers including cone–cone, double cone and plate, plate–plate, coni-cylindrical, and disk. These will not be discussed here because they are not widely used in the food field.

Other Rotational Viscometers

There are some empirical viscometers in which a paddle, a cylinder, or bars rotate in a container, usually with large clearances between the rotating member and the wall. The geometry of these viscometers is complex and usually not amenable to rigorous mathematical analysis. These instruments are generally rugged, moderate in cost, and fairly easy to manipulate. They have their place, particularly for quality control purposes in the plant where detailed mathematical analysis is not needed. They are widely used in industry. Examples of this type are the Brabender Viscocorder, the Brookfield Viscometer, and the FMC Consistometer.

The Brabender Viscocorder is designed to be a rugged and easy to operate instrument suitable for use in the laboratory or the factory. An electric motor causes a horizontal platform to rotate at any speed between 20 and 300 rpm, and an Eddy Current Tachometer indicates the revolutions per minute. A cup containing the liquid sample is placed on the platform. The cup is supplied by the investigator. An empty 303 × 406 can is a suitable size. A stationary paddle suspended from the top of the Viscocorder is immersed in the fluid when the cup is raised into the operating position. The paddle is connected to a torsion spring that moves a pointer over a chart that is divided into 1000 arbitrary units. Five torsion springs are available ranging in capacity from 125 to 2000 cm-g torque at full scale. Six paddles are available ranging from a narrow flag paddle for viscous products to a four-leaf wide paddle for less viscous products. When the motor is switched on, the drag of the fluid on the paddle causes the torsion spring to rotate and move the pointer over the chart. Chart speed is adjustable from 5 to 400 mm min^{-1}.

By selecting the size of paddle and stiffness of the torsion spring it is possible to handle materials ranging from about 0.1 to about 5000 P. The instrument is used for Newtonian and non-Newtonian fluids and can be used for studies of time-dependent effects. The Brabender Company has a good reputation for helping potential customers select the correct combination of paddle size and torsion spring stiffness for their own product line.

The Brabender Viscoamylograph is similar to the Viscocorder but with the additional feature of a controlled programmed heating system surrounding the cup. It is designed specifically to measure the apparent viscosity of starch suspensions and record how the viscosity changes as the temperature of the water–

starch slurry is raised past the gelatinization temperature, held at this elevated temperature for a period, and then cooled again.

The FMC Consistometer was originally designed by the Food Machinery Corporation to measure the consistency of cream-style corn. It is now distributed by C. W. Brabender Instruments and has been used for routine quality control purposes for catsup, tomato paste, strained baby foods, and other products that have a similar consistency. The product is placed in a stainless-steel cup, the paddle is lowered into the cup, the motor is switched on causing the cup to rotate at a single fixed speed of 78 rpm, and the torque on the paddle is read from a scale on top of the instrument. Four paddles with different dimensions are provided with this instrument. The instrument is approximately 38 cm high, 26 cm wide, 31 cm long, and weighs about 16 kg.

The Corn Industries Viscometer CIV was once widely used to measure changes in viscosity of corn syrup and starch pastes, but this instrument is no longer commercially available.

The Brookfield Synchro-Lectric Viscometer is an instrument that may be held in the hand or supported on a stand. A synchronous induction type motor gives a series of speeds of rotation that are constant. Various spindles that take the form of cylinders, disks, and T bars are attached to a small chuck. When the spindle is immersed in the liquid and the motor switched on, the viscous drag of the fluid on the spindle is registered as torque on a dial. A Factor Finder scale provided by the manufacturer enables the operator to quickly convert the dial reading into apparent viscosity.

Several models of the Brookfield Synchro-Lectric Viscometer are available covering a range of apparent viscosity from about 1 cP to about 60,000P. The company can supply a Helipath stand that automatically lowers the Brookfield Viscometer, thus ensuring that the spindle is continuously moving into previously undisturbed material. This accessory is useful when studying fluids that exhibit time effects or that have a tendency to settle.

The Brookfield Engineering Laboratories have over 40 years experience with viscosity measurement and have a good reputation for helping potential customers identify the precise model of instrument and optimum mode of operation for their own applications.

Brookfield Synchro-Lectric Viscometers are widely used in the food field. They have the advantages that they are of moderate cost, portable, simple to operate, well adapted to many viscosity problems, give results quickly, can be used to measure viscosity in almost any container ranging from a 200-ml beaker to a 1000-gal tank, can be used on Newtonian and non-Newtonian liquids, can be used to measure time dependency and hysteresis, not affected by large particles in suspension, and require minimum maintenance. The disadvantages are that there is a limited range of shear rates, the shear rate can only be changed stepwise, the shear rate varies across the fluid, there can be problems in obtaining shear rate and apparent viscosity for non-Newtonian liquids, and the geometry and flow pattern do not lend themselves to rigorous mathematical analysis.

Falling-Ball Viscometers

This type of viscometer operates on the principle of measuring the time for a ball to fall through a liquid under the influence of gravity. The falling ball reaches a limiting velocity when the acceleration due to the force of gravity is exactly compensated for by the friction of the fluid on the ball. Stokes (1819–1903) was one of the first to study the limiting velocity of falling balls and the following equation is named the "Stokes equation" in his honor:

$$\eta = [\tfrac{2}{9}(\rho_s - \rho_l)gR^2]/V,$$

where η is the viscosity; ρ_s, the density of the falling ball; ρ_l, the density of the fluid; R, the radius of the falling ball; g, gravity; and V, the limiting velocity.

This is a simple type of instrument that is useful for Newtonian fluids but has limited applicability to non-Newtonian fluids. It cannot be used for opaque fluids because the ball cannot be seen. Stokes' law applies when the diameter of the ball is so much smaller than the diameter of the tube through which it is falling that there is no influence of the wall on the rate of fall of the ball.

A falling-ball viscometer can be easily improvised in the laboratory (see Fig. 27). Fill a large graduated glass cylinder with the test fluid and gently drop a steel ball in the center of the cylinder. Allow sufficient distance of fall for the ball to reach the limiting velocity, then time the fall of the ball with a stopwatch. Steel ball bearings with a range of precisely controlled diameters can be obtained from engineering supply houses. The larger the ball, the faster it falls. Therefore it is necessary to select a diameter ball that is small enough to fall at a rate that can be measured with some degree of accuracy with a stopwatch. The lower the density of the ball, the slower it falls. It is possible to obtain balls of material other than steel that have a different density. For example, glass marbles have a density of about 2.6 compared with 7.8 for steel. A glass marble will fall more slowly than a steel ball of equal size.

The Gilmont Viscometer is a falling-ball viscometer in which a glass or stainless-steel ball falls down a vertical tube slightly larger than the ball. The interior of the tube is beaded to ensure that the ball stays centered as it falls. Gilmont (1963) used the theory of flow rotameters with spherical floats to derive the following two equations:

$$\eta = K(\rho_f - \rho)t,$$

where η is the viscosity, ρ_f is the density of ball (2.53 for glass and 8.02 for stainless steel). ρ is the density of liquid, t is the time for ball to fall between two

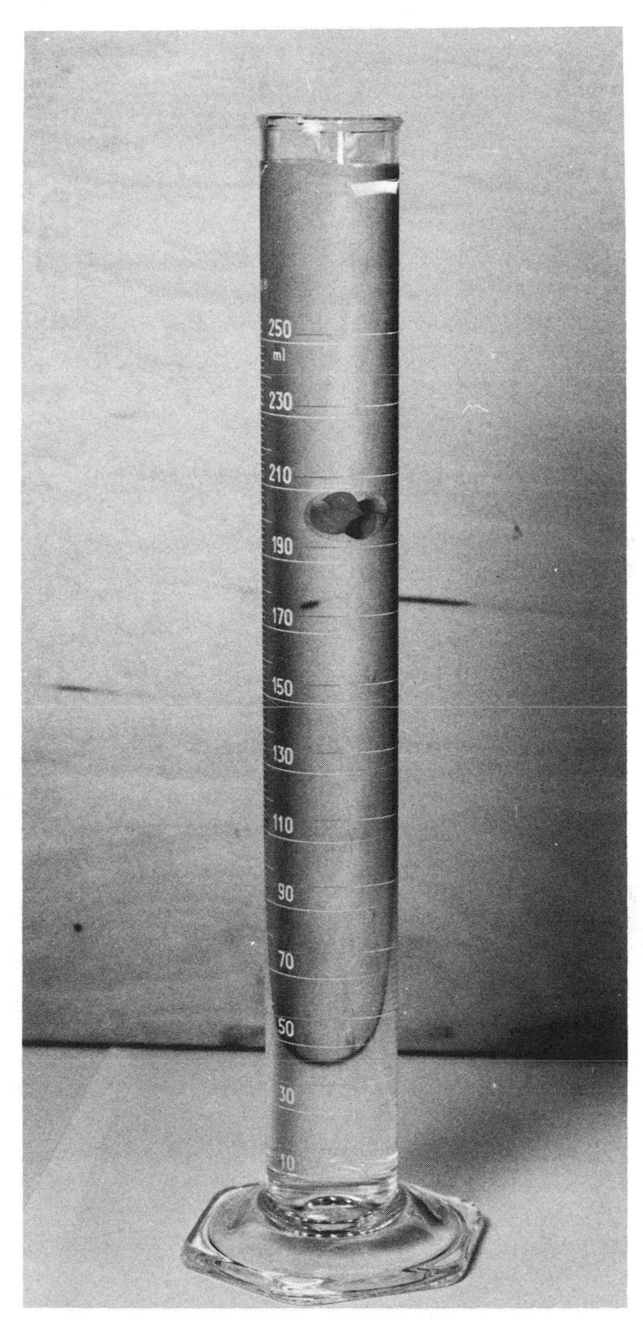

FIG. 27. A falling-ball viscometer improvised in the laboratory (note the glass marble falling through the liquid).

sets of fiduciary lines etched into the tube as measured by stopwatch, and K is the instrument constant; and

$$K = \frac{8.80\ D_f^2}{L} \cdot R^{5/2}\left(2 + \frac{R}{100}\right),$$

where D_f is the diameter of falling ball, L is the distance the ball falls between fiduciary marks, $R = 100(D_t - D_f)/D_f$, and D_t is the tube diameter.

In practice, the value of K is usually obtained by measuring the time of descent for a liquid of known viscosity and rearranging the viscosity equation into the form:

$$K = \frac{\eta_s}{\rho_f - \rho} \cdot t,$$

where η_s is the viscosity of liquid of known viscosity.

The Gilmont Viscometer uses a 10 ml sample. Two sizes of tubes and two balls (glass and stainless steel) are available. It is suitable for Newtonian liquids in the viscosity range of 0.25–300 cP.

A variation of the falling-ball viscometer is the rolling-ball viscometer in which a ball falls through the liquid in a tube inclined at an angle of about 10° from the vertical. The tube is only slightly larger in diameter than the ball, and there is a strong influence of the wall on the ball.

The best-known rolling-ball apparatus is the Hoeppler Viscometer (see Fig. 28). The instrument consists of a heat-resistant chemically inert 20-cm-long glass tube with a precision bore about 16 mm diam. It is enclosed in an 80-mm-diam glass tube through which water from a constant temperature bath is circulated. A screw cap at the top of the tube is removed, the tube is filled with sample (about 30 ml), a designated ball is placed in the tube, all air is removed, and the cap is replaced. When the system has reached equilibrium temperature, the tube assembly is inverted and the rate of fall of the ball between markings on the glass tube is measured with a stopwatch.

Hubbard and Brown (1943) developed general relations between the variables involved in the streamline region of fluid flow for rolling-ball viscometers which led to the equation

$$\eta = \frac{5\pi}{42} K \cdot \frac{d^2 \rho g \sin \theta}{V} \cdot \frac{\rho_s - \rho}{\rho} \cdot \frac{D + d}{d},$$

where K is a dimensionless correlation factor; d, the diameter of the ball; D, the internal diameter of the tube; V, the terminal rolling velocity of the ball; g, acceleration of gravity; ρ, the density of the liquid; ρ_s, the density of the ball; and θ, the angle of inclination of the tube to horizontal.

For a given instrument operating under standard conditions, D, d, θ, and K are constant and the above equation reduces to

$$\eta = C(\rho_s - \rho)/V,$$

where C is the instrument coefficient, which is equal to $(5\pi/42)Kg \sin\theta$ $D(D + d)$.

By selecting balls of different composition and different diameters it is possible to measure viscosities over the range of less than 1 cP to about 2000P. The

FIG. 28. The Hoeppler Viscometer.

Hoeppler viscometer can give results reproducible to 0.5% or better with Newtonian fluids.

Oscillation Viscometry

A vibrating surface in contact with a liquid experiences "surface loading" because the shear waves imparted to the liquid are damped at a rate that is a function of the viscosity of the liquid. The power required to maintain a constant amplitude of oscillation is proportional to the viscosity of the fluid. Oscillation viscometers usually take the form of a stainless-steel ball immersed in the fluid and vibrated at high frequency and low amplitude. This type of viscometer has the advantages of high precision, high sensitivity to small changes in viscosity, rapid accumulation of data, and the equipment is easy to clean. The disadvantages are that it operates at one shear rate only.

The size of the test sample is not critical so long as it exceeds that volume below which reflection from the walls of the container occur. This distance is usually less than 5 mm. Roth and Rich (1953) give the following equation for the propagation distance for the amplitude of the shear waves to fall to $1/e$ of their value in a Newtonian fluid:

$$\delta = (2\eta)^{1/2}/\omega\rho,$$

where δ is the propagation distance; η, the viscosity of the fluid; ρ, the density of the liquid; and ω, the vibrational frequency.

A commercial viscometer of this type is available from the Nametre Company (Fitzgerald and Matusik, 1976; Ferry, 1977). It consists of a 1¼-in.-diam polished stainless-steel ball attached to a stainless-steel rod. The ball is immersed in the liquid and vibrated at a frequency of 646 Hz and am amplitude of 25 μ. A digital readout dial displays the viscosity. A viscosity range from about 1 cP to 1000 P can be measured. Minimum sample size is 35 ml up to 100 P and 70 ml up to 1000 P. The author has not seen reports of the use of this instrument for foods, but it appears to have possibilities for many liquids, including in-line quality control (Oppliger *et al.*, 1975).

Imitative Viscometers

These empirical instruments imitate the flow of non-Newtonian fluid foods under practical conditions. They are simple instruments that usually give a one-point measurement. Although they have their limitations they can be useful for quality control purposes. Examples of this type of viscometer are the Bostwick Consistometer, the Grawemeyer and Pfund Consistometer (also known as the Adams Viscometer), and sag meters. These types were discussed in the previous chapter.

Use of One-Point Measurements for Non-Newtonian Fluids

Throughout this chapter we have emphasized the severe problems associated with attempts to describe a non-Newtonian fluid by means of a one-point measurement. However, having expressed these cautions, it is now time to point out that under certain conditions it is possible to use a one-point measurement as a quality control technique for non-Newtonian fluids. In some highly standardized systems the change in viscous properties during processing moves in a reproducible manner along a predetermined path. A one-point measurement may satisfactorily determine the endpoint in such a system.

An example of this can be found in the concentration of tomato juice to make catsup. Tomato catsup is essentially tomato puree that has been flavored with salt, sugar, vinegar, and spices. It is manufactured by adding these ingredients to tomato juice and boiling until a satisfactory consistency is obtained. Close con-

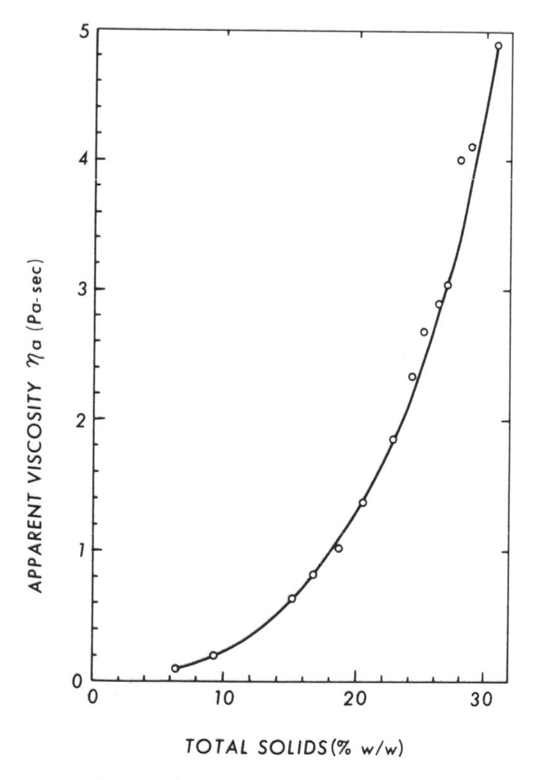

FIG. 29. Apparent viscosity of tomato juice and concentrates as a function of solids concentration measured at a constant shear rate of 100 sec^{-1}. Nova cultivar. (From Rao *et al.*, 1981.)

trol of this endpoint consistency is critical. If slightly too thin, the catsup gushes out of the bottle too fast, while if slightly too thick, it becomes difficult to make it flow from the bottle. Figure 29 plots the apparent viscosity of tomato puree as a function of the solids content in the puree. Although the flow properties of the puree at each concentration are complex (see Rao *et al.*, 1981), the viscous properties do move in a reproducible way along this complex path. Hence, a single-point apparent viscosity measurement can be used successfully to determine the finishing point for tomato catsup.

CHAPTER 6

Sensory Methods of Texture and Viscosity Measurement

Introduction

Sensory evaluation is the measurement of a product's quality based on information received from the five senses: sight, smell, taste, touch, and hearing. Sensory texture measurement is perceived primarily by touch (the tactile sense), although the eyes and ears can provide information on some important components of the total texture profile of a product. The signals generated at the nerve endings of the senses are transmitted via the central nervous system to the brain where they are integrated with past experience, expectations, and other conceptual factors before the opinion of the response is summarized (Amerine *et al.*, 1965; Larmond, 1970).

Sensory methods of measuring food quality appear to lack the precision that is desirable in scientifc research because of the variability from person to person and variability from hour to hour and day to day in likes and dislikes of each person. In spite of these obstacles, sensory measurement of texture is a very important aspect of food quality that cannot be ignored. Later in this chapter it will be shown that some sensory testing methodology can be as reproducible and precise as objective measurements.

Importance of Sensory Evaluation

Instruments are calibrated in absolute units such as newtons force, millimeters distance, pascal-seconds viscosity, and so on, but these readings mean little unless correlated with sensory judgments of quality. There is no point in measuring properties that are not perceived or not judged important by the human senses. People will not purchase or consume food unless it has high acceptability according to their perception of quality.

Sensory methods are the ultimate method of calibrating instrumental methods of texture measurement. Even though sensory methods are generally time consuming, expensive, and not subject to absolute standards, the fact remains that eventually all objective measurements have to be calibrated against the human senses. We have to face the fact that if the palate sends a value judgment message that says the food has undesirable textural properties, then the texture is undesirable regardless of the readings given by our instruments.

Sensory evaluation offers the opportunity to obtain a complete analysis of the textural properties of a food as perceived by the human senses. A number of processes occur while food is being masticated, including deformation, flow, comminution, mixing and hydration with saliva, and sometimes changes in temperature, size, shape, and surface roughness of the food particles. All of these changes are recorded with great sensitivity by the human senses, but many of them are difficult to measure by objective methods. The entire complex of events that occurs during mastication cannot be measured completely by instruments. There is no instrument available that has the sophistication, elegance, sensitivity, and range of mechanical motions as the mouth or that can promptly change the speed and mode of mastication in response to the sensations received during the previous chew.

Sensory evaluation is an important aspect of product development. It is the best method for evaluating texture of new types of foods in the early stages of development, especially fabricated foods, and for providing a basis on which instrumental methods might later be designed for use as a quality measure and production control.

Sensory Texture Profiling

The most complete system of sensory texture measurement is the General Foods Sensory Texture Profiling technique (Brandt *et al.*, 1963; Szczesniak *et al.*, 1963; Civille and Szczesniak, 1973; Civille and Liska, 1975). The following description is based on the material in these references. This technique is an extremely powerful tool and is highly recommended. It should be far more widely used than it is. Most other methods for sensory analysis of texture may be viewed as partial texture profile techniques. The best way to learn the procedure and have confidence in it is to do it; it is less satisfactory to describe it because verbal or written descriptions do not give the sense of the strength, accuracy, flexibility, and reproducibility of the technique that is obtained by actually doing it. One might liken this to learning to drive a car by sitting behind a steering wheel and actually driving the car versus learning to drive a car by reading about how to drive cars but never driving one.

The major steps in the operation of establishing a sensory texture profile are (1) selection of panel, (2) training the panel, (3) establishing standard rating scales, (4) establishing a basic texture profile analysis (TPA) score sheet, and (5) developing a comparative TPA score sheet for each commodity. These steps will now be described in sequence.

1. Selection of Panel Members

A properly trained panel leader is needed to start the sensory texture profiling. It is best for this person to have been trained in a formal training workshop conducted by people who are well experienced with the procedure. The panel leader should possess all the attributes needed for panel members described below and in addition should have (a) the type of personality that puts people at ease and encourages them to put forth their best efforts as a group; (b) some scientific training and understanding of the scientific method, although it is not necessary to have advanced training in these areas; (c) leadership qualities that will bring the panel to a consensus of opinion without imposing personal ideas upon them.

The general requirements for texture profile panelists are listed in Table 1. At least twice as many persons as needed should be chosen for preliminary selection because not all persons meeting the requirements in Table 1 will be found suitable. A panel normally comprises five to seven persons. In order to have a complete panel at all times, it is necessary to start with two to three times this number in order to allow for those who cannot pass the preliminary selection test and also to allow for attrition due to absences, relocations, and retirements.

In the preliminary selection process each candidate is given four consecutive samples from the hardness scale presented in random order and asked to grade them in order of increasing hardness. Peanuts, carrot slices, peanut brittle, and rock candy are easy to obtain, fairly standard in hardness, not highly perishable, and are a good set to present to the panel. Those persons who can rank these four commodities with complete success in increasing order of hardness are used for further training. Those who are unsuccessful in this preliminary test are excluded from further participation.

2. Training of the Panel

The panel should be located in surroundings that are conducive to concentration: a place that is well lighted (not glary), quiet, and free from odors and distractions that might lower the concentration of the panel from the task at hand. Successful sensory texture profiling requires much concentration. The temperature of the surroundings should be comfortable. The panel is seated around a

TABLE 1

REQUIREMENTS FOR TEXTURE PROFILE PANELISTS[a,b]

1. Ability to work cooperatively and harmoniously with a group and develop a feeling of team identity with the group.
2. Able to spare the time for training (2–3 hr a day for several weeks) and the regular operation of the panel for an indefinite period.
3. Their supervisor must approve this expenditure of time willingly, not reluctantly.
4. Panel members should be very interested in their work, and dedicated to developing a team that can give results with the precision and reproducibility of a scientific instrument.
5. Panel members must have common sense and reasonable intelligence. A high I.Q. is not essential. No special education is necessary. In fact, laboratory technicians and secretaries (for example) frequently make the best panel members because they can more readily spare the time; they are always available; they are less likely to be preoccupied with other matters (as are senior scientists and administrators), and hence are able to devote their whole interest to the work at hand.
6. Panel members should be able to discuss the tests with the other members of the panel and be able to reach a consensus. People with a domineering or bossy attitude, and people who are excessively timid or cannot express an opinion are unsuited for panel work.
7. They should be able to develop a professional attitude toward their work, and take pride in it.
8. They should not have dentures because false teeth may restrict the perception of some texture attributes.
9. People who are deeply involved in product development should not be on the panel because they tend to come to the panel with preconceived ideas of the textural quality of the products to be examined.
10. It is desirable to have members of both sexes represented on the panel, although the panel can be comprised predominantly of one sex.

[a]From Civille and Szczesniak (1973).

[b]The panel leader should possess the above attributes and in addition should have the following: (1) the type of personality that puts people at ease and encourages them to put forth their best efforts as a group; (2) leadership qualities that will bring the panel to a consensus opinion without imposing personal ideas upon them; and (3) some scientific training and understanding of the scientific method.

large table and provided with score sheets as needed, a glass of water for rinsing the mouth, and a paper cup for spitting out any material that is no longer needed for the test. There should be room in the center of the table to hold the samples that are currently being tested. A blackboard or a set of large flip sheets of paper should be available for recording scores and any other comments made by the panel.

The first step in the training is to familiarize the panel with the standard rating scales described in detail in the next section. The panel is presented with one complete standard rating scale at a time. The panel leader gives a full explanation of the scale and then the panel samples each item on the scale in ascending order of magnitude. This is followed by discussion of the scale and further sampling of the commodities on that scale until the panel feels they have mastered the scale. At that time, a food of unknown intensity on the scale being considered is presented and the panel is asked to rate it to the nearest quarter point on the scale. The scores are called out to the panel leader who writes them down on the blackboard. When all the scores have been written down, any differences in the scores are discussed, and sampling of the unknown and the standards is repeated until the entire panel gives a score within ±¼ point of the mean. The panel should work on each standard rating scale until they can obtain this degree of consistency between panelists. When this has been satisfactorily completed, the panel moves on to the next standard rating scale and repeats the procedure. This is continued until all of the scales have been covered, and the panel has a clear impression of the type of property being measured in each scale and the intensities that can be experienced in that scale using the standard items as anchor points.

When the panel has thoroughly grasped the standard scales, including the geometrical scales, they develop (as an exercise) a complete texture profile on a simple product such as soda crackers using the basic TPA score sheet, which is described in detail below. The complete texture profile is developed in one session without the presence of any of the food items on the standard scales. When it is found that panel members show substantial disagreement in some areas, the exercise is repeated in the following session with the items from the standard scales available for reference on the disputed points. The panel now repeats the evaluation of the disputed points on the scale using the standard scales for reference and continues to do this until they resolve their differences to within ±¼ point. This exercise generally makes the panel realize the value of having the standard scales for reference as anchor points.

Having successfully developed a reproducible texture profile for a simple product the panel turns its attention to the commodity of interest and develops a texture profile for it. The time required to develop a texture profile for the product of interest varies. With a simple product a good profile may be developed in two or three sessions. A difficult product may take a number of sessions before a complete, reproducible, and satisfactory profile is developed.

When the basic texture profile for the commodity of interest has been completed, the panel leader develops the comparative texture profile ballot, which is discussed on p. 262. The panel then uses the comparative texture profile ballot and perfects it by means of discussions between the panel and the leader, and by referring to the standard rating scales when questions or differences of opinion arise.

The panel has now been trained and is ready for routine work on the commodity for which the comparative texture profile ballot has been prepared.

Whenever a new commodity is to be studied, the panel utilizes the basic training it has already received. They first develop the basic texture profile for the new product and then move on to develop and perfect the comparative texture profile for that product.

3. Establishing Standard Rating Scales

Textural characteristics are divided into three classes:

a. *Mechanical Characteristics*

The mechanical characteristics are related to the reaction of the food to stress and are made quantitative by means of standard rating scales, analogous to Mohs scale of hardness used by mineralogists. The standard hardness scale consists of nine food products ranging from low hardness (Philadelphia cream cheese) to high hardness (rock candy).

Other standard scales are fracturability (originally termed "brittleness") (7 points), chewiness (7 points), gumminess (5 points), adhesiveness (5 points), and viscosity (8 points). The original standard scales are listed in Tables 2–7.

The items selected to be used for the standard rating scales are chosen on the basis of having that particular textural property as a dominant characteristic coupled with fairly uniform intervals between points in the desired characteristic. A panel of five to eight people with adequate training can rate the mechanical properties of a sample on each of the six standard scales to within about one fifth of a point with a high degree of reproducibility.

The items listed in the standard scales shown in Tables 2–7 were used to construct the original scales because they were available in eastern United States. Some of them may not be available in other areas of the United States and many of them are not available in other countries. Under these conditions substitute commodities must be selected to fill out these scales. Each scale should encompass the full range of intensity of that textural characteristic encountered in foods. Other factors to be considered in selection of the standard commodities are (a) select well-known brands that have good quality control and give a consistent quality of the product; (b) use products that require the minimum amount of preparation in order to eliminate recipe variables; and (c) use products that do not change greatly with small temperature variations or with short-term storage. The reference items should be standardized as much as possible with respect to size, temperature, brand name, and handling to ensure the stability of each scale point.

With these criteria in mind it is possible to change any commodity in the standard scales. An example of this is shown in Table 8 where the hardness and viscosity scales developed for a texture profile panel in Colombia are shown and contrasted with the original scales developed by the General Foods Group in eastern United States. Similar scales have been developed in Colombia for the

TABLE 2

STANDARD HARDNESS SCALE[a]

Panel rating	Product	Brand or type	Manufacturer	Sample size
1	Cream cheese	Philadelphia	Kraft foods	½-in.
2	Egg white	Hard-cooked, 5 min	—	½-in. tip
3	Frankfurters	Large, uncooked, skinless	Mogen David Kosher Meat Products Corp.	½-in.
4	Cheese	Yellow, American, pasteurized process	Kraft foods	½-in.
5	Olives	Exquisite, giant size, stuffed	Cresca Co.	1 olive
6	Peanuts	Cocktail type in vacuum tin	Planters Peanuts	1 nut
7	Carrots	Uncooked, fresh	—	½-in.
8	Peanut brittle	Candy part	Kraft foods	—
9	Rock candy	—	Dryden and Palmer	

[a]From Szczesniak *et al.* (1963); reprinted from *J. Food Sci.* **28,** 398, 1963. Copyright by Institute of Food Technologists.

TABLE 3

STANDARD FRACTURABILITY SCALE[a,b]

Panel rating	Product	Brand or type	Manufacturer	Sample size
1	Corn muffin	Finast	First National Stores	½-in.
2	Angel puffs	Dietetic, heated for 5 min at 190°F	Stella D'Oro Biscuit Co.	1 puff
3	Graham crackers	Nabisco	National Biscuit Biscuit Co.	½-in. cracker
4	Melba toast	Inside piece	Devonsheer Melba Corp.	½-in.
5	Jan Hazel cookies	—	Keebler Biscuit Co.	½-in.
6	Ginger snaps	Nabisco	National Biscuit Co.	½-in.
7	Peanut brittle	Candy part	Kraft foods	½-in.

[a]This was originally known as the "brittleness" scale.

[b]From Szczesniak *et al.* (1963); reprinted from *J. Food Sci.*)**28,** 399, 1963. Copyright by Institute of Food Technologists.

TABLE 4

STANDARD CHEWINESS SCALE[a]

Product rating	Average no. of chews	Product	Brand or type	Manufacturer	Sample size
1	10.3	Rye bread	Fresh, center cut	Pechter Baking Co.	½-in.
2	17.1	Frankfurter	Large, uncooked, skinless	Mogen David Kosher Meat Products Corp.	½-in.
3	25.0	Gum drops	Chuckle	Fred W. Amend Co.	½-in.
4	31.8	Steak	Round, ½-in.-thick broiled on each side for 10 min	—	½-in. square
5	33.6	Black crows candy	—	Mason Candy Corp.	1 piece
6	37.3	Peanut chews	—	Whitman Co.	1 piece
7	56.7	Tootsie rolls	Midget size	Sweets Co. of America	1 piece

[a]From Szczesniak *et al.* (1963); reprinted from *J. Food Sci.* **28,** 399, 1963. Copyright by Institute of Food Technologists.

other mechanical characteristics using different commodities than in eastern United States (Bourne *et al.*, 1975).

When necessary, the scales can be expanded in selected areas to allow for a more precise description of differences between closely related samples. For example, when working with semisolids such as puddings and whipped toppings the lower end of the scale may require the addition of softer standards than cream cheese, which ranks number one on the standard scale.

TABLE 5

STANDARD GUMMINESS SCALE[a]

Panel rating	Product	Brand or type	Manufacturer	Sample size
1	40% flour paste	Gold Medal	General Foods	1 tbs
2	45% flour paste	Gold Medal	General Foods	1 tbs
3	50% flour paste	Gold Medal	General Foods	1 tbs
4	55% flour paste	Gold Medal	General Foods	1 tbs
6	60% flour paste	Gold Medal	General Foods	1 tbs

[a]From Szczesniak *et al.* (1963); reprinted from *J. Food Sci.* **28,** 400, 1963. Copyright by Institute of Food Technologists.

TABLE 6
STANDARD ADHESIVENESS SCALE[a]

Panel rating	Product	Brand or type	Manufacturer	Sample size
1	Hydrogenated vegetable oil	Crisco	Procter and Gamble Co.	½ tsp
2	Buttermilk biscuit dough	—	Pillsbury Mills	¼ biscuit
3	Cream cheese	Philadelphia	Kraft Foods	½ tsp
4	Marshmallow topping	Fluff	Durkee-Mower	½ tsp
5	Peanut butter	Skippy, smooth	Best Foods	½ tsp

[a]From Szczesniak *et al.* (1963); reprinted by *J. Food Sci.* **28,** 400, 1963. Copyright by Institute of Food Technologists.

The scales can also be expanded between points by adding other selected foods to serve as intermediate anchor points. For example, if a given formulated food always has a hardness between 3 and 5, then a new hardness scale can be constructed just for that food using 5 to 10 anchor points. In this case there is no need to use points 1, 2, 6, 7, 8, and 9 of the standard hardness scale because they will never be used for this food.

TABLE 7
STANDARD VISCOSITY SCALE[a]

Panel rating	Product	Brand or type	Manufacturer	Sample size
1	Water	Spring	Crystal Springs Co.	½ tsp
2	Light cream	Sealtest	Sealtest Foods	½ tsp
3	Heavy cream	Sealtest	Sealtest Foods	½ tsp
4	Evaporated milk	—	Carnation Co.	½ tsp
5	Maple syrup	Premier 100%	Francis H. Leggett and Co.	½ tsp
6	Chocolate syrup	—	Hershey Chocolate Corp.	½ tsp
7	Mixture: ½ cup mayonnaise and 2 tbs heavy cream	Hellman's Sealtest	Best Foods Sealtest Foods	½ tsp
8	Condensed milk	Magnolia, sweetened	Borden Foods	½ tsp

[a]From Szczesniak *et al.* (1963); reprinted from *J. Food Sci.* **28,** 401, 1963. Copyright by Institute of Food Technologists.

TABLE 8

HARDNESS AND VISCOSITY SCALES IN NEW YORK AND BOGOTA

Scale value	New York[a]	Bogota[b]
Hardness		
1	Philadelphia cheese (Kraft)	Philadelphia cheese (Alpina)
2	Cooked egg white	Cooked egg white
3	Frankfurters (Mogen David)	Cream cheese (Ubaté)
4	Processed cheese (Kraft)	Frankfurters (Suiza)
5	Pickled olives (Cresca)	Mozzarella cheese (LaPerfecta)
6	Peanuts (Planters)	Peanuts (LaRosa)
7	Carrot (raw)	Carrot (raw)
8	Peanut brittle (Kraft)	Candied peanuts (Colombina)
9	Rock candy	Milk candy (Colombina)
Viscosity		
1	Water	Water
2	Light cream (Sealtest)	40% sucrose syrup
3	Heavy cream (Sealtest)	50% sucrose syrup
4	Evaporated milk	60% sucrose syrup
5	Maple syrup	Maple syrup
6	Chocolate syrup (Hershey)	96% sweetened condensed milk + 4% water
7	Mixture: ½ cup mayonnaise and 2 tbs heavy cream	Sweetened condensed milk
8	Sweetened condensed milk	

[a]From Szczesniak *et al.* (1963)
[b]From Bourne *et al.* (1975).

For panel use, the definitions of the mechanical characteristics are given in terms that are closely related to the actual perception.

Hardness is the force required to compress a substance between the molar teeth (in the case of solids) or between the tongue and palate (in the case of semisolids). To evaluate the hardness of solid foods, the item is placed between the molar teeth and the panelist bites down evenly, evaluating the force to compress the food. For semisolids, hardness is measured by compressing the food against the palate with the tongue.

Fracturability is a parameter that was initially called brittleness. It is the force with which a sample crumbles, cracks, or shatters; for example, peanut brittle shatters with greater force than graham crackers. Foods that exhibit fracturability are products that possess low cohesiveness and some degree of hardness. To evaluate fracturability the food is placed between the molar teeth and the panelist bites down evenly until the food crumbles, cracks, or shatters. The degree of fracturability of a food is measured as the horizontal force with which a food moves away from the point where the vertical force is applied. Another factor that helps determine fracturability is the suddenness with which the food breaks.

Chewiness is the length of time required to masticate a sample at a constant rate of force application to reduce it to a consistency suitable for swallowing. An alternative way to use this scale is to record the actual number of chews instead of using the numbers from the scale. To evaluate chewiness the standard is placed in the mouth and masticated at the rate of one chew per second. Chewiness is the number of chews required for a standard-sized piece before the product is swallowed. There may be a wide range in the number of chews from person to person, but the average number of chews for the whole panel represents a range for each scale value and adjacent ranges should not overlap.

Adhesiveness is the force required to remove material that adheres to the mouth (generally the palate) during the normal eating process. The technique for evaluating adhesiveness is to place the food in the mouth, press it against the palate, and evaluate the force required to remove it with the tongue. Since the amount of saliva in the mouth affects the degree of adhesiveness, it is desirable to rinse the mouth with water immediately prior to each evaluation.

Gumminess is the denseness that persists throughout mastication or the energy required to disintegrate a semisolid food to a state ready for swallowing. The technique for evaluating gumminess is to place the sample in the mouth and move it between the tongue and the palate. The degree of gumminess is judged as the extent of manipulation required before the food disintegrates.

Viscosity is the force required to draw a liquid from a spoon over the tongue. The technique for evaluating viscosity is to place the spoon containing the food directly in front of the mouth and draw the liquid from the spoon over the tongue by slurping. The degree of viscosity is measured as the force required to draw the liquid over the tongue.

b. *Geometrical Characteristics*

Geometrical characteristics are related to the arrangement of the physical constitutents of the food product such as size, shape, arrangement of particles within a food, surface roughness, etc.; they are qualitative and partly quantitative.

These characteristics relate to particle size, shape, orientation, and surface roughness. Some standards for geometrical characteristics are given in Table 9. Geometrical characteristics do not lend themselves to as clear-cut scaling as do the mechanical characteristics. They may be divided into two general groups of qualities: size and shape, and shape and orientation.

(1) Those related to size and shape are perceived as discrete particles that are relatively harder than the surrounding medium or the carrier. This group can be scaled in the same manner as the mechanical characteristics. For example, chalky, gritty, grainy, and coarse comprise a scale of increasing particle size.

TABLE 9

GEOMETRICAL CHARACTERISTICS OF TEXTURE[a]

Descriptive term	Example
A. Characteristics relating to particle size and shape:	
Powdery	Confectioner's sugar
Chalky	Raw potato
Grainy	Farina, Cream of Wheat
Gritty	Pear stone cells, sand
Lumpy	Cottage cheese
Beady	Tapioca pudding
B. Characteristics relating to shape and orientation:	
Flaky	Boiled haddock
Fibrous	Base of asparagus shoot, breast of chicken
Pulpy	Orange sections
Cellular	Raw apples, white cake
Aerated	Chiffon pie filling, milk shake
Puffy	Puffed rice, cream puffs
Crystalline	Granulated sugar

[a]From Brandt *et al.* (1963); reprinted from *J. Food Sci.* **28,** 405, 1963. Copyright by Institute of Food Technologists.

Note that this is particle size evaluation; the hardness of the particles must be evaluated independently.

(2) Characteristics related to shape and orientation represent highly organized structures of different geometrical arrangements within each product. For example, a puffy texture is an organization of hard or firm outer shells filled with large, often uneven, air pockets (e.g., puffed rice), while an aerated texture is a network of relatively small even cells filled with air and surrounded by cell walls (e.g., whipped egg white). The geometrical characteristics are sensed primarily by the tongue but may be sensed to some extent on the palate and on the teeth.

c. *Other Characteristics*

Other characteristics are properties related to the moisture and fat content of the food as perceived by the human senses; they are qualitative and partly quantitative. These are sometimes called chemical characteristics because they measure the factors of moistness, dryness, oiliness, and fattiness. No standard scales for these characteristics were published in the original texture profile method (Brandt *et al.*, 1963), but it should be possible to develop standard scales for these properties. These terms are not the same as moistness or fat content as determined by chemical analysis. For example, two apples may have the same moisture content as determined by chemical analysis but in a sensory test one might be found to be dry and mealy while the other is moist and juicy. It is possible to have two cuts of beef that have been shown to have equal moisture contents by chemical analysis, and yet one cut will be termed juicy because of the sensation of moisture in the mouth while the other cut will be determined dry because the sensation of moisture is lacking.

4. Developing the Basic TPA Score Sheet

A systematic method of recording all the texture characteristics of a given food is based upon the "order of appearance" principle, which relates to the time sequence in which the various attributes of the product appear. Unlike flavor, where the order of appearance of the notes cannot be anticipated, texture perception follows a definite pattern regarding the order in which the various characteristics are perceived. These are subdivided into initial (first bite), masticatory, and residual phases. The basic texture profile score sheet is shown in Fig. 1. This should be consulted frequently during the discussion that follows:

1. *Initial.* In the first bite the product is placed between the molars and a single bite is made. On this bite the mechanical properties of hardness, fracturability, and viscosity are measured and also geometrical properties and other properties (moistness, oiliness). The mechanical characteristics are graded to

BASIC TEXTURE PROFILE BALLOT

Product: ____________ Date: ____________ Name: ____________

I. *INITIAL (perceived on first bite)*
  (a) Mechanical
    Hardness (1–9 scale)
    Fracturability (1–7 scale)
    Viscosity (1–8 scale)
  (b) Geometrical
  (c) Other characteristics (moistness, oiliness)
II. *MASTICATORY (perceived during chewing)*
  (a) Mechanical
    Gumminess (1–5 scale)
    Chewiness (1–7 scale)
    Adhesiveness (1–5 scale)
  (b) Geometrical
  (c) Other characteristics (moistness, oiliness)
III. *RESIDUAL (changes induced during mastication and swallowing)*
  Rate of breakdown
  Type of breakdown
  Moisture absorption
  Mouth coating

FIG. 1. The basic texture profile score sheet. (Courtesy of Dr. A. S. Szczesniak.)

within 0.2 units on the standard scale, although some particularly sensitive people can grade to within 0.1 of a unit. The geometrical characteristics and other characteristics, if present, are listed without assigning a number to them, but adjectives such as slight, moderate, or strong may be appended to these characteristics at this time. It is worth noting that some characteristics may be absent and should be given a score of 0. For example, the fracturability scale and viscosity scale are mutually exclusive. If a food is fracturable, it is a brittle solid and is not a liquid. A food that has a fracturability component will have no viscosity component. Conversely, viscosity refers to liquid foods that have no fracturability, so if a product is given a score on the viscosity scale the fracturability score will be zero.

2. *Masticatory.* The second or masticatory phase is performed by placing a piece of food between the teeth and chewing at a standard rate, approximately 60 chews per minute, and determining the mechanical properties of gumminess, chewiness, and adhesiveness, and also assessing any geometrical and other characteristics that appear during chewing. As noted above, chewiness may be graded on the 1–7 scale or it can be listed simply as the total number of chews to swallowing. There will always be a score for chewiness of solid foods. If gumminess and adhesiveness are absent, they should be given a score of 0.

3. *Residual characteristics.* The third or residual phase measures the changes induced in chemical, mechanical, geometrical, and all the characteristics throughout the course of mastication up to the completion of swallowing. These are divided separately into the rate of breakdown, type of breakdown, moisture absorption, and mouth coating. In the two previous phases (initial and masticatory), numbers are used to describe the mechanical characteristics and words or phrases are used to describe geometrical and chemical characteristics. In contrast, in the residual phase, numbers are rarely used but phrases and short sentences are used to describe these residual characteristics. To the person who has only read about sensory profiling, the residual characteristics may seem to be of minor importance. In fact, the parameters developed in the residual characteristics are one of the most important aspects of texture profile analysis. The fact that these parameters cannot have numbers assigned to them should not be interpreted as a sign of minor importance. These characteristics are the ones that are most difficult to duplicate in instrumental tests. The residual characteristics section of the texture profile anlaysis is the section where textural parameters are least likely to be measured or detected by any instrumental method. It is an essential part of the total texture profile procedure.

When the panel has completed the basic texture profile ballot for a food, the leader asks each person in turn to call out the scores they have written on their sheet and these scores are written on the board for the rest of the panel to see. When the scores of all the panelists have been written on the board, the leader and the panel examine the scores together. Whenever the score for any parameter that has a standard rating scale varies by more than ±⅕ of a point, the leader and the panelists discuss the problems that were experienced. After discussing the situation with the other panelists and leader, the panelists repeat that section of the ballot and by means of discussion and repeated testing generally reach a consensus. For those parameters for which words, phrases, or sentences are used, the panelists discuss among themselves any differences and by tasting and discussion (led by the leader) reach a consensus. The sampling–discussion–reference to standard scales–sampling–discussion sequence continues until a consensus or near consensus has been reached. Occasionally one panelist will not agree with the rest of the panelists. The score of that panelist is rejected from the final report. At first glance this seems to be a nonscientific approach to reject some of the data, but since the texture profile is built around the concept of consensus following adequate discussion, an out-of-line datum must be rejected if it deviates far from the consensus.

Figure 2 shows the basic texture profile ballot for meatballs, and Fig. 3 shows the basic texture profile ballot for soda crackers.

The following similarities and differences between these two foods during the mastication sequence are found by comparing Figs. 2 and 3.

1. *Initial.* Mechanical characteristics show that soda crackers have a little more hardness than the Swedish meatballs and more fracturability, but viscosity is absent in both commodities. Geometrical properties of the crackers are flaky and puffy while the meatballs are lumpy with a grainy surface. Other characteristics show that the crackers are dry while the Swedish meatballs are moist. The surface of the meatball is slippery, but the uncut surface is not slippery.

2. *Masticatory.* Mechanical characteristics indicate the soda crackers have no gumminess while the meat balls have a gumminess of 1.2. The number of chews for mastication is approximately the same for both commodities and both have a small amount of adhesiveness. The geometrical characteristics of the soda crackers continue to be flaky while the meatballs are coarse and grainy and fibrous particles begin to be felt. Other characteristics show that the soda crackers continue to be dry and the meatballs continue to be moist.

3. *Residual sensations.* The rate of breakdown is high for soda crackers. The meatballs break down fast, forming grains that break down at a medium rate. In the type of breakdown, we find that the crackers break down into little rough sheets that change into a smooth dough, while the meatballs become a nonhomogenous paste that is grainy and the grain size steadily decreases; some stringy fibrous grains are present but become more noticeable toward the end and require more effort to chew. With moisture absorption we find that the soda crackers absorb a lot of saliva at a slow rate and gradually change into a moist dough; the saliva mixes easily with the Swedish meatballs to form a slurry, and

BASIC TEXTURE PROFILE BALLOT FOR MEATBALLS

Product: *Swedish meat balls* Date: ______ Name: ______

I. *INITIAL (perceived on first bite)*
  (a) Mechanical
    Hardness *3.4*
    Fracturability *0.7*
    Viscosity *Not applicable*
  (b) Geometrical *Lumps, with a grainy surface*
  (c) Other characteristics *Moist, uncut surface is slippery and cut surface is not slippery*

II. *MASTICATORY (perceived during chewing)*
  (a) Mechanical
    Gumminess *1.2*
    Chewiness *17.7 chews*
    Adhesiveness *1.2*
  (b) Geometrical *Coarse, grainy, some fibrous particles present*
  (c) Other characteristics *Moist*

III. *RESIDUAL (changes induced during mastication and swallowing)*
  Rate of breakdown—*Large lumps break down fast. Grains break down at a medium rate.*
  Type of breakdown—*Lumps turn into a nonhomogeneous paste that is grainy, and grain size decreases. Some stringy fibrous grains are present that become more noticeable towards the end and require more effort to chew.*
  Moisture absorption—*Initially moist. Saliva mixes easily with slurry and the bolus becomes progressively more moist. Residual grains feel dry.*
  Mouth coating—*Slight residual oiliness. A few fibrous particles stick between the teeth and around the mouth.*

FIG. 2. The basic texture profile score sheet for meatballs. (From unpublished data of M. C. Bourne.)

the bolus becomes progressively more moist, leaving residual grains that feel dry. With mouth coating we find that little pieces of cracker stick to the mouth and gums; there is some slight residual oiliness with the meatballs, and a few fibrous particles stick between the teeth and around the mouth.

These two foods, although very different in nature, have many similarities in the texture profile for the initial and masticatory phases. The major differences between these two commodities show up in residual sensations, illustrating their importance in the texture profile.

5. Developing the Comparative Texture Profile Analysis Ballot

The final step in texture profile analysis is to develop from the standard score sheets a comparative texture profile ballot for each commodity. The basic ballot can be used for any commodity. Each comparative ballot is especially designed for a particular commodity and it enables one to identify and quantify small differences in textural properties of similar materials caused by differences in

BASIC TEXTURE PROFILE BALLOT FOR SODA CRACKERS

Product: *Soda Crackers* Date: Name:

I. *INITIAL (perceived on first bite)*
 (a) Mechanical
  Hardness *4.0*
  Fracturability *2.5*
  Viscosity *Not applicable*
 (b) Geometrical *Flaky and puffy*
 (c) Other characteristics *Dry*

II. *MASTICATORY (perceived during chewing)*
 (a) Mechanical
  Gumminess *0*
  Chewiness *16*
  Adhesiveness *0.7*
 (b) Geometrical *Flaky*
 (c) Other characteristics *Dry*

III. *RESIDUAL (changes induced during mastication and swallowing)*
 Rate of breakdown—*High*
 Type of breakdown—*In the beginning it breaks down into little rough sheets, then it changes into a smooth dough*
 Moisture absorption—*It absorbs a lot of saliva slowly and changes into a moist dough*
 Mouth coating—*Little pieces stick to the mouth and gums*

FIG. 3. The basic texture profile score sheet for soda crackers. (From unpublished data of M. C. Bourne.)

quality of ingredients, formulation, storage, or processing. In the comparative texture profile ballot one material is selected as the "target" material whose textural properties are desirable to reproduce. It acts as the control and is assigned a score of zero for every textural parameter.

A basic texture profile ballot for arepa is shown in Fig. 4, and a comparative texture profile ballot for arepa is shown in Fig. 5. These should be referred to frequently in the following discussion. Arepa is a corn-based staple food that is widely used in Colombia and other countries in Latin America. The textural properties that have been identified during the initial, mastication, and residual phases are listed in the order of appearance and the experimental samples are graded equal to, less than, or greater than the control sample in that particular quality factor. The control in this case was arepa made fresh each day by the traditional village method (Bourne *et al.*, 1975). The grading is made semiquantitative by grading from 1 to 5 plus and 1 to 5 minus. One plus means that the sample is slightly greater than the control in that particular textural property; five plus means it is much greater than the control in that particular textural property. The minus score is used to indicate slightly less than to much less than the control sample.

The comparative texture profile ballot identifies those formulations and pro-

BASIC TEXTURE PROFILE SHEET FOR AREPA

Product: *Arepa de Peto* Date: Name:

I. *INITIAL (perceived on first bite)*
  (a) Mechanical
    Hardness *4.2*
    Fracturability *2.5*
    Viscosity *0*
  (b) Geometrical *Tough skin and coarse center. Sandwich-like structure with thin tough skin and doughy matrix. Black patches are crispy and located only on the surface*
  (c) Other characteristics *Dry surface and moist center*

II. *MASTICATORY (perceived during chewing)*
  (a) Mechanical
    Gumminess *0*
    Chewiness *16*
    Adhesiveness *0.5*
  (b) Geometrical *Surface is coarse and center is doughy with little pieces*
  (c) Other characteristics *Dry surface and moist center*

III. *RESIDUAL (changes induced during mastication and swallowing)*
  Rate of breakdown—*Moderate*
  Type of breakdown—*Skin breaks in little sheets. Center breaks in little pieces to form a nonuniform dough.*
  Moisture absorption—*Center absorbs moisture more quickly than the surface. Bolus has some little, hard and rough pieces.*
  Mouth coating—*Little pieces leave a coating in the mouth, especially on the gums and teeth. After swallowing, the mouth is dry.*

FIG. 4. The basic texture profile score sheet for arepa. (From Bourne *et al.*, 1975. Reprinted from *J. Texture Stud.*; with permission of Food and Nutrition Press.)

cessing variables that bring the experimental samples closer to the target. It is definitely the best technique for identifying desirable textural properties and eliminating undesirable textural properties in product formulation. A study of Figs. 4 and 5 shows that 28 different textural characteristics were derived from the basic texture profile ballot. These figures show how complete a texture analysis can be performed by a trained sensory panel. There is no instrumental method or group of methods available that could give as complete an analysis of texture profile as is seen in Fig. 5.

Figure 6 shows the comparative texture profile for meatballs that was developed from the basic ballot shown in Fig. 2. Initial sensations list the mechanical properties of hardness and fracturability while viscosity does not appear. Geometrical properties list lumpiness and scratchiness of the grains; and other characteristics list slipperiness of the uncut surface and moistness. Masticatory sensations list the mechanical properties of gumminess, chewiness, and adhesiveness; geometrical properties list coarseness, graininess, and amount of fibrous grains; and other characteristics list moistness. Residual sensations list under the rate of

COMPARATIVE TEXTURE PROFILE BALLOT FOR AREPA

	Control	
	0	+
I. Initial Sensation		
(a) Measure force to: (a) bite off with incisors		
(b) pull out with hand		
(b) Hardness		
(c) Sideways sliding of center		
(d) Toughness of skin		
(e) Doughiness of center		
(f) Dryness of skin		
(g) Moistness of center		
II. Mastication		
(a) Chewiness (no. of chews)		
(b) Adhesiveness		
(c) Pastiness of center		
(d) Graininess of center		
(e) Toughness of skin		
(f) Roughness of skin pieces		
(g) Dryness of skin pieces		
(h) Moistness of center		
III. Final Sensations		
(a) Rate of breakdown of skin		
(b) Rate of breakdown of center		
(c) Moistness of paste		
(d) Graininess of paste		
(e) Dryness of skin particles		
(f) Roughness of skin particles		
(g) Presence of coarse particles other than skin		
(h) Absorption of moisture by mass		
(i) Absorption of moisture by pieces of skin		
(j) Mouthcoating of mass		
(k) Presence of skin particles around mouth		
(l) Scratchiness of residual skin particles		

Instructions:
Put an X in "0" column if sample is equal to control.
Put 1 X to 5 X in (+) column if sample is more than control and in (−) column if less than control (X = slightly different; XXXXX = strongly different).

FIG. 5. Comparative texture profile score sheet for arepa. (From Bourne *et al.*, 1975. Reprinted from *J. Texture Stud.;* with permission of Food and Nutrition Press.)

Comparative Texture Profile for Meat Balls and Fish Balls

I. INITIAL SENSATIONS
(a) hardness
(b) fracturability
(c) lumpiness
(d) scratchiness of grains
(e) slipperiness (uncut surface)
(f) moistness

II. MASTICATORY
(g) gumminess
(h) chewiness
(i) adhesiveness
(j) coarseness
(k) graininess
(l) amount of fibrous grains
(m) moisture

III. RESIDUAL SENSATIONS
(n) rate of breakdown of lumps
(o) rate of breakdown of grains n.a.
(p) rate of loss of cohesiveness between particles
(q) homogeneity of bolus
(r) appearance of stringy, fibrous grains
(s) chewiness of fibrous grains x = n.a.
(t) dryness of residual grains
(u) ease of mixing of saliva and slurry
(v) oily mouthcoating
(w) residual fibrous particles
(x) residual sandy particles

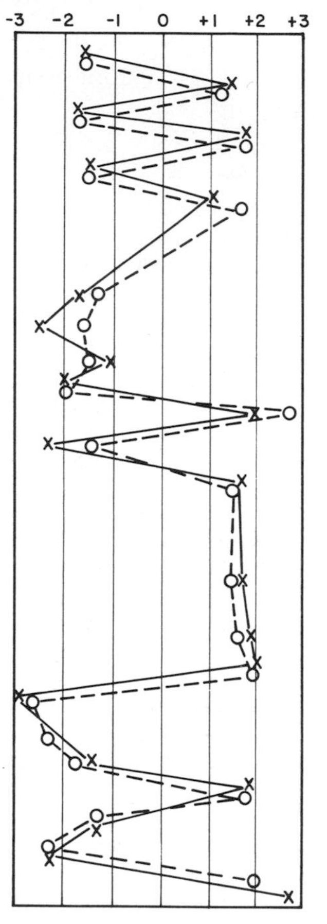

Swedish meat balls were used as control. Sample "X" is fish balls.
Sample "O" is fish balls with 10% replacement of fish with soy protein.

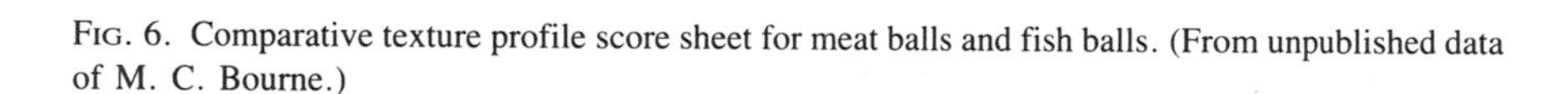

Fig. 6. Comparative texture profile score sheet for meat balls and fish balls. (From unpublished data of M. C. Bourne.)

breakdown heading, the rate of breakdown of the grains, and rate of loss of cohesiveness between the particles; under type of breakdown heading are listed the homogeneity of the bolus, appearance of stringy fibrous grains, and chewiness of the fibrous grains; under moisture absorption heading are listed dryness of residual grains, and ease of mixing of saliva and slurry; under the mouth coating heading are listed oily mouth coating, presence of residual fibrous particles, and presence of residual sandy particles.

The right-hand side of Fig. 6 plots compare the texture profile of fish balls against meatballs. Sample X is made from fish instead of beef. Sample O is fish balls in which 10% of the fish was replaced with textured soy protein. The figure shows that the fish balls have lower hardness, higher fracturability, less lumpiness, more scratchiness of the grains, less slipperiness, and more moistness than the meat balls in the initial sensations; less gumminess, chewiness, adhesiveness, and coarseness than the meatballs, more graininess, less amount of fibrous grains, and higher moistness in the masticatory phase; a higher rate of breakdown of the lumps, a higher rate of breakdown of the grains, a higher rate of loss of cohesiveness between the particles, and a higher homogeneity of the bolus in residual sensations. The fish balls have much fewer stringy fibrous grains, lower chewiness, and less dryness than the meatballs. The fish balls mix with saliva to form a slurry better than the meatballs, have less oily mouth coating, less residual fibrous particles, but have more residual sandy particles.

The product in which 10% of the fish was substituted with soy protein has a similar texture profile to the one made entirely from fish; however, there are a few exceptions, namely, the substitution of the soy gives a product that has higher moistness, chewiness, adhesiveness, graininess, and amount of fibrous grains, and less sandy particles. If this were a project in product development, various other formulations and processing variables would be tried; in each case the changes in the texture profile would be noted, particularly whether the change in formulation or processing brought the experimental sample closer to the target or further away from the target with certain textural properties.

Figure 7 shows a texture profile ballot that was developed for cooked rice. This is a particularly difficult food for TPA, probably because (a) rice is a staple and people become very sensitive to small differences in textural properties of foods that are consumed frequently and in large quantities; (b) rice has a bland flavor and bland flavors usually increase the attention given to textural properties.

The examples listed above show how the basic texture profile technique can be extended to cover all commodities by suitable adaptation.

Variations on the Texture Profile Technique

For certain purposes it may not be necessary to use the entire texture profile. The parameter of interest might be simply chewiness, in which case only the chewiness part of the texture profile would be performed. For example, Harrington and Pearson (1962) used a panel to measure chew count for measuring the tenderness of pork loins with various degrees of marbling and found a good correlation between chew count and Warner–Bratzler shear readings. In contrast, Cover *et al.* (1962) used a panel to evaluate the juiciness of beef and six

COMPARATIVE TEXTURE PROFILE BALLOT FOR COOKED RICE

Score

– 0 +

STAGE 1. Place a spoonful of rice in mouth, manipulate gently without breaking kernels. Evaluate kernel surface for:

Wetness	—*Degree of moisture on kernel surface and type of moisture (watery or starchy).*
Kernel stickiness	—*Degree of manipulation required to remove kernels adhering to tongue and roof of mouth.*
Roughness	—*Feel of the kernel surface on the tongue.*
Uniformity of size	—*Refers to size and shape of individual kernels.*
Clumpiness	—*Degree to which kernels adhere to one another.*
Plumpness	—*Degree to which the kernel is rounded and full.*

STAGE 2. Place a spoonful of rice in mouth. Chew twice with molars. Evaluate:

Hardness	—*Force required to penetrate kernels with the molar teeth on first chew.*
Crumbliness	—*Degree to which kernels fall apart when sheared with the teeth.*
Rubberiness	—*Resistance of the kernel to the teeth prior to shearing.*
Gluiness	—*Degree to which chewed kernels adhere to each other after being sheared and exposed to saliva due to starch paste.*
Inner moisture	—*Amount of moisture inside kernel that is released upon chewing.*

STAGE 3: Place a spoonful of rice in mouth. Chew with molars three or more times. Evaluate:

Kernel uniformity	—*Degree of texture similarity between inside and outside of kernel.*
Cohesiveness of mass	—*Denseness and cohesion of the mass of chewed kernels throughout mastication.*
Stickiness	—*Degree to which kernels adhere to and pack in the teeth during mastication.*
Describe breakdown	—*Includes rate, type, and uniformity. Also the geometrical characteristics observed during breakdown.*
Moisture absorption	—*Degree to which saliva is absorbed by and mixed with chewed kernels.*
Mouth or throat coating	—*Degree of coating perceived in the mouth or throat after swallowing.*

FIG. 7. Comparative texture profile score sheet for cooked rice. (Courtesy of Dr. A. S. Szczesniak.)

components of tenderness: softness to the tongue and cheek, softness to tooth pressure, ease of fragmentation across the grain, mealiness, apparent adhesions between fibers, and connective tissue.

Szczesniak (1979) reported that for fluid foods such as beverages a deeper analysis is required than what is obtained by the single parameter of viscosity that appears in the standard texture profile analysis. Table 10 lists the classification of sensory mouthfeel terms, typical descriptive words, and examples, as developed by Szczesniak (1979).

The previous discussion has dealt with the operation of an expert trained panel for texture profile analysis. Szczesniak *et al.* (1975) simplified this procedure to the point where it can be used by untrained consumer panels. The procedure is based on the original sensory texture profiling technique (Brandt *et al.*, 1963; Szczesniak *et al.*, 1963) and on popular texture terminology as determined in surveys (Szczesniak and Skinner, 1973).

A list of descriptive texture terms for the commodity of interest is compiled by a trained texture profile panel and used to prepare a ballot. A typical ballot is shown in Fig. 8. Texture terms are listed in random order in the left-hand column. The consumers check one of six boxes alongside each word to indicate the degree to which they feel this sample has the texture characteristic described by that term, ranging from "not at all" to "very much so." Some antonyms are included in the list of words as an internal check that the respondents understand the meaning of the words. For example, since "soft" and "hard" convey bipolarity, a sample that is rated high on softness should rate low on hardness, and vice versa. A few comparative terms such as "good" and "bad" are included in order to obtain an overall measure of textural quality. This technique has proved successful for a variety of foodstuffs in both central location and home use situations (Szczesniak *et al.*, 1975). Figure 9 plots the results of a consumer texture profile on whipped toppings done by two panels in two different locations. The excellent reproducibility of results between these two panels demonstrates the reliability of the technique.

The Texture Profile as an Objective Method

Texture profiling is, without question, a *sensory* method, but this does not necessarily mean that it is a *subjective* method. The word "subjective" has the connotation that the personal feelings, biases, and previous experiences of the judge play a major role in the results that are obtained. A subjective method measures an individual's response to the test material; that is, the data are some complex combination of the properties of the test material and the personal characteristics of the judge, and both factors carry considerable weight in the results.

TABLE 10

CLASSIFICATION OF SENSORY MOUTHFEEL TERMS OF BEVERAGES[a]

Category	Responses (% total)	Typical words	Beverages that have this property	Beverages that do not have this property
Viscosity-related terms	30.7	Thin	Water, iced tea, hot tea	Apricot nectar, milk shake, buttermilk
		Thick	Milk shake, eggnog, tomato juice	Club soda, champagne, drink made from dry mix
Feel on soft tissue surfaces	17.6	Smooth	Milk, liqueur, hot chocolate	—
		Pulpy	Orange juice, lemonade, pineapple juice	Water, milk, champagne
		Creamy	Hot chocolate, eggnog, ice cream soda	Water, lemonade, cranberry juice
Carbonation-related terms	11.2	Bubbly	Champagne, ginger ale, club soda	Prune juice, iced tea, lemonade
		Tingly	Ginger ale, champagne, club soda	Instant orange, hot tea, coffee
		Foamy	Beer, root beer, ice cream soda	Cranberry juice, lemonade, water
Body-related terms	10.2	Heavy	Milk shake, eggnog, liqueur	Water, lemonade, ginger ale
		Watery	Bouillon, iced tea, hot tea, drink made from dry mix	Milk, V-8 juice, apricot nectar
		Light	Water, iced tea, canned fruit drink	Buttermilk, hot chocolate, V-8 juice
Chemical effect	7.3	Astringent	Hot tea, iced tea, lemonade	Water, milk, milk shake
		Burning	Whiskey, liqueur	Milk, tea, drink made from dry mix
		Sharp	Prune juice, pineapple juice	Water, hot chocolate, canned fruit drink

Coating of oral cavity	4.5	Mouth coating	Milk, eggnog, hot chocolate	Water, apple cider, whiskey
		Clinging	Milk, milk shake, ice cream soda, liqueur	Water, ginger ale, bouillon
Resistance to tongue movement	3.6	Slimy	Prune juice, milk, light cream	Water, ginger ale, champagne
		Syrupy	Liqueur, apricot nectar, root beer	Water, milk, club soda
Afterfeel–mouth	2.2	Clean	Water, iced tea, wine	Buttermilk, beer, canned fruit drink
		Drying	Hot chocolate, cranberry juice	Water
		Lingering	Hot chocolate, light cream, milk	Water, iced tea, club soda
		Cleansing	Water, hot tea	Milk, pineapple juice, V-8 juice
Afterfeel–physiological	3.7	Refreshing	Water, iced tea, lemonade	Buttermilk, prune juice, hot chocolate
		Warming	Whiskey, liqueur, coffee	Lemonade, champagne, iced tea
		Thirst quenching	Coca-Cola, water, drink made from dry mix	Milk, coffee, cranberry juice
Temperature-related	4.4	Cold	Ice cream soda, milk shake, iced tea	Liqueur, hot tea
		Cool	Iced tea, water, milk	Eggnog
		Hot	Hot tea, bouillon, whiskey	Ginger ale, lemonade, iced tea
Wetness-related	1.3	Wet	Water	Milk, coffee, apple cider
		Dry	Lemonade, coffee	Water

[a]From Szczesniak (1979); reprinted with permission of Academic Press Inc. (London) Ltd.

Instructions: Here is a list of terms commonly used to describe texture, that is, how foods feel in the mouth. Using these terms, we would like you to describe the texture of this sample. To do this, please check one of the six boxes along the side of each term to indicate the degree to which you feel this sample has the texture characteristic described by that term. It is very important to our test that you make a choice for each term.

	Not at all					Very much so
Crisp	☐	☐	☐	☐	☐	☐
Soft	☐	☐	☐	☐	☐	☐
Airy	☐	☐	☐	☐	☐	☐
Brittle	☐	☐	☐	☐	☐	☐
Chunky	☐	☐	☐	☐	☐	☐
Flaky	☐	☐	☐	☐	☐	☐
Soggy	☐	☐	☐	☐	☐	☐
Dry	☐	☐	☐	☐	☐	☐
Bad	☐	☐	☐	☐	☐	☐
Chewy	☐	☐	☐	☐	☐	☐
Crunchy	☐	☐	☐	☐	☐	☐
Hard	☐	☐	☐	☐	☐	☐
Slippery	☐	☐	☐	☐	☐	☐
Doughy	☐	☐	☐	☐	☐	☐
Good	☐	☐	☐	☐	☐	☐
Gritty	☐	☐	☐	☐	☐	☐

FIG. 8. Typical consumer texture profile score sheet for cold cereals. (From Szczesniak *et al.*, 1975. Reprinted from *J. Food Sci.* **40,** 1253, 1975. Copyright by Institute of Food Technologists.)

In contrast, an *objective* method is usually thought of as an instrumental or chemical method. This concept of an objective method may not be always correct. The true characteristics of an objective method are (1) that the data obtained are independent of the individual observer; that is, the result is fair, impartial, factual, and unprejudiced by the personal characteristics of the observer; (2) that the results are repeatable and verifiable by others; that is, other laboratories can obtain the same results within the limits of experimental error.

The author believes that a properly trained texture profile panel is objective, not subjective, because the texture profile method complies with the two criteria of objectivity enunciated in the previous paragraph:

(1) Freedom from personal bias. The data obtained are partly quantitative and partly descriptive, but always objective because the panel is trained to take an analytical approach and use intensity scaling, not acceptability scaling. The members of the panel are trained to observe and record data, not allowing their personal likes and dislikes to influence their results.

(2) Repeatability. Results from different panels are reproducible to a high degree. The author has seen a panel produce a texture profile on a product one

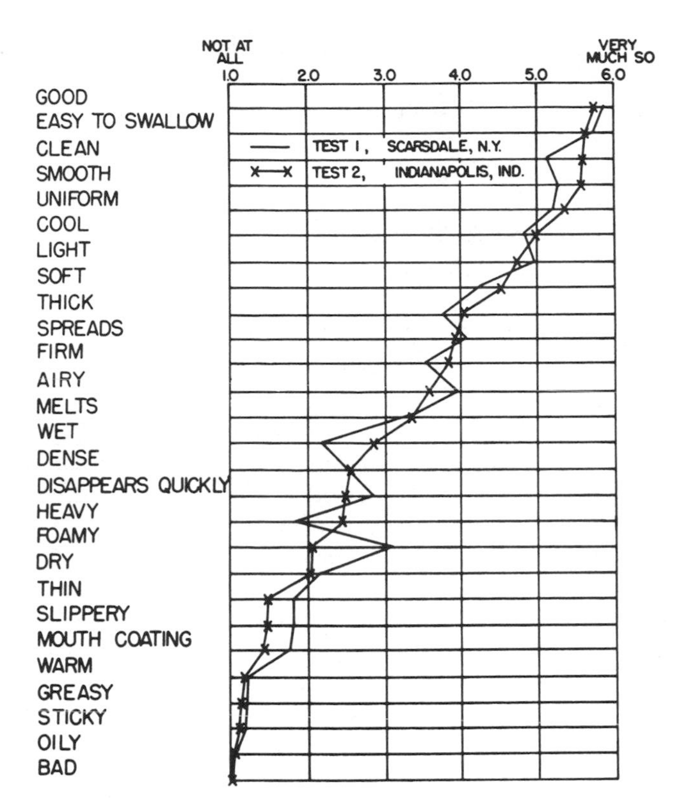

FIG. 9. Consumer texture profile of whipped toppings done by two separate panels in two locations. Note the high degree of reproducibility of the test. (From Szczesniak *et al.*, 1975. Reprinted from *J. Food Sci.* **40,** 1256, 1975. Copyright by Institute of Food Technologists.)

week and then reproduce that profile a week later in the absence of the previous data, on a product they were led to believe was different but was not. He has personal experience with a panel in Ithaca, New York, and another panel in Bogota, Colombia, who gave substantially identical texture profiles for soda crackers. Figure 9 (see above) shows how well even two consumer texture profile groups, one in Scarsdale, New York, and the other in Indianapolis, Indiana, can reproduce data. Szczesniak *et al.* (1975) show that a group can reproduce their score on the same product even when the second test is conducted 16 months after the first test.

The texture profile technique passes the tests of impartiality and reproducibility and should, therefore, be considered as an objective method. The texture profile technique trains a small group of people to use their mouth as a scientific instrument, similar to a balance or a pH meter. The advantages of the texture profile technique over other objective methods are that this particular scientific instrument (the trained mouth) can measure a number of textural parameters that can be measured by no other objective method at the present time, and, in many instances, it can measure a given textural parameter with greater sensitivity than an instrument.

Correlations between Subjective and Objective Measurements

The correlation of objective measurements with subjective perceptions is a complex matter. All that can be given here is a simplified brief summary of the situation.

Sensory testing methodology may be divided into two broad classes: (1) The first class is intensity scaling, which is how much of some property is present in the test material. For this scale there should be a direct relationship between the sensory score and the objective measurement within the limits of sensitivity of the panel. (2) The second class is acceptability scaling, which is how much a person likes the food. This scale usually takes the form of an inverted U when plotted against an objective scale, although the location of the peak in the liking scale may vary widely between various groups. This kind of test is best performed by large-scale consumer testing and will not be covered any further here. The difference between these two types of sensory scaling is shown schematically in Fig. 10.

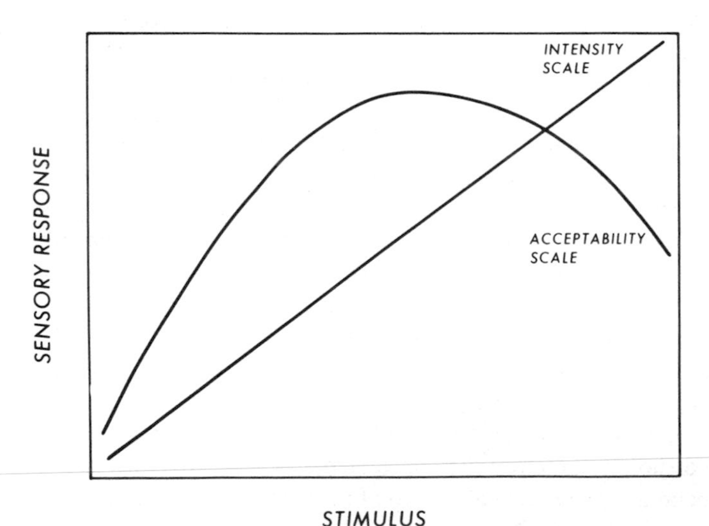

FIG. 10. Schematic representation of sensory intensity scores and sensory acceptability scores versus objective test.

The correlation between an intensity scale and an objective measurement usually follows one of three psychophysical models.

1. The linear model. There is a direct linear relationship between the stimulus (measured by some objective method) and the response which is the sensory measurement. It can be described by the equation

$$R = AS + B,$$

where R is the response to stimulus S, and A and B are constants.

(2) The Weber–Fechner (semilog) relationship. The sensory response makes a linear relationship when plotted against the logarithm of the stimulus. It is described by the equation

$$R = A \log S + B.$$

(3) The Power model (log–log relationship). This model is described by the equation

$$R = CS^n,$$

which may be rearranged into the form

$$\log R = n \log S + \log C,$$

where n and C are constants.

Each of these psychophysical models have been successfully applied to certain systems. There has been a long debate by psychologists as to which is the most suitable model. The present consensus seems to be that the power model is the correct one because it satisfactorily describes most situations that arise. In other words, a plot of the logarithm of the objective measurement versus the logarithm of the subjective measurement will be linear in most circumstances.

The numerical value of the exponent n in the power model is an index of the degree of compression or expansion of the physical scale by the senses.

(a) When $n < 1.0$, there is compression of the physical scale; that is, a tenfold increase in the stimulus will give a less than tenfold increase in the sensory response. This allows a wide stimulus range to be compressed into a smaller and more manageable response range for the senses and brain to process.

(b) When $n = 1.0$, there is no compression or expansion of the scale; a tenfold increase in stimulus gives a tenfold increase in response. In this special case the power model becomes almost identical with the linear model.

(c) When $n > 1.0$, there is expansion of the scale; that is, a tenfold increase in stimulus gives a more than tenfold increase in response. Some experimentally measured values for the exponent n for human subjects are given in Table 11. Note that pressure on the palm has a value of 1.1 for the exponent n, indicating slight expansion, while tactual roughness has a value of 1.5 (great expansion) and tactual hardness has a value of 0.8 (moderate compression).

TABLE 11

MEASURED EXPONENTS AND THEIR POSSIBLE FRACTIONAL VALUES FOR POWER FUNCTIONS RELATING TO SUBJECTIVE MAGNITUDE TO STIMULUS MAGNITUDE[a]

Continuum	Measured exponent	Stimulus condition
Loudness	0.67	3000-Hz tone
Brightness	0.33	5° target in dark
Brightness	0.5	Very brief flash
Smell	0.6	Heptane
Taste	1.3	Sucrose
Taste	1.4	Salt
Temperature	1.0	Cold on arm
Temperature	1.5	Warmth on arm
Vibration	0.95	60 Hz on finger
Vibration	0.6	250 Hz on finger
Duration	1.1	White-noise stimuli
Finger span	1.3	Thickness of blocks
Pressure on palm	1.1	Static force on skin
Heaviness	1.45	Lifted weights
Force of handgrip	1.7	Hand dynamometer
Vocal effort	1.1	Vocal sound pressure
Electric shock	3.5	Current through fingers
Tactual roughness	1.5	Rubbing emery cloths
Tactual hardness	0.8	Squeezing rubber
Visual length	1.0	Projected line
Visual area	0.7	Projected square
Angular acceleration	1.41	5-sec stimulus

[a]From Stevens (1970); reprinted with permission from *Science*. Copyright 1970 by the American Association for the Advancement of Science.

Nonoral Methods of Sensory Measurement

Although most of the sensing of texture occurs in the mouth and with the lips, it is possible to measure textural properties outside the mouth, most commonly with the fingers and the hand. It is a common practice to hold foods in the hand or squeeze them, and this frequently gives a good method of measuring the textural quality of the food. The food may be squeezed between the forefinger and the opposed thumb or between two, three, or four fingers and the opposed thumb. It may be squeezed by pressing with the whole palm on top of the food, which is resting on a firm surface such as a table, or the two palms may be placed at opposite ends of the food and squeezed. The size of the object frequently determines the method that is used. The forefinger and opposed thumb are generally used for small objects while the entire hand or two hands are used on large objects such as a loaf of bread.

In the squeeze test the fingers sense the distance they move as they apply a force to that food. The fingers are well suited to perform the squeeze test because they are able to sense small distances quite accurately. When the fingers move a greater distance, the food is considered to be soft, and vice versa. Whether firmness is a desirable or undesirable characteristic depends upon the food being squeezed. The simple hand squeeze tells a potential customer that there are more leaves in a firm head of lettuce than in a soft head of equal size, that the soft marshmallow is fresh while the firm marshmallow is older and probably stale. The squeeze test also enables one to determine the ripeness of many fruits and some vegetables.

Bourne (1967b) measured how firmly people squeeze foods by hand. Some of his results are shown in Fig. 11, which plots the force exerted in successive squeezes on the same product by four individuals. The group of lines marked A were obtained by a young lady squeezing a large fresh cucumber. Notice how uniformly she squeezes each time. This degree of uniformity is unusual. This

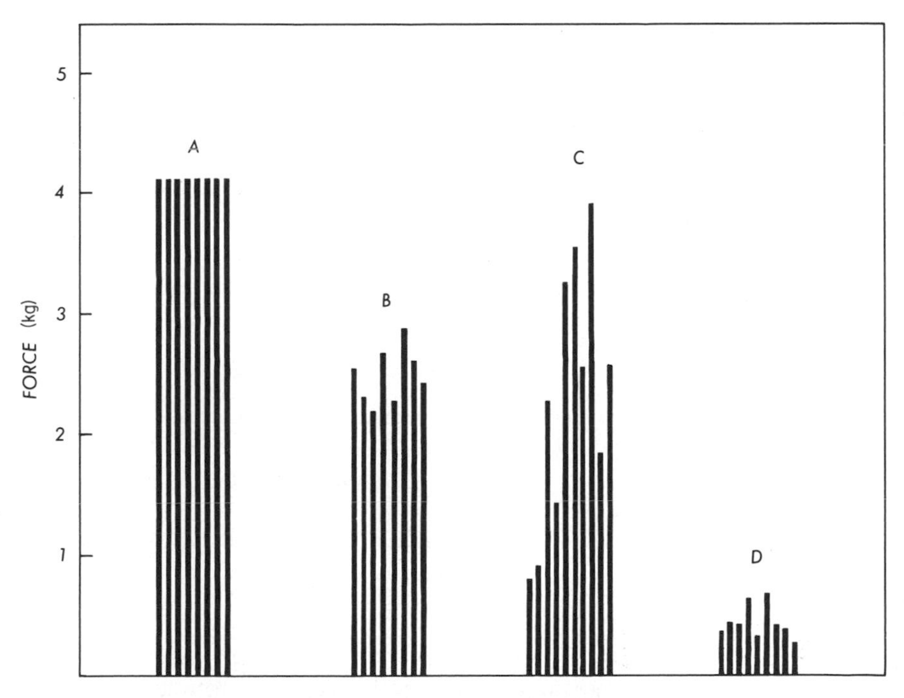

FIG. 11. Firmness of successive hand squeezes of foods: A, B, C, from three individuals squeezing a whole cucumber; D is individual B squeezing a loaf of fresh bread. (From Bourne, 1967b; reprinted with permission from New York State Agricultural Experiment Station.)

lady squeezes quite hard: a little over 4 kg at each squeeze. The group of curves marked B shows how hard another person squeezes the same cucumber. There is some change in force exerted from one squeeze to the next. This amount of variation in force exerted from one squeeze to the next is about normal for most people. Operator B squeezes the cucumber at an average force of about 2.5 kg. Operator C squeezed the same cucumber but the force exerted in successive squeezes fluctuates widely. The C type of squeezing pattern is less common than the B pattern.

The series of lines marked D were obtained from operator B squeezing a fresh loaf of bread. There is still about the same amount of variation from one squeeze to the next but the average squeezing force drops from about 2.5 for the cucumber to about 0.5 kg for the fresh bread. The fresh bread is much softer than the fresh cucumber. People generally squeeze soft and spongy foods more gently than harder foods. The force exerted by the hand in the squeeze test is therefore partly dependent upon the person making the test and partly dependent upon the nature of the food.

The measurement of firmness by an objective deformation test was discussed on p. 87 where it was shown that a small deforming force gives a better resolution between similar samples than a high deforming force. This principle should apply to the sensory deformation test: a *gentle* squeeze should discriminate better between the firmness of two samples of food than a hard squeeze. Squeezing gently has another point in its favor—there is less damage to the food. All the advantages lie with the gentle squeeze.

Peleg (1980) studied the sensitivity of the human tissue in squeeze tests and pointed out that in these types of tests there can be significant deformation of the human tissues (e.g., the balls of the fingers) in addition to the deformation of the specimen. He pointed out that the combined mechanical resistance in a squeezing test is given by the equation

$$M_c = M_1 M_x (M_1 + M_x), \qquad (1)$$

where M_c is the combined mechanical resistance of the sample and the fingers; M_1, the resistance of the human tissue; and M_x, the resistance to deformation of the test specimen. This equation provides a simple explanation as to why there are differences in the sensing range between the fingers and the jaws and why the human senses are practically insensitive to hardness beyond certain levels.

There are three different types of responses that can be drawn from this equation:

Case No. 1: $M_1 >> M_x$. This case occurs when a soft material is deformed between hard contact surfaces (e.g. a soft food is deformed between the teeth). Under these conditions Eq. 1 becomes $M_c = M_x$ (since $M_1 + M_x \simeq M_1$). In this situation the sensory response is primarily determined by the properties of the test specimen.

Case No. 2: M_x and M_1 are of comparable magnitude. In this case the response is regulated by both the properties of the test material and the tissue applying the stress, as given in Eq. 1.

Case No. 3: $M_x >> M_1$. This case occurs when a very firm product is compressed between soft tissues. For example, pressing a nut in the shell between the fingers. Under these conditions the equation becomes $M_c = M_1$ (because $M_1 + M_x \simeq M_x$). In this situation the response is due to the deformation of the tissue and is insensitive to the hardness of the specimen. This appears to be interpreted as "too hard to detect" or "out of range."

Voisey and Crête (1973) measured the amount of force and the rate at which force is applied to fruits and vegetables by the hands of consumers who were judging firmness. They found that males generally squeeze harder than females and applied the force more quickly. In squeezing an onion the mean force for females was 3910 g and for males 5670 g, while for tomatoes the mean force was 1522 g for females and 1705 g for males. The rate of force application on onions for females was 11,900 g sec^{-1} and for the males 17,560 g sec^{-1}, while for tomatoes the rate of force application for females was 14,040 g sec^{-1} and for males 5470 g sec^{-1}.

Stirring a fluid or semifluid food with a spoon or a finger is frequently used to measure viscosity or consistency. It is possible to use other parts of the anatomy such as cheeks, elbows, and feet to obtain some index of the textural qualities of foods.

A visual manifestation of texture can be found according to the rate and degree that foods spread or slump. For example, one observes the fluidity of a food by the ease with which it pours from a container or flows across the plate. With more solid foods we can see how far the food slumps; for example, a firm jelly holds its shape well while a soft jelly sags to a greater degree.

CHAPTER 7

Selection of a Suitable Test Procedure

Introduction

The previous chapters have described a large number of methods for measuring texture or viscosity of foods. A food technologist can easily become bewildered when first faced with the problem of developing a suitable procedure for measuring the textural properties of a particular food. Where does one begin? The following discussion is intended as a guide for selecting and establishing a texture measurement, particularly for those who are just entering the field.

A number of factors should be considered before setting up a new test procedure, otherwise a good deal of money and time can be wasted. The following recommended steps are based on the author's experience on a wide range of problems.

Factors to be Considered

Objective or Subjective

The first decision is whether to use an objective or a subjective test. Objective tests are generally preferred because they are generally believed to be more reproducible, use less time, and utilize a minimum of labor. On the other hand, there are times when subjective methods are the only way in which adequate information can be obtained. If the decision is made to use subjective methods, the reader is referred back to the previous chapter because the remainder of this chapter is directed to the selection and use of instrumental methods.

Nature of Product and Purpose of Test

The kind of material (liquid, solid, brittle, plastic, homogeneous, heterogeneous) affects which type of instrument will be selected. Is the test to be used for quality control, for setting legally binding official standards, for product development, or for basic research? These questions should be answered because they are an essential feature of the selection process. The previous chapters discussed a wide range of instruments ranging from simple and inexpensive to highly sophisticated. Each of them has its place. In some cases a single-point measurement is adequate; in other cases a multipoint measurement is needed. One is usually prepared to sacrifice sophistication for the sake of rapidity for routine quality control purposes where a rapid test is essential. On the other hand, difficult problems that are handled in the research laboratory will need more sophisticated instrumentation.

The difference between simple and sophisticated instruments might be likened to the difference between a $50 and a $500 camera. A $50 camera usually has a fixed focus, is simple to operate, and almost foolproof. The quality of the picture is not as good as that obtained with an expensive camera that has been properly operated, and it is restricted in the conditions of lighting and movement of the subject under which satisfactory pictures can be taken. Nevertheless, a great number of low-cost cameras are sold because the simplicity of operation and low cost of the camera are of paramount consideration. In contrast, the $500 camera has a better lens, it provides better-quality pictures, and it can be used under a wide range of conditions. However, the person operating the $500 camera needs to know something about its operation because of the complexity of the adjustments that need to be made. Novices frequently take poorer pictures with an expensive camera than with a cheap camera because they do not know how to set the adjustments on the expensive camera. A similar situation occurs with texture-measuring instruments. Sophisticated instrumentation has it place, but there is also room for the simple low-cost instruments.

Accuracy Required

Another question that should be resolved is the required accuracy of the results. Greater accuracy is obtained as the number of times the test is replicated is increased. Generally, a larger sample size gives a result closer to the true mean than a small sample size, and hence fewer replicate tests are needed to obtain a given degree of accuracy. But a larger sample size usually means that higher forces are needed, and the force capacity of the instrument may be exceeded. When the "spread" of values between individual units is needed, it is preferable to use a small sample size and run a large number of replicates in order to increase the probability of obtaining the full spread of values.

It comes as a surprise to some researchers to find a large inherent variability from unit to unit in the same sample lot. This is especially noticeable on most native foods where coefficients of variation of 10%, 20%, or higher are common. This variation is inherent in the commodity and it is to be expected. It is not a defect of the instrument, provided the instrument is correctly operated. When working with a new commodity, it is advisable to run a preliminary test to ascertain the degree of inherent variability in the product and to establish how large a sample size and number of replicates is necessary to give the desired degree of confidence in the data. An example of this type of exercise is shown in Table 1. Fairly large numbers of apples are required for reliable reproducible results at harvest time because of the large variation in firmness readings within the same lot of apples. A smaller number suffices after 4 months storage because of the reduced fruit-to-fruit variation.

Since this inherent wide variability is the norm for most foods, the primary consideration in most texture work is to look for an instrument that can perform tests rapidly, thus allowing a number of replicate tests to be made. A high degree of precision is a secondary consideration because there is little point in attempting to measure some textural parameter to a precision of 0.1% or better when the replicate samples may vary by 20% or more, especially when considerable time is required to obtain the high degree of precision. It is usually preferable to replicate a 1-min test five times than to run a more precise 5-min test only once.

TABLE 1

DIFFERENCES IN MAGNESS–TAYLOR MEASUREMENTS (LB) ON FRESH APPLES REQUIRED FOR EVALUATING SIGNIFICANCE BETWEEN TREATMENT MEANS FOR VARIOUS SAMPLE SIZES[a]

	95% Confidence level[b]				99% Confidence level[b]			
	10	20	100	200	10	20	100	200
At harvest								
Red delicious	1.9	1.3	0.6	0.4	2.6	1.8	0.8	0.6
Golden delicious	1.1	0.8	0.4	0.3	1.5	1.1	0.5	0.3
Rome	1.1	0.8	0.3	0.2	1.4	1.0	0.5	0.3
York	2.0	1.4	0.7	0.5	2.8	1.9	0.9	0.6
Stored 4 months at 31°F								
Red delicious	0.8	0.5	0.2	0.2	1.0	0.7	0.3	0.2
Golden delicious	0.8	0.6	0.3	0.2	1.1	0.8	0.3	0.2
Rome	0.9	0.6	0.3	0.2	1.2	0.8	0.4	0.3
York	1.7	1.2	0.5	0.4	2.3	1.6	0.7	0.5

[a]Taken from Worthington and Yeatman (1968); reprinted with permission from the *Proc. Am. Soc. Hort. Sci.* See also Schultz and Schneider, (1955).

[b]Columns are set up according to number of apples.

Destructive or Nondestructive?

Destructive tests ruin the structure and organization of the sample, rendering it unsuitable for repeating the test. Nondestructive tests should leave the food in a condition so close to its original state that the test can be repeated and give the same result as the first time. Both destructive and nondestructive tests have had their successes and failures (Bourne, 1979a). Because the majority of the textural parameters of foods are sensed in the mouth and mastication is a destructive process, it seems logical that destructive tests should be the predominant type to be used on foods. Nevertheless, nondestructive tests are sometimes effective, and they offer the advantage that the same piece of food can be repeatedly tested, thus eliminating variations in geometry from piece to piece.

Costs

How much money can be spent on this test? This includes the initial cost of the instrument, and maintenance and operating costs. An instrument that uses chart paper has an operating cost that is not found with an instrument in which a dial reading is taken. The maintenance cost should be considered. Does the instrument need spare parts and what is their availability and cost? Is the instrument used occasionally or frequently? Another element is the labor cost. A simple instrument can be operated by unskilled or semiskilled personnel whereas sophisticated instruments need to be operated by a person with higher qualifications. An automatic instrument costs more than a simple instrument but may cost less per test because of its reduced labor requirement and less chance of making errors.

Time

How much time can be spent on the test? Routine quality control tests need an instrument that gives results rapidly. In contrast, some tests in the research laboratory may be so sophisticated that the amount of time required to obtain reliable data is not of great consequence. Research needs may require measuring a number of textural parameters, which will take more time than a one-point measurement.

Location

Where will the instrument be operated? Any instrument can be used in a clean, dry laboratory. Instruments used in the plant may need to withstand steam, water, dust, vibration, and other hazards that render some instruments unsuitable. Instruments using a chart or complex electronic systems are likely to suffer damage in the steamy atmosphere of a processing plant unless specially designed to withstand the poor environment.

Eliminate Unsuitable Tests

Some test principles are obviously unsuitable for the commodity that needs to be tested and should be eliminated from consideration. For example, an extrusion test is unsuited for crackers and bread because these products do not flow; the puncture test works poorly on most brittle foods because they crumble or fracture before penetration; a cone penetrometer test is unsuitable for fibrous materials such as meat or raw vegetables; a snapping test will not be effective for flexible or fluid materials. Sometimes the geometry of the sample (size and shape) may impose limitations. For example, a large item cannot be tested in an instrument that has a small compartment for holding the sample.

Preliminary Selection

The steps described above will reduce the number of instruments under consideration. The next step is to narrow the field to the most promising two or three test principles. It is advisable to observe what kind of test principle people use in the sensory evaluation of textural quality because one can usually get good clues for the type of objective test to select by observing how people test the commodity. For example, if people judge textural quality by gently squeezing in the hand, consideration should be given to a test that works on the deformation principle. If people use a bending or snapping test, then this test principle should be given a high priority. If people bite the product between the incisors, the cutting-shear principle should be included among the preliminary tests.

The test principles that should be considered are

Puncture	Viscosity–consistency
Deformation	Crushing
Extrusion	Indirect methods (such as chemical analysis)
Penetration	Tensile
Cutting shear	Texture profile analysis
Snapping–bending	

All researchers should be warned about persevering with an instrument just "because it is there." By all means, try out an instrument if it is available and continue to use it if it gives satisfactory results. However, if it fails to give satisfaction after adequate testing, it use should be abandoned and one should look for another instrument that uses a different principle. One can easily spend far more than the cost of another instrument in labor costs by persevering with a test that uses the wrong principle for that particular application.

Sometimes none of the established procedures give satisfactory results. In these cases the researcher should have the confidence to develop a new test procedure or apparatus that is suitable for the purpose.

Final Selection

By this time, the number of principles should have been reduced to a small number. It is now time to test each of the remaining principles over the full range of textures that will normally be encountered with the food (i.e., excellent to poor) and identify the most suitable one. If any principle proves to be ineffective after being given a fair try, do not persevere with it; abandon it and try some other principle. For example, if the Magness–Taylor puncture test fails to give satisfactory results after a fair trial, then other instruments that work on the puncture principle will probably be unsatisfactory also. Therefore, abandon the puncture test principle and look at instruments that use another principle such as deformation or extrusion. The author has seen instances where a laboratory has persevered with a single test principle for a long time hoping that it will eventually give satisfactory results when in fact an unsuitable test principle was being used that would never be satisfactory for the commodity under study. In these cases, refining the test is not going to help because an inappropriate test principle is being used.

The selection among several principles to identify that principle that gives the best results can be done rather quickly. For example, the author was once faced with measuring the firmness of whole potatoes. Having gone through the preliminary selection it was agreed that the most suitable test would be either a puncture test or a deformation test. Three groups of potatoes (soft, medium, hard) were selected by hand with about 10 potatoes in each group. Each of these potatoes was then tested in the Instron using first a deformation test and then a puncture test. The mean values were calculated and are plotted in Fig. 1. It is obvious from this simple test, which only needed a few hours to perform, that the puncture principle is unsuitable for measuring the kind of firmness that was being sensed in the hand, but the deformation test showed promise. Therefore, we concentrated on refining the deformation test and wasted no more time trying to perfect the puncture test principle for this particular application (Bourne and Mondy, 1967).

Refine Test Conditions

The final step is to standardize the test conditions such as sample size, test cell dimensions, force range, speed of travel of moving parts, chart speed, temperature, and perhaps other factors. Several variations of the test conditions should be studied to find which gives the best resolution between different samples. For example, a small deforming force generally gives a better resolution in deformation tests than a high force (see p. 87). The test conditions finally selected should then be recorded for future use.

The textural properties of some foods change in unison and in the same

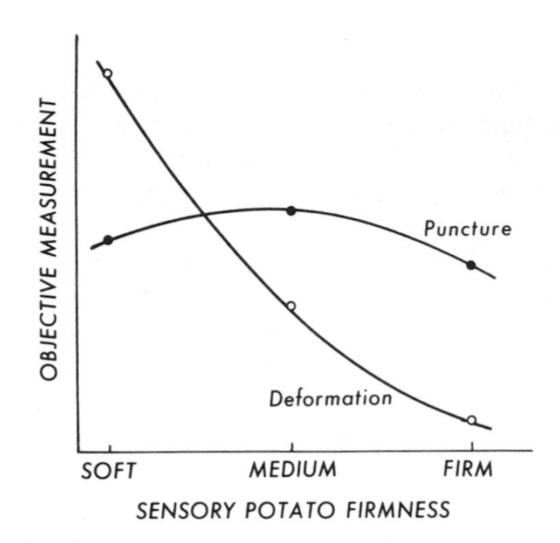

FIG. 1. Two objective methods for measuring firmness of whole potatoes versus sensory evaluation of firmness of the same potatoes.

direction during processing and storage; in these cases several types of texture measurement will correlate well with other texture test principles and with a panel. An example of this is fruit that softens greatly as it ripens (pears, peaches, bananas). Measuring the changes in firmness of these commodities is fairly straightforward. Each of several different tests will give satisfactory results (see Fig. 2). In this situation one can measure the wrong parameter and still get the right answer because of the nature of the interrelationships between the different parameters. In these cases the most convenient instrument and easiest to perform test principle should be selected. These foods are easy to measure by means of a simple parameter ("one-point" measurement) because each textural property correlates highly with all the other textural properties.

The textural properties of other foods change in different directions; it may be necessary to make several different kinds of tests to adequately describe the changes in textural properties of these foods. Under these conditions one can select several test principles or use texture profile analysis or an abbreviated version of texture profile analysis. For nonhomogeneous foods that contain more than one textural component, it is necessary to test separately each component. For example, in a candy containing nuts it is necessary to measure both the texture of the nut and the texture of the candy surrounding the nut. It may be necessary to use a different test principle on each component.

It is useful to make a scatter diagram of the preliminary subjective and objective measurements before calculating the correlation coefficient because this enables one to see certain aspects of the correlation that may otherwise be overlooked. Some of the possibilities are shown schematically in Fig. 3, which

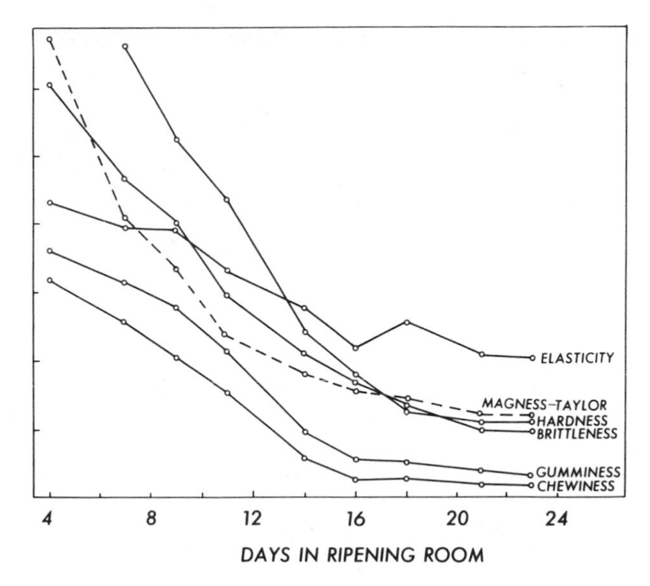

FIG. 2. Changes in texture profile parameters and Magness–Taylor puncture test on pears as they ripen. Notice how all parameters change in the same direction at approximately the same rate. (From Bourne, 1968. Reprinted from *J. Food Sci.* **33,** 225, 1968. Copyright by Institute of Food Technologists.)

shows nine potential relationships between instrumental tests (I) and sensory scores (S).

The first column on the left-hand side of Fig. 3 shows good correlations. The top graph shows a rectilinear relationship with the two desirable factors of a steep slope over the range of interest and a small degree of scatter. It is a very satisfactory relationship. The middle graph is just as satisfactory as the one above it, the only difference being that it has a negative slope. Three curvilinear relationships are shown in the bottom graph, each having the desirable features of low degree of scatter, and a steep slope. (For the sake of clarity, the scatter points are shown for only one line in this graph.) The curve may be concave or convex and may have a positive or negative slope, but it is very satisfactory. The simple correlation coefficient for any one of these three curves does not adequately reflect the goodness of fit of the experimental points to the line because it measures the goodness of fit to a straight line, not to a curve. Under these conditions it is advisable to transform the data in some way to straighten the curve (e.g., by taking logarithms on one or both axes) before calculating the correlation coefficient.

The three graphs in the center column of Fig. 3 show a marginal predictive relationship. They can be used to correlate instrumental tests with sensory judgments but not with the degree of certainty that is desirable. It is worth some effort

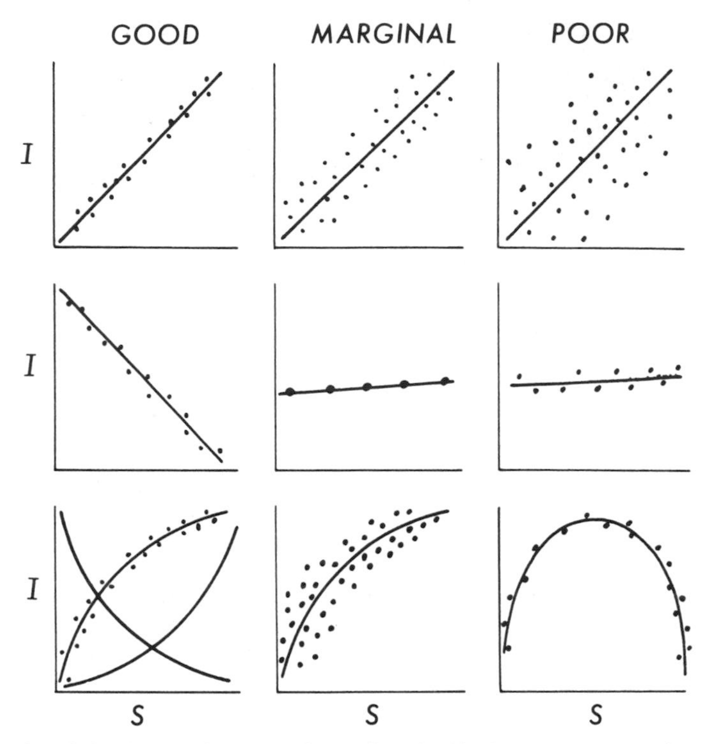

FIG. 3. Correlation between instrumental tests (I) and subjective (sensory) tests (S) on foods.

to improve these relationships before using them. The top graph in the center column of Fig. 3 has the desirable steep slope, but the degree of scatter of the points is greater than the top curve in the left-hand column. The center graph in the center column has a good fit of the points to the line, but the line has a shallow slope which limits the usefulness of the correlation. The bottom graph in the center column of Fig. 3 has a desirable steep slope, even with its curvature, but an undesirably wide degree of scatter.

The right-hand column in Fig. 3 shows relationships between instrument tests and sensory scores that are so poor that they should not be used for predictive purposes. In the top curve the degree of scatter from the line is too great, even though the slope of the line is steep. In the center curve the low slope of the line coupled with a moderate degree of scatter makes this relationship unsatisfactory for predictive purposes. The bottom right-hand graph in Fig. 3 is unsatisfactory because the relationship changes slope. It does not matter whether the slope changes from positive to negative or from negative to positive; any relationship in which the direction of the slope changes is unsatisfactory, even when there is a good fit of the data points to the curve.

The use of scatter diagrams as recommended above should not replace adequate statistical analysis of the data. Statistical analysis is definitely needed. Since it is outside the scope of this book to cover the analysis of the data, the reader should refer to a good book on statistical analysis or consult with a qualified statistician. The function of the scatter diagrams shown in Fig. 3 is to enable one to have a better understanding of the relationship before embarking on statistical analysis. They can also save unnecessary effort in computation. For example, if any of the relationships shown in the right-hand column of Fig. 3 are obtained, it would be better to continue to look for a better test procedure than to put a lot of effort into sophisticated statistical analysis of data that is so obviously unsatisfactory.

A useful guide of the suitability of a correlation for quality control purposes, provided that representative samples and adequate sample size have been used, was given by Kramer (1951). When the simple correlation coefficient between the instrument test and sensory score is ± 0.9 to ± 1.0, the instrument test is a good one and it can be used with confidence as a predictor of sensory score. When the correlation coefficient lies between ± 0.8 and ± 0.9, the test can be used as a predictor but with less confidence; it is worth some effort to improve the test to bring the correlation coefficient above ± 0.9. Extending this concept further, when the correlation coefficient lies between ± 0.7 and ± 0.8, the test is of marginal use as a predictor; when it is less than ± 0.7, it is practically worthless for predictive purposes.

A statistically significant relationship between an instrument test and sensory score may be found even with a low correlation coefficient if the sample size is large enough. For example, a correlation coefficient of 0.3 may be statistically significant, but it is far from adequate for predictive purposes. One needs to distinguish between statistical significance and predictive reliability.

The full textural range that will be encountered under reasonable circumstances should be used when setting up the preliminary tests. When a restricted range is used, the correlation coefficient will be lower than when the full range is used. Conversely, the correlation coefficient will be spuriously high when an excessively wide range is used.

The effect of too narrow or too broad a range on the numerical value of the correlation coefficient is demonstrated in Fig. 4, which is a hypothetical example of comparisons between instrumental and sensory testing of firmness of a food. The normal range of variability in this commodity is A—B and the correlation coefficient r over this range is 0.828. When a narrower range C—D is covered, the correlation coefficient drops to 0.695. This is a spuriously low figure because an unnecessarily narrow range was studied. On the other hand, when an extremely wide range E—F is covered, the correlation coefficient increases to 0.910. But this is a spuriously high figure because an abnormally wide range was taken.

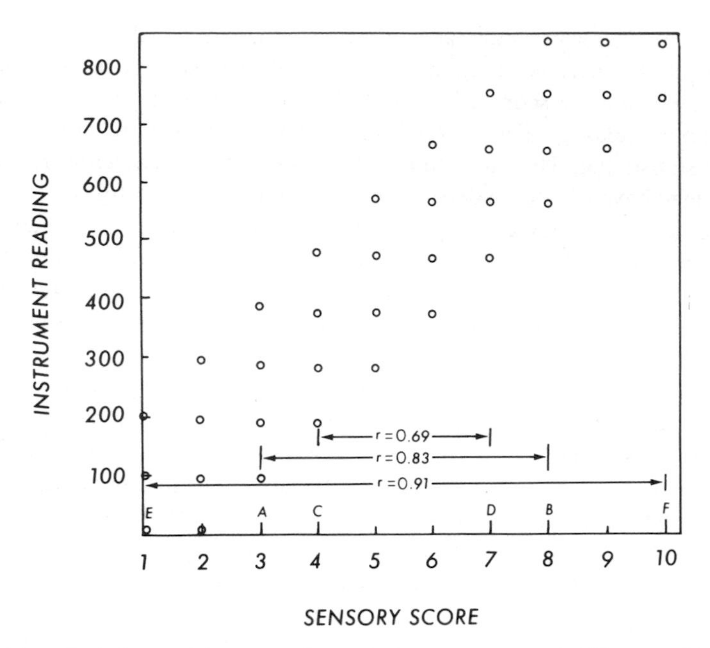

FIG. 4. A hypothetical case of instrumental versus sensory measurement of firmness showing how the range of firmness examined affects the correlation coefficient *r:* A–B, normal range; C–D, narrow range; E–F, excessively wide range.

These simple precautions in looking at the data enable the researcher to avoid some serious pitfalls.

Preparation of the Sample

Adequate sample preparation is an important element in performing food texture measurements. Problems with sample preparation sometimes impel the researcher to use a particular test principle or instrument. The sample selected for testing should be representative of the lot from which it was drawn. This point is so well known and so important that it should not require an extended discussion. Some other points in sample preparation are discussed below.

Temperature

The effect of temperature on viscosity of fluids is well known but many researchers are not aware that the temperature can affect the textural properties of many solid foods. Feuge and Guice (1959) measured the hardness of completely hydrogenated cottonseed oil with a ball penetrometer, obtaining a hardness index of 250 units at 10°C, which decreased to 10 units at 57°C. Miyada and Tappel (1956) attempted to calibrate a motorized Christel Texturemeter with beeswax but found a maximum force of 237 lb at 80°F, which continuously decreased to 59 lb at 110°F. Simon *et al.* (1965) in puncture tests on frankfurters obtained a value of 5.0 lb at 32°F and 2.3 lb at 70°F, but found almost no change from 70 to 120°F. Szczesniak (1975a) characterized the texture–temperature relationship of dessert gels and whipped toppings and showed a large variation in temperature effects on a number of different textural parameters of these foods. Caparaso *et al.* (1978) reported that the Warner–Bratzler Shear force on a sample of cooked beef was 6.3 lb at 50°C and 7.2 lb at 22°C.

Bourne (1982) found that firmness of raw fruits and vegetables as measured by puncture, extrusion, and deformation usually decreased with increasing temperature over the range 0–45°C. The change in firmness was highly variable, ranging from 0 to 7% change in firmness per degree temperature change. He defined the firmness–temperature coefficient as

$$\frac{\text{firmness at } T_2 - \text{firmness at } T_1}{\text{firmness at } T_1 \cdot (T_2 - T_1)} \cdot 100\% \text{ change in firmness per degree temperature change,}$$

where T_1 and T_2 are the lowest and highest temperatures at which firmness was measured. This definition assumes linearity between the texture parameter and temperature. For foods in which this relationship is not linear, this definition can still be used if the temperature range is narrowed to an approximately linear segment and the temperature range over which the coefficient applies is specified.

The above definition could be used to describe the temperature relationship of textural parameters other than firmness. A suitable general term would be the "texture–temperature coefficient."

Food samples should be tested at approximately the same temperature in order to obtain consistent results unless it has been determined that the texture– temperature coefficient is small. Some latitude in test temperature is acceptable for those commodities that have a low texture–temperature coefficient, but the temperature must be controlled for those foods that have a high coefficient.

Great care should be given to temperature control of those foods that are close to a temperature-induced solid–liquid phase change because the texture–temperature coefficient is usually high in that range. For example, fats usually have a high coefficient. In addition, the previous temperature history of storage of a fat sample can have a profound effect on its texture.

Sample Geometry

The size and shape of the sample may or may not be a factor in a texture test, depending upon the test and the food. The size of the sample is not important in puncture tests and penetrometer tests provided that it is above the minimum size

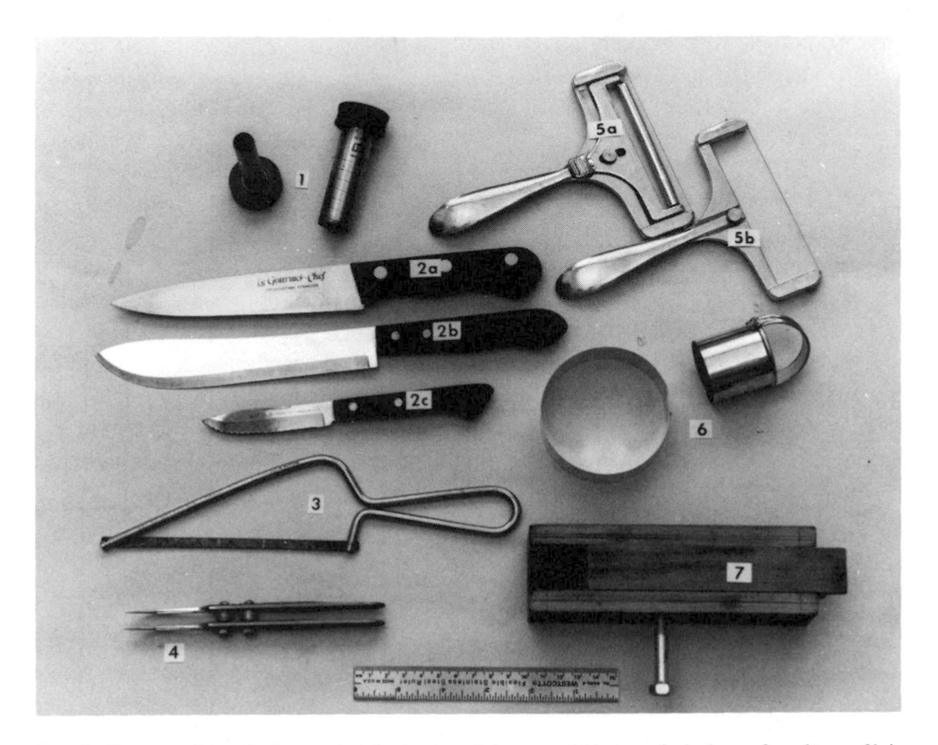

Fig. 5. Some useful tools for shaping foods: 1, cork boreres; 2, knives; 2a is the preferred type; 2b is undesirable because it is hollow ground and has a thick blade; 2c is undesirable because it is too small, is hollow ground, and has a serrated edge; 3, small fine-tooth saw; 4, a pair of scalpels bolted together; 5a, household cheese cutter as purchased; 5b, cheese cutter with roller removed leaving space to cut samples up to 25 mm thick; 6, a large and a small cookie cutter; 7, a miter box with an adjustable slide.

necessary to avoid breaking away at the edges or breaking through the bottom of the sample. The fill of container in the back extrusion test often has little effect on maximum force provided it is above a certain minimum (Bourne and Moyer, 1968). The fill of the standard FTC Texture Press cell has a great effect on maximum force for some foods while for other foods it has little effect above a certain fill level (Szczesniak *et al.*, 1970; see also p. 140).

For many tests, including deformation, cutting-shear, bending, crushing, tensile, and texture profile analysis, the geometry of the sample has a profound effect on the results, and therefore it becomes necessary to standardize the dimensions of the test piece.

The shaping of foods to standard measurements is a practical problem that is often frustrating and time consuming. Some practical tips that the author has found useful are the following (see Fig. 5):

1. A cork borer is useful for cutting out cylinders. A motorized borer is preferred to a hand-operated borer. Make sure the borer is kept sharp. Apply a light uniform pressure when cutting because an uneven diameter is obtained if the pressure is not held steady. Continuous heavy pressure will give an hourglass shape instead of a uniform cylinder especially on highly deformable foods such as meat.

2. A sharp knife is useful for cutting many foods. We find that a fairly long thin blade that is not hollow ground and not serrated gives the best results. A back-and-forth sawing motion under gentle pressure gives better control of dimensions than applying a heavy downward cutting action. It is difficult to get surfaces flat and parallel when using a thick-bladed or hollow ground knife.

3. A small saw with very fine teeth is useful for cutting hard fracturable materials to size. We use a saw blade 6 in. long with 32 teeth per inch.

4. Two scalpels bolted together with spacers between them make an implement that is useful for some applications.

5. A wire cutter is good for shaping adhesive foods such as soft cheese. An easy way to get one is to buy a household cheese cutter and remove the roller bar.

6. A circular cookie cutter is helpful for cutting dumbbell shapes suitable for a tensile test.

7. A miter box used in conjunction with a sharp kinife, a small saw, or a wire cutter helps in cutting samples to a standard length and cutting uniform cubes for texture profile analysis.

Food technologists have to face the fact that some foods cannot be shaped. It is impossible to cut a head of lettuce, a peanut, or a potato chip to a standard geometry without destroying the integrity of the sample as a whole. In these cases the best one can do is to select units of as uniform shape and size as possible and be realistic about the fact that the data points will show more scatter than if pieces of standard size and shape had been available. A nondestructive test should be used for these foods if it is suitable because the same unit of food can be repeatedly tested as it undergoes the experimental treatments while the geometry factor remains constant.

APPENDIX

Suppliers of Texture and Viscosity Measuring Instruments[a]

Instrument	Supplier	Price range
Adams Consistometer (see TUC Meter)	National Manufacturing Co. P.O. Box 30226 Lincoln, NE 68503	B
Albumen Height Gauge	(see Haugh Meter)	
Amylograph	C. W. Brabender Instruments, Inc. 50 E. Wesley St. South Hackensack, NJ 07606	D
AVS/N Viscometer[b]	Jenaer Glaswerk Schott & Gen Inc.[c] D 6238 Hofheim Q.T.S. West Germany	C
Baker Compressimeter	F. Watkins Corporation P.O. Box 445 Caldwell, NJ 07006	B
Ballauf Pressure Tester	(see Magness–Taylor Pressure Tester)	
Biscuit Texture Meter (B.B.I.R.A.)	Baker Perkins (Exports) Ltd. Westwood House, 13 Stanhope St. Park Lane, London W1 England	C

(*continued*)

[a]Although this compilation represents the best information available to the author at the time of writing, the author takes no responsibility for the accuracy of the information. The reader should contact the manufacturer directly for the latest information on availability, delivery, and price. The price range designations are as follows: A, less than \$200; B, \$200–\$1000; C, \$1000–\$5000; D, \$5000–\$25,000; and E, more than \$25,000.

[b]The AVS/N is a glass capillary viscometer.

[c]United States branch office: 11 East 26th Street, New York, NY 10010

Instrument	Supplier	Price range
Bloom Gelometer	G.C.A. Precision Scientific Group 3737 W. Cortland St. Chicago, IL 60647 (see also Stevens LFRA Texture Analyzer)	C
Bostwick Consistometer	Central Scientific Co. 2600 South Kostner Ave. Chicago, IL 60623	B
Brookfield Viscometer	Brookfield Engineering Laboratories, Inc. 240 Cushing St. Stoughton, MA 02072	B
Butter Consistometer	Accurate Manufacturing Co. 945 King Ave. Columbus, OH 43212	C
Butter Consistency Meter	B. V. Apparatenfabriek van Doorn Utrechtseweg 364 Postbus 17, DeBilt Holland	C
Capillary Glass Viscometers[d]	Cannon Instrument Co.[e] P.O. Box 16 State College, PA 16801 Fisher Scientific Co. 711 Forbes Ave. Pittsburg, PA 15219 V.W.R. Scientific P.O. Box 8188 Philadelphia, PA 19101 Curtin-Matheson Scientific Inc. P.O. Box 1546 Houston, TX 77001 Sargent-Welch 7300 North Linder Avenue Skokie, IL 60077 SGA Scientific Inc. 735 Broad St. Bloomfield, NJ 07003 Arthur H. Thomas Co. Vine St. at Third P.O. Box 779 Philadelphia, PA 19105	A
Cell Fragility Tester	(see Torry Brown Homogenizer)	

(continued)

[d]Most laboratory supply houses carry capillary viscometers.

[e]This company also supplies oils of standard viscosity for standardization of kinematic viscometers.

APPENDIX Continued

Instrument	Supplier	Price range
Chatillon Testers	John Chatillon and Sons, Inc. 83-30 Kew Gardens Road Kew Gardens, NY 11415	A
Cheese Curd Torsiometer	(see Plint Cheese Torsiometer)	
Cherry–Burrell Curd Tester	(see Marine Colloids Gel Tester)	
Consistometer	(see Adams, Bostwick, USDA, FMC Consistometers; TUC Cream Corn Meter)	
Contraves Viscometer	Contraves AG Zurick[f] Schaffhauserstrasse 580 P.O. Box CH-8052 Zurich Switzerland	D
Corn Breakage Tester	Fred Stein Laboratories 121 North Fourth St. Atchison, KS 66002	B
Deer Rheometer	Rheometer Marketing Ltd. Crown House, Armley Road Leeds LS12 2EJ England	D
Effi-Gi Tester	Effi-G[g] Corso Garibaldi 102 48011 Alfonsine Ravenna Italy	A
Extensigraph	C. W. Brabender Instruments Co. 50 E. Wesley St. South Hackensack, NJ 07606	D
Fann Viscometer	Curtin-Matheson Scientific Inc. P.O. Box 1546 Houston, TX 77001	C
Farinograph	C. W. Brabender Instruments Co. 50 E. Wesley St. South Hackensack, NJ 07606	D
Ferranti–Shirley Viscometer	Ferranti Ltd.[h] Instrument Department Moston, Manchester M10 OBE England	E

(*continued*)

[f]The United States distributor is Tekmar Company, P.O. Box 37202, Cincinnati, OH 45222.

[g]The United States agent is McCormick Fruit Tree Inc., 1315 Fruitvale Blvd., Yakima, WA 98902.

[h]The United States agent is Ferranti Electric Inc., 87 Modular Ave., Commack, NY 11725.

Instrument	Supplier	Price range
F.I.R.A. Jelly Tester	H. A. Gaydon Co., Ltd. 93 Lansdowne Road Croydon England	B
FMC Consistometer	C. W. Brabender Instruments Co. 50 E. Wesley St. South Hackensack, NJ 07606	C
Food Technology Corporation (FTC) Texture Test System (Kramer Shear Press)	Food Technology Corporation 12300 Parklawn Drive Rockville, MD 20852	D
General Foods Texturometer	Zenken Company Ltd. Kyodo Bldg., No. 5, 2-Chome Honcho Nihonbashi, Chuo-Ku Tokyo 103 Japan	D
Gilmont Viscometer	Gilmont Instruments 401 Great Neck Road Great Neck, NY 11021	A
Haake Viscometers (Rotovisco)	Haake Inc.[i] Dieselstrasse 6, D-7500 Karlsruhe 41 West Germany	C
Haugh Meter	B. C. Ames Co. Lexington St. Waltham, MA 02154	A
	Mattox and Moore Inc. 1503 E. Riverside Dr. Indianapolis, IN 46207	
	VAL-A Company 700 West Root St. Chicago, IL 60609	A
Hilker–Guthrie Plummet	Whitman Laboratories Inc. 7 Wedgewood Drive New Hartford, NY 13413	A
Hoeppler Viscometer	Sargent-Welch 7300 North Linder Ave. Skokie, IL 60077	C
	SGA Scientific Inc. 735 Broad St. Bloomfield, NJ 07003	
Instron Universal Testing Machine	Instron Corporation 2500 Washington St. Canton, MA 02021	D

(continued)

[i]The United States agent is Haake Buchler Instruments, Inc., 244 Saddle River Road, Saddle Brook, NJ 07662.

APPENDIX Continued

Instrument	Supplier	Price range
Kramer Shear Press	(see Food Technology Corporation Texture Test System)	
Lamb–Lewis Viscometer	John Dimick 526 Clayton El Cerrito, CA 94530	A
	National Food Processors Association 1950 Sixth St. Berkeley, CA 94710	
Lauda Automatic Viscometer (Viscoboy)	Brinkman Instruments Inc. (subsidiary of Sybron Corporation) Cantiague Road Westbury, NY 11590	C
Loaf Volumeter	National Manufacturing Co. P.O. Box 30226 Lincoln, NE 68503	B
Magness–Taylor Pressure Tester (see Puncture testers)	D. Ballauf Co. 619 H Street N.W. Washington, D.C. 20001	A
Marine Colloids Gel Tester	Marine Colloids Inc. 2 Edison Place Springfield, NJ 07081	C
Maturometer	Sardik Engineering Pty Ltd. 31 Higginbotham Road Gladesville N.S.W. 2111 Australia	C
Mixograph	National Manufacturing Co. P.O. Box 30226 Lincoln, NE 68503	C
Nametre Viscometer	Nametre Company 1778 State Highway 27 Edison, NJ 08817	C
Ottawa Texture Measuring System (OTMS)	Canners Machinery Ltd. P.O. Box 190 Simcoe, Ontario N3Y 4LI Canada	
	without recorder	C
	with recorder	D
	Queensboro Instruments[j] 645 Brierwood Avenue Ottawa, Ontario K2A 2J3 Canada	

(*continued*)

[j]Additional accessories and custom systems for the OTMS are available from this company.

Instrument	Supplier	Price range
Pabst Texture Tester	P.E.P. Inc. 8928 Spring Branch Drive Bldg A3 Houston, TX 77080	D
Penetrometer[k]	G.C.A. Precision Scientific Group 3737 West Cortland St. Chicago, IL 60647	B
	Lab-Line Instruments 15th and Bloomingdale Ave. Melrose Park, IL 60160	B
Plint Cheese Torsiometer (N.I.R.D. Cheese Torsiometer)	Plint and Partners Ltd.[l] Fishponds Road Wokingham Berkshire RG11 2QG England	C
Puncture Testers[m]	(see Ballauf, Chatillon, Effi-Gi, Magness–Taylor, Marine Colloids, Maturometer, Stevens, UC Fruit Firmness Testers)	
Rheometrics Fluids Rheometer	Rheometrics Inc. 2438 US Highway #22 Union, NJ 07083	E
Ridgelimeter	Sunkist Growers Inc. Products Sales Division Ontario, CA 91764	A
Stevens LFRA Texture Analyzer	C. Stevens and Sons Ltd.[n] Dolphin Yard Holywell Hill St. Albans, Hertshire AL1 1EX England	D
Stormer Viscometer	Fisher Scientific Co. 711 Forbes Ave. Pittsburgh, PA 15219	B
	Sargent-Welch 7300 North Linder Ave. Skokie, IL 60077	B
	Arthur H. Thomas Co. Vine St. at Third P.O. Box 779 Philadelphia, PA 19105	B

(continued)

[k]Many laboratory supply houses carry penetrometers in stock.

[l]The United States agent is C.A.A. Scientific, P.O. Box 1234, Darien, CT 06820.

[m]Food Technology Corporation, Instron Corporation, and Ottawa Texture Measuring System supply various puncture probes as accessories to their basic machine.

[n]The United States agent is Voland Corp., 27 Centre Ave., New Rochelle, NY 10802.

APPENDIX Continued

Instrument	Supplier	Price range
	SGA Scientific Inc. 735 Broad St. Bloomfield, NJ 07003	B
Structograph	C. W. Brabender Instruments Inc. 50 E. Wesley St. South Hackensack, NJ 07606	D
Succulometer[o]	The United Company TUC Road Westminster, MD 22157	B
SURRD Hardness Tester	Custom Scientific Instruments P.O. Box A Whippany, NJ 07981	C
Tensipresser	Taketomo Electronic Co. Ltd. 1-55 Wakamatsu-Cho Shinjuku-Ku Tokyo 162 Japan	D
Torry Brown Homogenizer	A. G. Brown Electronics Ltd. Morden House 10-11 Royal Crescent Glasgow C3 Scotland	D
TUC Cream Corn Meter (see also Adams Consistometer)	The United Company TUC Road Westminster, MD 21157	B
UC Fruit Firmness Tester	Western Industrial Supply Inc. 236 Clara St. San Francisco, CA 94107	B
Universal Testing Machines[p]	Baldwin-Lima-Hamilton Corp. 42 4th Avenue Waltham, MA 02154 John Chatillon and Sons, Inc. 83-30 Kew Gardens Road Kew Gardens, NY 11415 Custom Scientific Instruments P.O. Box A Whippany, NJ 07981 Dillon and Co. 14620 Keswick St. Van Nuys, CA 91407	

(*continued*)

[o]The FTC Texture Test System offer a succulometer as an accessory to their basic machine.

[p]See also Food Technology Corporation Texture Test System, General Foods Texturometer, Instron Universal Testing Machine, Ottawa Texture Measuring System.

Instrument	Supplier	Price range
	Labquip Corp. 4520 West North Avenue Chicago, IL 60639	
	The Marqordt Corp. 16555 Saticoy St. Van Nuys, CA 91406	
	Olson Tinius Testing Machine Co. 2100 Easton Road Willow Grove, PA 19090	
	Riehle Testing Machine Div. American Machine and Metals 48 Thomas St. East Moline, IL 61244	
	Scott Testers Inc. 77 Blackstone St. Providence, RI 02905	
	Soil Test Inc. 2205 Lee St. Evanston, IL 60202	
	Testing Machines Inc. 400 Bayview Ave. Amityville, NY 11701	
	Zwick and Co.[q] Eisingen Germany	
USDA Consistometer	J. H. Broetzman Hydraulics and Plastics P.O. Box 625 Vienna, VA 22180	A
Van Doorn Tester	(see Butter Consistency Meter)	
Vettori–Manghi Tenderometer (Pea Tenderometer)	Vettori-Manghi C s.p.a. via Spezia 54 - C/P 31P 43100 Parma Italy	C
Viscocorder	C. W. Brabender Instruments Inc. 50 E. Wesley St. South Hackensack, NJ 07606	D
Viscometers	(see AVS/N, Brookfield, Capillary Glass, Contraves, Deer, Fann, Ferranti–Shirley, FMC Consistometer, Gilmont, Haake, Hoeppler, Lamb–Lewis, Lauda, Nametre, Rheometrics, Stormer, Viscocorder, Weissenberg, Zahn)	

(continued)

[q]The United States agent is William J. Hacker and Co., Inc., P.O. Box 646, West Caldwell, NJ 07007.

APPENDIX Continued

Instrument	Supplier	Price range
Warner–Bratzler Shear[r]	G-R Electric Mfg. Co. Route 2 Manhattan, KS 66502	B
Weissenberg Rheogoniometer	Sangamo Schlumberger[s] Rheology Division North Bersted, Bognor Regis West Sussex PO 22 9B5 England	E
Zahn Viscosimeter	Fisher Scientific Co. 711 Forbes Avenue Pittsburgh, PA 15219	A
	Sargent-Welch 7300 North Linder Ave. Skokie, IL 60077	A
	Curtin-Matheson Scientific Inc. P.O. Box 1546 Houston, TX 77001	
	VWR Scientific P.O. Box 8188 Philadelphia, PA 19101	A

[r]Food Technology Corporation, Instron Corporation, and the Ottawa Texture Measuring System supply a Warner–Bratzler shear blade as an accessory fitting to their basic machine.

[s]The United States distributor is Sangamo Transducers, 60 Winter Street, Weymouth, MA 02188.

References

Abbott, J. A., N. F. Childers, G. S. Bachman, J. V. Fitzgerald, and F. J. Matusik. 1968a. Acoustic vibration for detecting textural quality of apples. *Proc. Am. Soc. Hortic. Sci.* **93,** 725–737.

Abbott, J. A., G. S. Bachman, R. F. Childers, J. V. Fitzgerald, and F. J. Matusik. 1968b. Sonic techniques for measuring texture of fruits and vegetables. *Food Technol.* **22,** 635–646.

Adams, M. C., and E. L. Birdsall. 1946. New Consistometer measures corn consistency. *Food Ind.* **18,** 844–846, 992, 994.

Alexander, J. 1906. The grading and use of glues and gelatin. *J. Soc. Chem. Ind., London* **25,** 158–161.

Alexander, J. 1908. Method of and apparatus for testing jelly. U.S. Patent 882, 731.

Allen, F. W. 1932. Physical and chemical changes in the ripening of deciduous fruits. *Hilgardia* **6,** 381–441.

American Association of Cereal Chemists. 1969. "Cereal Laboratory Methods." American Association of Cereal Chemists, St. Paul, Minnesota.

Amerine, M. A., R. M. Pangborn, and E. B. Roessler. 1965. "Principles of Sensory Evaluation of Foods." Academic Press, New York.

Andreotti, R., and R. Agosti. 1965. Perfezionamento del Tenderometro modello stazione sperimentale conserve alimentari. *Ind. Conserva* **2,** 103–106.

Ang, J. K., F. M. Isenberg, and J. D. Hartman. 1960. Measurement of firmness of onion bulbs with a shear press and potentiometric recorder. *Proc. Am. Soc. Hortic. Sci.* **75,** 500–509.

Anonymous. 1959. Pectin standardization. Final report of IFT Committee. *Food Technol.* **13,** 496–500.

Anonymous. 1963. New research tool acquired by Station. *Farm Res.* **29**(2), 16.

Anonymous. 1964. Sensory testing guide for panel evaluation of foods and beverages. Prepared by the Committee on Sensory Evaluation of the Institute of Food Technologists. *Food Technol.* **18,** 1135–1141.

Anonymous. 1966. A new instrument for measuring the texture of potato french fries. *Am. Potato J.* **43,** 175.

Anonymous. 1973. "A Study of Consumer Attitudes Toward Product Quality." A. C. Nielsen Co., Northbrook, Illinois.

Arnold, P. C., and N. N. Mohsenin. 1971. Proposed techniques for axial compression tests on intact agricultural products of convex shape. *Trans.ASAE* **14,** 78–84.

Babcock, C. J. 1922. The whipping quality of cream. *U.S. Dep. Agric. Bull.* No. 1075.

Bailey, C. H. 1934. An automatic shortometer. *Cereal Chem.* **11,** 160–163.

Ball, C. O., W. E. Claus, and E. F. Stier. 1957. Factors affecting quality of prepackaged meats. 1B. Loss of weight and study of texture. *Food Technol.* **11,** 281–283.

Bechtel, W. G., D. F. Meisner, and W. B. Bradley. 1953. The effect of the crust on the staling of bread. *Cereal Chem.* **30,** 160–168.

Benbow, J. J. 1971. The dependence of output rate on die shape during catalyst extrusion. *Chem. Eng. Sci.* **26**(14), 67–73.

Benbow, J. J. 1981. Letter to the Editor. *J. Texture Stud.* **12,** 91–92.

Bice, C. W., and W. F. Geddes. 1949. Studies on bread staling. IV. Evaluation of methods for the measurement of changes which occur during bread staling. *Cereal Chem.* **26,** 440–465.

Bingham, E. C. 1914. A new viscometer for general scientific and technical purposes. *Ind. Eng. Chem.* **6,** 233–237.

Bingham, E. C. 1922. "Fluidity and Plasticity." McGraw-Hill, New York

Bingham, E. C. 1930. Fundamental definitions of rheology. *J. Rheol.* **1,** 507–516.

Bloom, O. T. 1925. Machine for testing jelly strength of glues, gelatins and the like. U.S. Patent 1,540,979.

Bookwalter, G. N., A. J. Peplinski, and V. F. Pfeiffer. 1968. Using a Bostwick Consistometer to measure consistencies of processed corn meals and their CSM blends. *Cereal Sci. Today* **13,** 407–410.

Borker, E., and K. G. Sloman. 1969. Dessert gel (gelatin) strength testing. II. Collaborative study. *J. Assoc. Off. Anal. Chem.* **52,** 657–659.

Borker, E., A. Stefanucci, and A. A. Lewis. 1966. Dessert gel strength testing. *J. Assoc. Off. Anal. Chem.* **49,** 528–33.

Bourne, M. C. 1965a. Studies on punch testing of apples. *Food Technol.* **19,** 413–415.

Bourne, M. C. 1965b. How skin affects apple pressure tests. *Farm Res.* **31**(2), 10–11.

Bourne, M. C. 1966a. A classification of objective methods for measuring texture and consistency of foods. *J. Food Sci.* **31,** 1011–1015.

Bourne, M. C. 1966b. Measurement of shear and compression components of puncture tests. *J. Food Sci.* **31,** 282–291.

Bourne, M. C. 1967a. Deformation testing of foods. I. A precise technique for performing the deformation test. *J. Food Sci.* **32,** 601–605.

Bourne, M. C. 1967b. Squeeze it gently. *Farm Res.* **32**(4), 8–9.

Bourne, M. C. 1968. Texture profile of ripening pears. *J. Food Sci.* **33,** 223–226.

Bourne, M. C. 1972. Standardization of texture measuring instruments. *J. Texture Stud.* **3,** 379–384.

Bourne, M. C. 1973. Use of the Penetrometer for deformation testing of foods. *J. Food Sci.* **38,** 720–721.

Bourne, M. C. 1974. Textural changes in ripening peaches. *J. Can. Inst. Food Sci. Technol.* **7,** 11–15.

Bourne, M. C. 1975a. Is rheology enough for food texture measurement? *J. Texture Stud.* **6,** 259–262.

Bourne, M. C. 1975b. Method for obtaining compression and shear coefficients of foods using cylindrical punches. *J. Texture Stud.* **5,** 459–469.

Bourne, M. C. 1975c. Texture measurements in vegetables. *In* "Theory, Determination and Control of Physical Properties of Food Materials" (C. Rha, ed.), pp. 131–162. Reidel Publ., Dordrecht, Netherlands.

Bourne, M. C. 1976. Interpretation of force curves from instrumental texture measurements. *In* "Rheology and Texture in Food Quality" (J. M. deMan, P. W. Voisey, V. F. Rasper, and D. W. Stanley, eds), pp. 244–274. Avi, Westport, Connecticut.

Bourne, M. C. 1977. Compression rates in the mouth. *J. Texture Stud.* **8,** 373–376.

Bourne, M. C. 1979a. Rupture tests vs. small strain tests in predicting consumer response to texture. *Food Technol.* **33**(10), 67–70.

Bourne, M. C. 1979b. Theory and application of the puncture test in food texture measurement. *In* "Food Texture and Rheology" (P. Sherman, ed.), pp. 95–142. Academic Press, New York/London.

Bourne, M. C. 1980. Texture evaluation of horticultural crops. Hortscience **15**(1), (7)–(13).

Bourne, M. C. 1982. Effect of temperature on firmness of raw fruits and vegetables. *J. Food Sci.* **47,** 440–444.

Bourne, M. C., and N. Mondy. 1967. Measurement of whole potato firmness with a universal testing machine. *Food Technol.* **21,** 1387–1406.

Bourne, M. C., and J. C. Moyer. 1968. The extrusion principle in texture measurement of fresh peas. *Food Technol.* **22,** 1013–1018.

Bourne, M. C., J. C. Moyer, and D. B. Hand. 1966. Measurement of food texture by a universal testing machine. *Food Technol.* **20,** 522–526.

Bourne, M. C., A. M. R. Sandoval, M. Villalobos C., and T. S. Buckle. 1975. Training a sensory texture profile panel and development of standard rating scales in Colombia. *J. Texture Stud.* **6,** 43–52.

Brabender, C. W. 1965. Physical dough testing, past, present, future. *Cereal Sci. Today* **10**(6), 291–304.

Brandt, M. A., E. Z. Skinner, and J. A. Coleman. 1963. Texture Profile Method. *J. Food Sci.* **28,** 404–409.

Bratzler, L. J. 1932. Measuring the tenderness of meat by means of a mechanical shear. M.S. Thesis, Kansas State Coll., Manhattan.

Bratzler, L. J. 1949. Determining the tenderness of meat by use of the Warner–Bratzler method. *Proc. Annu. Reciprocal Meat Conf.* **2,** 117–120.

Breene, W. M. 1975. Application of texture profile analysis to instrumental food texture evaluation. *J. Texture Stud.* **6,** 53–82.

Brennan, J. G., R. Jowitt, and A. Williams. 1975. An analysis of the action of the General Foods Texturometer. *J. Texture Stud.* **6,** 83–100.

Brinton, R. H., and M. C. Bourne. 1972. Deformation testing of foods. 3. Effect of size and shape of the test piece on the magnitude of deformation. *J. Texture Stud.* **3,** 284–297.

Briskey, E. J., R. W. Bray, W. G. Hoekstra, P. H. Phillips, and R. H. Grummer. 1959. The chemical and physical characteristics of various pork ham muscle classes. *J. Anim. Sci.* **18,** 146–152.

Brulle, R. 1893. Méthode générale pour l'analyse des beurres. C. R. Hebd. Seances Acad. Sci. **116,** 1255–1277.

Bruns, A. J., and M. C. Bourne, 1975. Effects of sample dimensions on the snapping force of crisp foods. Experimental verification of a mathematical model. *J. Texture Stud.* **6,** 445–458.

Bryan, W. L., B. J. Anderson, and J. M. Miller, 1978. Mechanically assisted grading of oranges for processing. *Tran. A. S. A. E.* **21,** 1226–1231.

Burnett, J., and G. W. Scott Blair. 1963. A speed-compensated torsiometer for measuring the setting of milk by rennet. *Dairy Ind.* **28,** 220–223.

Burnett, J., and G. W. Scott Blair. 1964. Note on the Torsiometer for measuring setting of milk by rennet. *Dairy Ind.* **29,** 97.

Burr, G. S. 1949. Servo-controlled tensile strength tester. *Electronics* **22,** 101–105.

Campion, D. R., J. D. Crouse, and M. E. Dikeman. 1975. The Armour Tenderometer as a predictor of cooked meat tenderness. *J. Food Sci.* **40,** 886–887.

Caparaso, F., A. L. Cortovarria, and R. W. Mandigo. 1978. Effects of post-cooking sample temperature on sensory and shear analyses of beef steaks. *J. Food Sci.* **43,** 839–841.

Carpenter, Z. L., G. C. Smith, and O. D. Butler. 1972. Assessment of beef tenderness with the

Armour Tenderometer. *J. Food Sci.* **37,** 126-129.

Carpi, S. 1884. Die Entdeckung fremder Oele im Olivenöl. *Arch. Pharm. (Weinheim, Ger.)* **221,** 963; from *Ann. Chim. Appl. Farm. Med.* **77,** No. 3.

Casimir, D. J., R. S. Mitchell, L. J. Lynch, and J. C. Moyer. 1967. Vining procedures and their influence on yield and quality of peas. *Food Technol.* **21,** 427–432.

Casimir, D. J., G. G. Coote, and J. C. Moyer. 1971. Pea texture studies using a single puncture Maturometer. *J. Texture Stud.* **2,** 419–430.

Casson, N. 1959. A flow equation for pigment oil suspensions of the printing ink type. *In* ''Rheology in Disperse Systems'' (C. C. Mill, ed.), pp. 84–104. Pergamon, Oxford.

Cathcart, W. H., and L. C. Cole. 1938. Wide-range volume measuring apparatus for bread. *Cereal Chem.* **15,** 69–79.

Cavigelli, W., and P. Schnyder. 1980. Messung rheologischer Eigenschaften bei Lebensmitteln. *Alimenta* **4,** 87–93.

Chevalley, J. 1975. Rheology of chocolate. *J. Texture Stud.* **6,** 177–196.

Christel, W. F. 1938. Texturemeter, a new device for measuring the texture of peas. *Cann. Trade* **60**(34), 10.

Civille, G. V., and I. H. Liska, 1975. Modifications and applications to foods of the General Foods Sensory Texture Profile Technique. *J. Texture Stud.* **6,** 19–32.

Civille, G. V., and A. S. Szczesniak. 1973. Guidelines to training a texture profile panel. *J. Texture Stud.* **4,** 204–223.

Claassens, J. W. 1958. The adhesion–cohesion, static friction and macrostructure of certain butters. I. A method of measuring the adhesion-cohesion of butter. *S. Afr. Tydskr. Landbouwet.* **1,** 457–463.

Claassens, J. W. 1959a. The adhesion–cohesion, static friction and macrostructure of certain butters. II. Factors affecting the values recorded with the hesion balance. *S. Afr. Tydskr. Landbouwet.* **2,** 89–118.

Claassens, J. W. 1959b. The adhesion–cohesion, static friction and macrostructure of certain butters. V. Friction and hesion measurements on the same test surfaces. *S. Afr. Tydskr. Landbouwet.* **2,** 551–571.

Clark, R. L., and P. S. Shackelford, J. 1976. Resonance and optical properties of peaches as related to firmness. *In* ''Quality Detection in Foods'' (J. J. Gaffney, ed.), ASAE Publ. No. 1–76, pp. 143–145. Am. Soc. Agric. Eng., St. Joseph, Michigan.

Clausi, A. S. 1973. Improving the nutritional quality of food. *Food Technol.* **27**(6), 36–40.

Cobb, N. A. 1896. The hardness of the grain in the principal varieties of wheat. *Agric. Gaz. N. S. W.* **7,** 279–298.

Cooke, J. R. 1972. An interpretation of the resonant behavior of intact fruits and vegetables. *Trans. ASAE* **15**(6), 1075–1080.

Cooke, J. R. and R. H. Rand. 1973. A mathematical study of resonance in intact fruits and vegetables using a 3-media elastic sphere model. *J. Agric. Eng. Res.* **18,** 141–157.

Corey, H. 1970. Texture in foodstuffs. *CRC Crit. Rev. Food Technol.* **1,** 161–198.

Couette, M. M. 1890. Etudes sur le frottement des liquides. *Ann. Chim. Phys.* **21,** 433–510.

Coulter, S. T., and W. B. Combs. 1936. A study of the body and texture of butter. *Minn. Agric. Exp. Stn., Tech. Bull.* No. 115.

Cover, S., S. J. Ritchey, and R. L. Hostetler. 1962. Tenderness of beef. I. The connective tissue component of tenderness. *J. Food Sci.* **27,** 469–475.

Cox, R. E., and R. H. Higby. 1944. A better way to determine the setting power of pectins. *Food Ind.* **16,** 441–443.

Crean, D. E. C., and D. R. Haisman. 1965. Elasticity coefficients and texture of cooked dried peas. *J. Sci. Food Agric.* **16,** 469–474.

Crossland, L. B., and H. H. Favor. 1950. A study of the effects of various techniques on the measurement of firmness of bread by the Baker Compressimeter. *Cereal Chem.* **27,** 15–25.

Culioli, J., and P. Sale. 1981. Spinnability of fababean protein II. Spinnability of dopes and textural properties of protein fibers. *J. Texture Stud.* **12,** 335–350.

Cumming, D. B., J. M. deMan, A. G. Lynch, W. G. Mertens, and M. Tanaka. 1971. Food texture measurements in Canada. *J. Texture Stud.* **2,** 441–450.

Dagbjartsson, B., and M. Solberg. 1972. A simple method to determine the water-holding capacity of muscle foods. *J. Food Sci.* **37,** 499–500.

Davey, C. L., and K. V. Gilbert. 1969. The effect of sample dimensions on the cleaving of meat in the objective assessment of tenderness. *J. Food Technol.* **4,** 7–15.

Davis, C. E. 1921. Shortening: Its definition and measurement. *Ind. Eng. Chem.* **13,** 797–799.

Davis, J. G. 1937. The rheology of cheese, butter, and other milk products. *J. Dairy Res.* **8,** 245–264.

DeBeaukelaer, F. L., J. R. Powell, and E. F. Bahlmann. 1930. Standard methods (revised) for determining viscosity and jelly strength of glue. *Ind. Eng. Chem., Anal. Ed.* **2,** 348–351.

DeBeaukelaer, F. L., O. T. Bloom, and J. T. DeRose. 1945. Determining jelly strength of gelatin-dessert jellies. *Ind. Eng. Chem., Anal. Ed.* **17,** 64.

Decker, R. W., J. N. Yeatman, A. Kramer, and A. P. Sidwell. 1957. Modifications of the shear press for electrical indicating and recording. *Food Technol.* **11,** 343–347.

deMan, J. M. 1969. Food texture measurements with the penetration method. *J. Texture Stud.* **1,** 114–119.

deMan, J. M. 1975. Texture of foods. *Lebensm.-Wiss. Technol.* **8,** 101–107.

deMan, J. M., and B. S. Kamel. 1981. Instrumental methods of measuring texture of poultry meat. *In* "Quality of Poultry Meat," Spelderholt Jubilee Symposia, pp. 157–164. Apeldoorn, Netherlands.

Diehl, K. C., D. D. Hamann,and J. K. Whitfield. 1979. Structural failure in selected raw fruits and vegetables. *J. Texture Stud.* **10,** 371–400.

Dikeman, M. E., J. H. Tuma, H. A. Glimp, R. E. Gregory, and D. M. Allen. 1972. Evaluation of the Tenderometer for predicting bovine muscle tenderness. *J. Anim. Sci.* **34,** 960.

Dixon, G. D., and J. V. Parekh. 1979. Use of the cone penetrometer for testing the firmness of butter. *J. Texture Stud.* **10,** 421–434.

Drake, B. K. 1961. An attempt at a geometrical classification of rheological apparatus. Unpublished manuscript, Göteborg, Sweden.

Drake, B. K. 1963. Food crushing sounds. An introductory study. *J. Food Sci.* **28,** 233–241.

Drake, B. K. 1965. Food crushing sounds: Comparison of objective and subjective data. *J. Food Sci.* **30,** 556–559.

Duke, R. W., and L. L. Chapoy. 1976. The rheology and structure of lecithin in concentrated solution and the liquid crystalline state. *Rheol. Acta* **15,** 548–557.

Edelman, E. C., W. H. Cathcart, and C. B. Berquist. 1950. The effect of various ingredients on the rate of firming of bread crumb in the presence of polyoxyethylene (mono) stearate and glycerylmonostearate. *Cereal Chem.* **27,** 1–14.

Ferry, J. D. 1977. Oscillation viscometry—effects of shear rate and frequency. *Meas. Control.* **11**(5), 89–91.

Feuge, R. D., and W. A. Guice. 1959. Effect of composition and polymorphic form on the hardness of fats. *J. Am. Oil. Chem. Soc.* **36,** 531–534.

Finney, E. E., Jr. 1970. Mechanical resonance within Red Delicious apples and its relation to fruit texture. *Trans. ASAE* **13,** 177–180.

Finney, E. E., Jr. 1971a. Dynamic elastic properties and sensory quality of apple fruit, *J. Texture Stud.* **2,** 62–74.

Finney, E. E., Jr. 1971b. Random vibration techniques for non-destructive evaluation of peach firmness. *J. Agric. Eng. Res.* **16,** 81–87.

Finney, E. E., Jr. 1971c. Vibration technique for testing fruit. *Pap.—Am. Soc. Agric. Eng.* No. 71–802.

Finney, E. E., Jr. 1972. Vibration techniques for testing fruit firmness. *J. Texture Stud.* **3,** 263–283.

Finney, E. E., Jr., and J. A. Abbott. 1972. Sensory and objective measurements of peach firmness. *J. Texture Stud.* **3,** 372–378.

Finney, E. E., Jr., and C. W. Hall. 1967. Elastic properties of potatoes. *Trans. ASAE* **10,** 4–8.

Finney, E. E., Jr., and K. H. Norris. 1968. Instrumentation for investigating dynamic mechanical properties of fruits and vegetables. *Trans. ASAE* **11,** 94–97.

Finney, E. E., Jr., I. Ben-Gera, and D. R. Massie. 1968. An objective evaluation of changes in firmness of ripening bananas using a sonic technique. *J. Food Sci.* **32,** 642–646.

Finney, E. E., Jr., J. A. Abbott, A. E. Watada, and D. R. Massie. 1978. Nondestructive sonic resonance and the texture of apples. *J. Am. Soc. Hortic. Sci.* **103,** 158–162.

Finney, K. F., and M. D. Shogren. 1972. A ten-gram Mixograph for determining and predicting functional properties of wheat flours. *Bakers Dig.* **46**(2), 32–35, 38–42, 77.

Fisher, J. D. 1933. Shortening value of plastic fats. *Ind. Eng. Chem.* **25,** 1171–1173.

Fitzgerald, J. V., and F. J. Matusik, 1976. Applications of a vibratory viscometer. *Am. Lab.* **8**(6), 63–68.

Freundlich, H., and W. Seifriz. 1923. Über die Elastizität von Solen und Gelen. *Z. Phys. Chem., Stoechiom. Verwandschaftsl.* **104,** 233–261.

Friedman, H. H., J. E. Whitney, and A. S. Szczesniak. 1963. The Texturometer—a new instrument for objective texture measurement. *J. Food Sci.* **28,** 390–396.

Fukushima, K. 1968. The texture of kamaboko. *Shokuhin Kogyo* **11**(16), 37.

Funk, K., M. E. Zabik, and D. A. Elgidaily. 1969. Objective measurements for baked products. *J. Home Econ.* **61,** 119–123.

Gardner, H. A., and A. W. VanHeuckeroth. 1927. Some further applications of a Mobilometer. *Ind. Eng. Chem.* **19,** 724–726.

Gillett, T. A., C. L. Brown, R. L. Leutzinger, R. D. Cassidy, and S. Simon. 1978. Tensile strength of processed meats determined by an objective Instron technique. *J. Food Sci.* **43,** 1121–1124, 1129.

Gilmont, R. 1963. A falling-ball viscometer. *Instru. Control Sys.* **36**(9), 121–122.

Goldthwaite, N. E. 1909. Contribution on the chemistry and physics of jelly-making. *Ind. Eng. Chem.* **1,** 333.

Goldthwaite, N. E. 1911. The priciples of jelly making. *Univ. Ill. Bull.* **3,** No. 7.

Gould, W. A. 1949. Instruments to quickly reveal quality of snap and wax beans. Food Packer **30**(12), 26–27.

Grawemeyer, E. A., and M. C. Pfund. 1943. Line-spread as an objective test for consistency. *Food Res.* **8,** 105–108.

Green, H. 1949 "Industrial Rheology and Rheological Structures." Wiley, New York.

Guthrie, E. S. 1952. A study of the body of cultured cream. *Cornell Agric. Exp. Stn. Bull. (Ithaca)* No. 880.

Guthrie, E. S. 1963. Further studies of the body of cultured cream. *Cornell Agric. Exp. Stn. Bull. (Ithaca)* No. 986.

Hagberg, S. 1960. A rapid method for determining alpha-amylase activity. *Cereal Chem.* **37,** 218–222.

Hagberg, S. 1961. Note on a simplified rapid method for determining alpha-amylase activity. *Cereal Chem.* **38,** 202–203.

Hagen, G. 1839. Ueber die Bewegung des Wassers in engen zylindrischen Roehren. *Pogg. Ann.* **46,** 423.

Haighton, A. J. 1959. The measurement of hardness of margarine and fats with cone penetrometers. *J. Am. Oil Chem. Soc.* **36,** 345–348.

Haller, M. H. 1941. Fruit pressure testers and their practical applications. *U.S. Dep. Agric., Circ.* No. 627.
Halmos, A. L., and C. Tiu. 1981. Liquid foodstuffs exhibiting yield stress ahd shear-degradability. *J. Texture Stud.* **12,** 39–46.
Halton, P., and G. W. Scott Blair. 1937. A study of some physical properties of flour doughs in relation to their bread-making properties. *Cereal Chem.* **14,** 201–219.
Hamann, D. D., and D. E. Carroll. 1971. Ripeness sorting of muscadine grapes by use of low-frequency vibrational energy. *J. Food Sci.* **36,** 1049–1051.
Hamann, D. D., L. J. Kushman, and W. E. Ballinger. 1973. Sorting blueberries for quality by vibration. *J. Am. Soc. Hortic. Sci.* **98,** 572–576.
Hamm, R. 1960. Biochemistry of meat hydration. *Adv. Food Res.* **10,** 355–463.
Hammerle, J. R., and W. F. McClure. 1971. The determination of Poisson's ratio by compression tests of cylindrical specimens. *J. Texture Stud.* **2,** 31–39.
Hansen, L. J. 1971. Meat tenderness testing. U.S. Patent 3,602,038 (W. Armour & Co.).
Hansen, L. R. 1972. Development of the Armour Tenderometer for tenderness evaluation of beef carcases. *J. Texture Stud.* **3,** 146–164.
Harrington, G., and A. M. Pearson. 1962. Chew count as a measure of tenderness of pork loins with various degrees of marbling. *J. Food Sci.* **27,** 106–110.
Harris, P. V. 1975. Meat chilling. *CSIRO Food Res. Q.* **35,** 49–56.
Hashimoto, Y., T. Fukazawa, R. Niki, and T. Yasui. 1959. Effects of storage conditions on some of the biochemical properties of meats and on the physical properties of an experimental sausage. *Food Res.* **24,** 185–197.
Haugh, R. R. 1937. The Haugh unit for measuring egg quality. *U.S. Egg Poult. Mag.* **43,** 552–555, 572.
Henrickson, R. L., J. L. Marsden, and R. D. Morrison. 1972. An evaluation of the Armour Tenderometer for an estimation of beef tenderness. *J. Food Sci.* **37**, 857–859.
Henry, W. F., and M. H. Katz. 1969. New dimensions relating to the textural quality of semi-solid foods and ingredient systems. *Food Technol.* **23,** 822–825.
Henry, W. F., M. H. Katz, F. J. Pilgrim, and A. J. May. 1971. Texture of semi-solid foods: sensory and physical correlates. *J. Food Sci.* **36,** 155–161.
Herschel, W. H., and C. Bergquist. 1921. The consistency of starch and dextrin pastes. *Ind. Eng. Chem.* **13,** 703–706.
Herschel, W. H., and R. Bulkley. 1926. Konsistenzmessungen von gummibenzollösungen. *Kolloid-Z.* **39,** 291–300.
Hilker, L. D. 1947. A method for measuring the body of cultured cream. *J. Dairy Sci.* **30,** 161–164.
Hill, R. L. 1923. A test for determining the character of the curd from cow's milk and its application to the study of curd and variance as an index to the food value of milk for infants. *J. Dairy Sci.* **6,** 509–526.
Hill, R. L. 1933. A decade and a half of soft-curd milk Studies. *Utah Agric. Exp. Stn. Circ. No.* 101.
Hindman, H., and G. S. Burr. 1949. The Instron tensile tester. *Trans. ASME* **71,** 789–796.
Hogarth, J. 1889. A new or improved mode of and means or appliance for testing and recording the characteristics or properties of flour and dough. U.K. Patent 16,389.
Holdsworth, S. D. 1971. Applicability of rheological models to the interpretation of flow and processing behavior of fluid food products. *J. Texture Stud.* **2,** 393–418.
Hoseney, R. C., K. H. Hsu, and R. C. Junge. 1979. A simple spread test to measure the rheological properties of fermenting dough. *Cereal Chem.* **56,** 141–143.
Hostetler, R. L., and S. J. Ritchey. 1964. Effect of coring methods on shear values determined by Warner–Bratzler shear. *J. Food Sci.* **29,** 681–685.
Houwink, R. 1958. "Elasticity, Plasticity and Structure of Matter." Dover, New York.
Howe, P. E., and S. Bull. 1927. A study of factors which influence the quality and palatability of

meat. *U.S. Dep. Agric., Natl. Coop. Proj., Bur. Anim. Ind.* Rev. ed.
Hubbard, R. M., and G. G. Brown. 1943. The rolling ball viscometer. *Ind. Eng. Chem., Anal. Ed.* **15,** 212–218
Huffman, D. L. 1974. An evaluation of the Tenderometer for measuring beef tenderness. *J. Anim. Sci.* **38,** 287–294.
Hulbert, E. 1913. Improved apparatus for testing the jelly strength of glues. *Ind. Eng. Chem.* **5,** 235.
Hunziker, O. E., H. C. Mills, and G. Spitzer. 1912. Moisture control of butter. *Indiana Agric. Exp. Stn., Stn. Bull.* No. 159, p. 347, 359.
Jansen, K. 1961. Hesion and consistency of butter. *J. Dairy Res.* **28,** 15–20.
Jauregui, C. A., J. M. Regenstein, and R. C. Baker. 1981. A simple centrifugal method for measuring expressible moisture, a water-binding property of muscle food. J. Food Sci **46,** 1271, 1273.
Jenkins, G. N. 1978. The Physiology and Biochemistry of the Mouth,'' 4th ed. Blackwell, Oxford.
Jerome, N. W. 1975. Flavor preferences and food patterns of selected U.S. and Caribbean blacks. *Food Technol.* **29**(6), 46.
Jowitt, R. 1974. The terminology of food texture. *J. Texture Stud.* **5,** 351–358.
Kapur, K. K., S. Soman, and A. Yurkstas. 1964. Test food for measuring masticatory performance of denture wearers. *J. Prosthet. Dent.* **14,** 483–491.
Kapur, K. K., H. H. Chauncey, and I. M. Sharon. 1966. Oral physiological factors concerned with ingestion of food. *In* ''Nutrition in Clinical Dentistry'' (A. E. Nizel, ed.), 2nd ed., pp. 296–304. Saunders, Philadelphia, Pennsylvania.
Karacsonyi, L. P., and A. C. Borsos. 1961. An apparatus for measuring the torsional strength of macaroni. *Cereal Chem.* **38,** 14–21.
Karmas, E., and K. Turk. 1976. Water binding of cooked fish in combination with various proteins. *J. Food Sci.* **41,** 977–979.
Kastner, C. L., and R. L. Henrickson. 1969. Providing uniform meat cores for mechanical shear force measurement. *J. Food Sci.* **34,** 603–605.
Katz, R., A. B. Caldwell, N. D. Collins, and A. E. Hostetter. 1959. A new grain hardness tester. *Cereal Chem.* **36,** 393–401.
Katz, R., N. D. Collins, and A. B. Caldwell. 1961. Hardness and moisture content of wheat kernels. *Cereal Chem.* **38,** 364–368.
Kertesz, Z. I. 1935. The quality of canned whole kernel corn as determined by the simplified method for alcohol insoluble solids. *Canner* **80**(11), 12–13.
Khan, N., and M. Elahi. 1980. A rapid batter expansion method for testing the baking quality of wheat flours. *J. Food Technol.* **15,** 43–46.
Kilborn, R. H., and C. J. Dempster. 1965. Power-input meter for laboratory dough mixers. *Cereal Chem.* **42,** 432.
Kissling, R. 1893. Zur Prüfung des Tatelleimes. *Chem.-Ztg.* **17,** 726–727.
Kissling, R. 1898. Zur Prüfung der Leimsorten des Handels. *Chem.-Ztg.* **22,** 171–172.
Klatsky, M. 1942. Masticatory stresses and their relation to dental caries. *J. Dent. Res.* **21,** 389.
Knaysi, G. 1927. Some factors other than bacteria that influence the body of artificial buttermilk. *J. Agric. Res. (Washington, D.C.)* **34,** 771–784.
Kosutány, T. 1907. ''Der ungarische Weizen und das ungarische Mehl.'' Budapest. Cited by D. W. Kent-Jones and A. J. Amos in ''Modern Cereal Chemistry,'' 6th ed., 1967. Food Trade Press, Ltd. London.
Kramer, A. 1951. Objective testing of vegetable quality. *Food Technol.* **5,** 265–269.
Kramer, A. 1973. Food texture—definition, measurement and relation to other food quality attributes. *In* ''Texture Measurements of Foods'' (A. Kramer and A. S. Szczesniak, eds.), pp. 1–9. Reidel Publ., Dordrecht, Netherlands.
Kramer, A., and H. R. Smith. 1946. The succulometer, an instrument for measuring the maturity of

raw and canned whole kernel corn. *Food Packer* **27**(8), 56–60.

Kramer, A., and B. A. Twigg. 1959. Principles and instrumentation for the physical measurement of food quality with special reference to fruit and vegetable products. *Adv. Food Res.* **9,** 153–220.

Kramer, A., R. B. Guyer, and L. E. Ide. 1949. Factors affecting the objective and organoleptic evaluation in sweet corn. *Proc. Am. Soc. Hortic. Sci.* **54,** 342–356.

Kramer, A., K. Aamlid, R. B. Guyer, and H. Rogers. 1951. New shear-press predicts quality of canned limas. *Food Eng.* **23**(4), 112–113, 187.

Krieger, I. M. 1979. Golden anniversary for rheology. *Phys. Today* **32**(10), 128.

Kruisheer, C. I. 1939. An analytical study on the quality and specially the firmness of butter. *Chem. Ind. (London)* **58,** 735–740.

Kruisheer, C. I., and P. C. den Herder. 1938. Onderzoekingen betreffende de consistentie van boter. *Chem. Weekbl.* **35,** 719.

Krumel, K. L., and Sakar, N. (1975). Flow properties of gums useful to the food industry. *Food Technol.* **29**(4), 36, 38, 40, 41, 43, 44.

La Belle, R. L. 1964. Bulk density—a versatile measure of food texture and bulk. *Food Technol.* **18,** 879–884.

La Belle, R. L., E. E. Woodams, and M. C. Bourne. 1964. Recovery of Montmorency cherries from repeated bruising. *Proc. Am. Soc. Hortic. Sci.* **84,** 103–109.

Lamb, F. C. 1967. Collaborative study of capillary viscometer method for consistency of fruit nectars and fruit juice products. *J. Assoc. Off. Anal. Chem.* **50,** 288–292.

Lamb, F. C., and L. D. Lewis. 1959. Consistency measurement of fruit nectars and fruit juice products. *J. Assoc. Off. Agric. Chem.* **42,** 411–416.

Larmond, E. 1970. Methods of sensory evaluation of food. *Publ.—Can. Dep. Agric.* No. 1284.

Larmour, R. K., E. B. Working, and C. W. Ofelt. 1939. Quality tests on hard red winter wheats. *Cereal Chem.* **16,** 733–752.

Lehmann, K. B. 1907a. Studien über die Zähigkeit des Fleisches und ihre Ursachen. *Arch. Hyg.* **63,** 134–179.

Lehmann, K. R. 1907b. Die Festigkeit (Zähigkeit) vegetabilischer Nahrungsmittel und ihre Veränderung durch das Kochen. *Arch. Hyg.* **63,** 180–182.

Leick, A. 1904a. Über Künstliche Doppelbrechung und Elastizität von Gelatineplatten. *Ann. Phys. (Leipzig)* **14,** 139–152.

Leick, A. 1904b. Über Künstliche Doppelchrechung und Elastizität von gelatineplatten. *J. Phys. Theor. Appl.* **3,** 866, 877.

Leighton, A., A. Leviton, and O. E. Williams. 1934. The apparent viscosity of ice cream. I. The sagging beam method of measurement II. Factors to be controlled III. The effects of milk fat, gelatin and homogenization temperature. *J. Dairy Sci.* **17,** 639–650.

Lewis, C. I., A. E. Murneek, and C. C. Cate. 1919. Pear harvesting and storage investigations in Rogue River Valley. *Ore. Agric. Exp. Stn., Bull.* No. 162.

Lindsay, J. B. 1901. Effect of feed on the composition of milk, butter fat, and on the consistency or body of butter. *13th Annu. Rep., Mass. Hatch Exp. Stn.* p. 29.

Lindsay, J. B., E. B. Holland, and P. H. Smith. 1909. Effect of soybean meal and soybean oil upon the composition of milk and butter fat, and upon the consistency or body of butter. *21st Annu. Rep., Mass. Agric. Exp. Stn.* Part II, pp. 66–110.

Lipowitz, A. 1861. "Neue Chemisch-Technische Abhandlungen." Bosselman, Berlin.

Locken, L., S. Loska, and W. Shuey. 1960. "The Farinograph Handbook." Am. Assoc. Cereal Chem., St. Paul, Minnesota.

Love, R. M., and E. Mackay. 1962. Protein denaturation in frozen fish. V. Development of the cell fragility method for measuring cold storage changes in muscle. *J. Sci. Food Agric.* **13,** 200–212.

Love, R. M., and M. Muslemuddin. 1972a. Protein denaturation in frozen fish, XII. The pH effect

and cell fragility determinations. *J. Sci. Food Agric.* **23,** 1224–1238.
Love, R. M., and M. Muslemuddin. 1972b. Protein denaturation in frozen fish, XIII. A modified cell fragility method insensitive to the pH of the fish. *J. Sci. Food Agric.* **23,** 1239–1251.
Low, W. H. 1920. Testing the strength of glue jellies. *Ind. Eng. Chem.* **12,** 355.
Lynch, L. J., and R. S. Mitchell. 1950. The physical measurement of quality in canning of peas. *CSIRO Aust. Bull.* No. 254.
Lynch, L. J., and R. S. Mitchell. 1952. Short term prediction of maturity and yield of canning peas. *Food Technol.* **6**(9), 24, 28, 30.
MacFarlane, P. G., and J. M. Marer. 1966. An apparatus for determining the tenderness of meat. *Food Technol.* **20,** 838–839.
McKennell, R. 1956. The cone-plate viscometer. *Anal. Chem.* **28,** 1710–1714.
MacMichael, R. F. 1915. A new direct reading viscometer. *Ind. Eng. Chem.* **7,** 961–963.
Magness, J. R., and G. F. Taylor. 1925. An improved type of pressure tester for the determination of fruit maturity. *U.S. Dep. Agric., Circ.* No. 350.
Maleki, M., and W. Siebel. 1972. Uber das Altbackenwerden von Brot. I. Beziehungen zwischen Sensorik Penetrometer und Panimeter beim Messen des Altbacken werdens von Weizenbrot. *Getreide Mehl Brot* **26,** 58.
Margules, M. 1881. Ueber die Bestimmung des Reibungs- und Gleitungscoefficienten aus ebenen Bewegungen einer Fluessigkeit. *Wien Sitzungsberger, Abt. 2A* **83,** 588; see also **84,** 491 (1882).
Martin, W. McK. 1937. The Tenderometer, an apparatus for evaluating tenderness in peas. Canner **84**(12), 108–112.
Martin, W. McK., R. H. Lueck, and E. D. Sallee. 1938. Practical application of the Tenderometer in grading peas. *Cann. Age* **19**(3), 146–149.
Matz, S. A. 1962. "Food Texture." Avi, Westport, Connecticut.
Meyer, G. S. 1929. Some critical comments on the methods employed in the expression of leaf saps. *Plant Physiol.* **4,** 103–111.
Meyeringh, W. 1911. Studie over het factoren op het vochgehalte der botter van invloed. 51–55. 's Gravenhage: De Gebroeders van Cleef. Diss. Delft. (From Mulder, 1953.)
Miller, B. S., J. W. Hughes, R. Rousser, and Y. Pomeranz. 1981a. Measuring the breakage susceptibility of shelled corn. *Cereal Foods World* **26,** 75–80.
Miller, B. S., J. W. Hughes, R. Rousser, and G. D. Booth. 1981b. Effects of modifications of a model CK2 Stein Breakage Tester on corn breakage susceptibility. *Cereal Chem.* **58,** 201–203.
Miller, B. S., J. W. Hughes, Y. Pomeranz, and G. D. Booth. 1981c. Measuring the breakage susceptibility of soybeans. *Cereal Foods World* **26,** 174–177.
Mitchell, R. S., D. J. Casimir, and L. J. Lynch. 1961. The Maturometer—instrumental test and redesign. *Food Technol.* **15,** 415.
Miyada, D. S., and A. L. Tappel. 1956. Meat tenderization I. Two mechanical devices for measuring texture. *Food Technol.* **10,** 142–145.
Mohsenin, N. N., and J. P. Mittal. 1977. Use of rheological terms and correlation of compatible measurements in food texture research. *J. Texture Stud.* **8,** 395–408.
Mohsenin, N. N., H. E. Cooper, and L. D. Tukey. 1963. Engineering approach to evaluating textural factors in fruits and vegetables. *Trans. ASAE* **6,** 85–88, 92.
Morris, O. M. 1925. Studies in apple storage. *Bull.—Wash. Agric. Exp. Stn.* No. 193.
Morrow, C. T., and N. N. Mohsenin. 1966. Consideration of selected agricultural products as viscoelastic materials. *J. Food Sci.* **31,** 686–698.
Morrow, C. T., and N. N. Mohsenin. 1976. Mechanics of a multiple needle Tenderometer for raw meat. *J. Texture Stud.* **7,** 115–128.
Mottram, F. J. 1961. Evaluation of pseudoplastic materials by cone penetrometers. *Lab. Prac.* **10,** 767–770.
Mueller, W. S. 1935. A method for the determination of the relative stiffness of cream during the

whipping process. *J. Dairy Sci.* **18,** 177–180.
Mulder, H. 1953. Consistency of butter. *In* "Foodstuffs, Their Plasticity, Fluidity and Consistency" (G. W. Scott Blair, ed.), p. 107. Wiley (Interscience), New York.
Muller, H. G. 1969a. Mechanical properties, rheology and haptaesthesis of food. *J. Texture Stud.* **1,** 38–42.
Muller, H. G. 1969b. Routine rheological tests in the British food industry. *J. Food Technol.* **4,** 83–92.
Munz, E., and C. W. Brabender. 1940. Prediction of baking value from measurements of plasticity and extensibility of dough I. Influence of mixing and molding treatments upon physical dough properties of typical American wheat varieties. *Cereal Chem.* **17,** 79–100.
Murneek, A. E. 1921. A new test for maturity of the pear. *Ore. Agric. Exp. Stn. Bull.* No. 186.
Murray, D. C., and L. R. Luft. 1973. Low DE corn starch hydrolysates. *Food Technol.* **27**(3), 33.
Nemitz, G. 1963. A new apparatus for measuring the course of rigor mortis in fish and mammalian muscle. (In Ger.) *Fleischwirtschaft* **15,** 18, 21.
Nemitz, G., W. Portman, and D. Sharra. 1960. Eine neue method zur Messung der Totenstarre. (A new method for measuring death rigor.) *Z. Lebensm.-Unters. -Forsch.* **112**(4), 261–272.
Oakes, E. F., and C. E. Davis. 1922. Jell strength and viscosity of gelatin solutions. *J. Ind. Eng. Chem.* **14,** 706–710.
Oka, S. 1960. The principles of rheometry. *In* "Rheology, Theory and Applications" (F. R. Eirich, ed.), pp. 17–82. Academic Press, New York.
Okabe, T. 1971. Studies on the measurement of food texture IV. Rheological analysis of mastication curves. (In Jpn.) *Sci. Cookery* **4,** 232.
Okabe, T. 1979. Texture measurement of cooked rice and its relationship to eating quality. *J. Texture Stud.* **10,** 131–152.
Oldfield, R. C. 1960. Perception in the mouth. *In* "Texture in Foods," Monograph, No. 7, pp. 3–9. Soc. Chem. Ind., London.
Oppliger, H. R., F. J. Matusik, and J. V. Fitzgerald. 1975. New technique accurately measures low viscosity on-line. *Control Eng.* **22**(7), 39–40.
Parrish, E. C., Jr., D. G. Olson, B. E. Miner, R. B. Young, and R. L. Small. 1973. Relationship of tenderness measurements made by the Armour Tenderometer to certain objective, subjective and organoleptic properties of bovine muscle. *J. Food Sci.* **38,** 1214–1219.
Pavlov, I. P. 1927. "Conditioned Reflexes. An investigation of the Physiological Activity of the Cerebral Cortex." Oxford Univ. Press, London. (Eng. trans.)
Pearson, C. A., and D. Raynor. 1975. Comments on article by Peter Voisey. *J. Texture Stud.* **6,** 393–394.
Peleg, M. 1980. A note on the sensitivity of fingers, tongue and jaws as mechanical testing instruments. *J. Texture Stud.* **10,** 245–251.
Perkins, A. E. 1914. An apparatus and method for determining the hardness of butterfat. *Ind. Eng. Chem.* **6,** 136–141.
Personius, C. J., and G. F. Sharp. 1938. Adhesion of potato tuber cells as influenced by temperature. *Food Res.* **3,** 513–514.
Perten, H. 1964. Application of the falling number method for evaluating alpha-amylase activity. *Cereal Chem.* **41,** 127–140.
Perten, H. 1967. Factors influencing falling number values. *Cereal Sci. Today* **12,** 516, 518, 519.
Pierson, A., and J. LeMagnen. 1970. Study of food textures by recording and chewing and swallowing movements. *J. Texture Stud.* **1,** 327–337.
Pitman, G. 1930. Further comparison of California and imported almonds. *Ind. Eng. Chem.* **22,** 1129–1131.
Platt, W., and P. D. Kratz. 1933. Measuring and recording some characteristics of test sponge cakes. *Cereal Chem.* **10,** 73–90.

Platt, W., and R. Powers. 1940. Compressibility of bread crumb. *Cereal Chem.* **17,** 601–621.

Poiseuille, J. L. M. 1846. Recherches expérimentales sur le mouvement des liquides dans les tubes de très petits diamètres. *Mem. Acad. R. Sci. Inst. Fr. Sci., Math. Phys.* **9,** 433–545. [Eng. trans., Rheol. Mem. **1,** No. 1 (1940).]

Pool, M. F. 1967. Objective measurement of connective tissue tenacity of poultry meat. *J. Food Sci.* **32,** 550–553.

Pool, M. F., and A. A. Klose. 1969. The relation of force to sample dimensions in objective measurements of tenderness of poultry meat. *J. Food Sci.* **34,** 524.

Poole, H. J. 1925. The elasticity of gelatin jellies and its bearing on their physical structure and chemical equilibria. *Trans. Faraday Soc.* **21,** 114–137.

Porst, E. G., and Moskowitz. 1922. Comparison of the various corn products starches as shown by the Bingham–Greene Plastometer. *Ind. Eng. Chem.* **14,** 49–52.

Potter, N. 1968. "Food Science." Avi, Westport, Connecticut.

Prentice, J. H. 1972. Rheology and texture of dairy products. *J. Texture Stud.* **3,** 415–458.

Priel, Z., M. Sasson, and A. Silverberg. 1973. Method for determining the time of flow in a capillary viscometer with an absolute accuracy of 3×10^{-6}. *Rev. Sci. Instrum.* **44,** 135.

Rao, M. A., and M. C. Bourne. 1977. Analysis of the Plastometer and correlation of Bostwich Consistometer data. *J. Food Sci.* **42,** 261–264.

Rao, M. A., M. C. Bourne, and H. J. Cooley. 1981. Flow properties of tomato pastes of high concentrations. *J. Texture Stud.* **12,** 521–538.

Rasper, V. F. 1975. Dough rheology at large deformations in simple tensile mode. *Cereal Chem.* **52,** 24r–41r.

Rasper, V. F., and J. M. deMan. 1980. Effect of granule size of substituted starches on the rheological character of composite doughs. *Cereal Chem.* **57,** 331–340.

Rasper, V. F.,J. Rasper, and J. M. deMan. 1974. Stress strain relationships of chemically improved unfermented doughs. *J. Texture Stud.* **4,** 438–466.

Reiner, M., and G. W. Scott Blair. 1967. Rheological terminology. *In* "Rheology: Theory and Applications" (F. R. Eirich, ed.), Vol. 4, pp. 461–488. Academic Press, New York.

Reynolds, D. 1883. An experimental investigation of the circumstances which determine whether the motion of water shall be direct or sinuous, and of the law of resistance in parallel channels. *Philos. Trans. R. Soc. London* **174,** 935–982.

Rha, C. 1980. Food rheology principles and practice. Lecture notes, Food Mater. Sci. Fabr. Lab., Mass. Inst. Technol., Cambridge, Massachusetts.

Rhodes, D. N., R. C. D. Jones, B. B. Chrystal, and J. M. Harries. 1972. Meat texture II. The relationship between subjective assessments and a compressive test on roast beef. *J. Texture Stud.* **3,** 298–309.

Roark, R. J. 1965. "Formulas for Stress and Strain." McGraw-Hill, New York.

Roberts, H. F. 1910. A quantitative method for the determination of hardness in wheat. *Kans. Agric. Exp. Stn., Bull.* No. 167, pp. 371–390.

Robertson, G. H., and S. H. Emani. 1974. Liquid jet penetrometry for physical analysis. Application to bread aging. *J. Food Sci.* **39,** 1247–1253.

Rogers, L. A., and G. P. Sanders. 1942. Devices for measuring physical properties of cheese. *J. Dairy Sci.* **25,** 203–210.

Ross, L. R., and W. L. Porter. 1968. Interpretation of multiple-peak shear force curves obtained on French fried potatoes. *Am. Potato J.* **45,** 461–471.

Ross, L. R., and W. L. Porter. 1969. Objective measurements of French fried potato quality. Laboratory techniques for research use. *Am. Potato J.* **46,** 192–200.

Ross, L. R., and W. L. Porter. 1971. Objective measurement of texture variables in raw and processed French fried potatoes. *Am. Potato J.* **48,** 329–343.

Ross, L. R., and W. L. Porter. 1976. Textural quality prediction of reheated frozen French fried potatoes by objective testing of raw stock. *J. Food Sci.* **41,** 206–208.

Rostagno, W. 1974. Rheological properties of chocolate. *In* ''Lebensmittel-Einfluss der Rheologie'' (D. Behrens and K. Fischbeck, eds.), DECHEMA Monograph, Vol. 77, No. 1505–1536, pp. 283–293. Verlag Chemie, Weinheim.

Roth, W., and S. R. Rich. 1953. A new method for continuous viscosity measurement. General theory of the Ultra-Viscoson. *J. Appl. Phys.* **24,** 940–950.

Rutgers, G. 1958. The consistency of starch-milk puddings. I. Preparation of puddings and measurement of consistency. *J. Sci. Food Agric.* **9,** 61–68.

Sayre, J. D., and V. H. Morris. 1931. Use of expressed sap in physiologic studies of corn. *Plant Physiol.* **6,** 139–148.

Sayre, J. D., and V. H. Morris. 1932. Use of expressed sap in determining the composition of corn tissue. *Plant Physiol.* **7,** 261–271.

Schultz, E. F., Jr., and G. W. Schneider. 1955. Sample size necessary to estimate size and quality of fruit growth of trees, and per cent fruit set of apples and peaches. *Proc. Am. Soc. Hortic. Sci.* **66,** 36–44.

Schwedoff, T. 1889. Recherches experimentales sur la cohesion des liquides. *J. Phys. Theor. Appl.* **8,** 341–359.

Scott Blair, G. W., and J. Burnett. 1963. A simple method for detecting an early stage in coagulation of rennetted milk. *J. Dairy Res.* **30,** 383–390.

Scott Blair, G. W., and F. M. V. Coppen. 1940. An objective measure of the consistency of cheese curd at the pitching point. *J. Dairy Res.* **11,** 187–195.

Scott Blair, G. W., G. Watts, and H. J. Denham. 1927. Effect of concentration on viscosity of flour suspensions. *Cereal Chem.* **4,** 63–67.

Segars, R. A., R. G. Hamel, and J. G. Kapsalis. 1977. Use of Poisson's ratio for objective–subjective texture correlations in beef. An apparatus for obtaining the required data. *J. Texture Stud.* **8,** 433–447.

Shama, F., and P. Sherman. 1973a. Evaluation of some textural properties of foods with the Instron Universal Testing Machine. *J. Texture Stud.* **4,** 344–353.

Shama, F., and P. Sherman. 1973b. Identification of stimuli controlling the sensory evaluation of viscosity. II. Oral methods. *J. Texture Stud.* **4,** 111–118.

Shama, F., C. Parkinson, and P. Sherman. 1973. Identification of stimuli controlling the sensory evaluation of viscosity. I. Non oral methods. *J. Texture Stud.* **4,** 102–110.

Sharma, M. G., and N. N. Mohsenin. 1970. Mechanics of deformation of a fruit subjected to hydrostatic pressure. *J. Agric. Eng. Res.* **15**(1), 65–74.

Sheppard, S. E., and S. S. Sweet. 1923. A preliminary study of a plunger type of jelly-strength tester. *Ind. Eng. Chem.* **15,** 571–576.

Sheppard, S. E., and S. S. Sweet, and J. W. Scott, Jr. 1920. The jelly strength of gelatins and glues. *Ind. Eng. Chem.* **12,** 1007–1011.

Sherman, P. 1970. ''Industrial Rheology.'' Academic Press, New York.

Shiffman, S. S. 1973. The dietary rehabilitation clinic and a multi-aspect, dietary, and behavioral approach to the treatment of obesity. b) Taste and smell of foods. *Meet. Assoc. Adv. Behav. Ther., Miami Beach, Fla.*

Simon, S., J. C. Field, W. E. Kramlich, and F. N. Tauber. 1965. Factors affecting frankfurter texture and a method of measurement. *Food Technol.* **14,** 410–413.

Sindall, R. W., and W. Bacon. 1914. The examination of commercial gelatins with regard to their suitability for papermaking. *Analyst (London),* **39,** 20–27.

Slattery, J. C. 1961. Analysis of the cone-plate viscometer. *J. Colloid Sci.* **16,** 431–437.

Smeets, H. S., and H. Cleve. 1956. Determination of conditioning by measuring softness. *Milling Prod.* **21**(4), 5, 12–13, 16.

Smith, C. R. 1920. Determination of jellying power of gelatins and glue by the polariscope. *Ind.*

Eng. Chem. **12,** 878–881.
Smith, E. S. 1909. Glue-Tester. U.S. Patent 911,277.
Smith, G. C., and Z. L. Carpenter. 1973. Mechanical measurements of meat tenderness using the Nip Tenderometer. *J. Texture Stud.* **4,** 196–203.
Sohn, C. E. 1893. Simple appliances for testing the consistency of semi-solids, with a note on a new method of examining butter. *Analyst (London)* **18,** 218–221.
Sone, T. 1972. "Consistency of Foodstuffs." Reidel Publ., Dordrecht, Netherlands.
Starita, J. M. 1980. A new research rheometer. *In* "Rheology, Vol. 2, Fluids" (G. Astarita, G. Marrucci, and L. Nicolais, eds.), pp. 229–234. Plenum, New York.
Sterling, C., and M. Simone. 1954. Crispness in almonds. *Food Res.* **19,** 276–281.
Stevens, S. S. 1970. Neural events and the psychophysical law. *Science* **170,** 1043–1050.
Stewart, F. C. 1923. The relation of moisture content and certain other factors to the popping of popcorn. *Bull.—N.Y. Agric. Exp. Stn. (Geneva)* No. 505.
Stone, H., and S. Oliver. 1966. Effect of viscosity on the detection of relative sweetness intensity of sucrose solutions. *J. Food Sci.* **31,** 129–134.
Sucharipa, E. 1923. Experimental data on pectin–sugar acid gels. *J. Assoc. Off. Agric. Chem.* **7,** 57–68.
Swanson, C. O., and E. B. Working. 1933. Testing the quality of flour by the recording dough mixer. *Cereal Chem.* **10,** 1–29.
Szczesniak, A. S. 1963a. Classification of textural characteristics. *J. Food Sci.* **28,** 385–389.
Szczesniak, A. S. 1963b. Objective measurements of food texture. *J. Food Sci.* **28,** 410–420.
Szczesniak, A. S. 1971. Consumer awareness of texture and of other food attributes. *J. Texture Stud.* **2,** 196–206.
Szczesniak, A. S. 1972. Consumer awareness of and attitudes to food texture. II. Children and teenagers. *J. Texture Stud.* **3,** 206–217.
Szczesniak, A. S. 1975a. Textural characterization of temperature sensitive foods. *J. Texture Stud.* **6,** 139–156.
Szczesniak, A. S. 1975b. General Foods texture profile revisited—ten years perspective. *J. Texture Stud.* **6,** 5–17.
Szczesniak, A. S. 1979. Classification of mouthfeel characteristics of beverages. *In* "Food Rheology and Texture" (P. Sherman, ed.), pp. 1–20. Academic Press, New York.
Szczesniak, A. S., and E. Farkas. 1962. Objective characterization of the mouthfeel of gum solutions. *J. Food Sci.* **27,** 381–385.
Szczesniak, A. S., and B. J. Hall. 1975. Application of the General Foods Texturometer to specific food products. *J. Texture Stud.* **6,** 117–138.
Szczesniak, A. S., and E. L. Kahn. 1971. Consumer awareness of and attitudes to food texture. I. Adults. *J. Texture Stud.* **2,** 280–295.
Szczesniak, A. S., and D. H. Kleyn. 1963. Consumer awareness of texture and other foods attributes. *Food Technol.* **17,** 74–77.
Szczesniak, A. S., and E. Z. Skinner. 1973. Meaning of texture words to the consumer. *J. Texture Stud.* **4,** 378–384.
Szczesniak, A. S., and K. Torgeson. 1965. Methods of meat texture measurement viewed from the background of factors affecting tenderness. *Adv. Food Res.* **14,** 53–165.
Szczesniak, A. S., M. A. Brandt, and H. H. Friedman. 1963. Development of standard rating scales for mechanical parameters of texture and correlation between the objective and sensory methods of texture evaluation. *J. Food Sci.* **28,** 397–403.
Szczesniak, A. S., P. R. Humbaugh, and H. W. Block. 1970. Behavior of different foods in the standard shear compression cell of the shear press and the effect of sample weight on peak area and maximum force. *J. Texture Stud.* **1,** 356–378.
Szczesniak, A. S., M. Einstein, and R. E. Pabst. 1974. The texture tester, principles and selected

applications. *J. Texture Stud.* **5,** 299–316.

Szczesniak, A. S., B. J. Loew, and E. Z. Skinner. 1975. Consumer texture profile technique. *J. Food Sci.* **40,** 1253–1256.

Tanaka, M. 1975. General Foods Texturometer applications to food texture research in Japan. *J. Texture Stud.* **6,** 101–116.

Tarr, L. W. 1926. Fruit Jellies III. Jelly strength measurements. *Del. Agric. Exp. Stn., Bull.* No. 142.

Tauti, M., I. Hirose, and H. Wada. 1931. A physical method of testing the freshness of raw fish. *J. Imp. Fish. Inst. (Jpn.)* **26,** 59–66.

Templeton, H. L., and H. H. Sommer. 1933. Studies on whipping cream. *J. Dairy Sci.* **16,** 329–345.

Thompson, J. B., and D. F. Meisner. 1950. The lever system of the Baker compressimeter. *Cereal Chem.* **27,** 71–73.

Tracy, A. F. 1928. A new jelly strength tester and some experiments on gelatin gels. *J. Soc. Chem. Ind., London, Trans. Commun.* **47,** 94T–96T.

Tressler, D. K., and W. T. Murray. 1932. Tenderness of meat. II. Determination of period of aging of grade A beef required to produce a tender quick-frozen product. *Ind. Eng. Chem.* **24,** 890–892.

Tressler, D. K., C. Birdseye, and W. T. Murray. 1932. Tenderness of meat. *Ind. Eng. Chem.* **24,** 242–245.

Tschoegl, N. W., J. A. Rinde, and T. L. Smith. 1970. Rheological properties of wheat flour doughs. I. Method for determining the large deformation and rupture properties in simple tension. *J. Sci. Food Agric.* **21,** 65–70.

Tsuji, S. 1981. Texture measurement of cooked rice kernels using the multiple point mensuration method. *J. Texture Stud.* **12,** 93–105.

Tsuji, S. 1982. Texture profile analysis of processed foods using the Tensipresser and the multi-point mensuration method. *J. Texture Stud.* **13,** (in press).

Underwood, J. C., and G. J. Keller. 1948. A method for measuring the consistency of tomato paste. *Fruit Prod. J. Am. Food Manuf.* **28**(4), 103–105.

Valenta, E. 1909. Neuer Apparat zur Bestimmung der Druckfestigkeit von Leimgallerten. *Chem.-Ztg.* **33,** 94.

Vanderheiden, G. J. 1970. An instrument for objective characterization of the consistency and texture of cheese. *Aust. J. Dairy Technol.* **25**(1), 37, 38.

VanWazer, J. R., J. W. Lyons, K. Y. Kim, and R. E. Colwell. 1963. "Viscosity and Flow Measurement. A Laboratory Handbook of Rheology." Wiley (Interscience), New York.

Vas. K. 1928. Koagulometer zur Beurteilung des Kasebruches. *Milchwirtsch. Forsch.* **6,** 231–235.

Vernon Carter, E. J., and P. Sherman. 1980. Rheological properties and applications of mesquite tree (*Prosopis juliflora*) gum. I. Rheological properties of aqueous mesquite gum solutions. *J. Texture Stud.* **11,** 339–344.

Vickers, Z. M., and M. C. Bourne. 1976a. Crispness in foods—a review. *J. Food Sci.* **41,** 1153–1157.

Vickers, Z. M., and M. C. Bourne. 1976b. A psychoacoustical theory of crispness. *J. Food Sci.* **41,** 1158–1164.

Voisey, P. W. 1971a. Modernization of texture instrumentation. *J. Texture Stud.* **2,** 129–195.

Voisey, P. W. 1971b. The Ottawa texture measuring system. *Can. Inst. Food Technol. J.* **4,** 91–103.

Voisey, P. W. 1972. Updating the shear press. *Can. Inst. Food Sci. Technol. J.* **5,** 6–12.

Voisey, P. W. 1974. Design and operational details of the Ottawa Tenderometer. *Agric. Can., Eng. Res. Serv. Rep.* No. 6820–8.

Voisey, P. W. 1975. Reply to C. A. Pearson and D. Raynor. *J. Texture Stud.* **6,** 394–398.

Voisey, P. W. 1976. Engineering assessment and critique of instruments used for meat tenderness evaluations. *J. Texture Stud.* **7,** 11–48.

Voisey, P. W. 1977a. Examination of operational aspects of fruit pressure tests. *Can. Inst. Food Sci. Technol. J.* **10,** 284–294.

Voisey, P. W. 1977b. Effect of blade thickness on readings from the FTC shear compression cell. *J. Texture Stud.* **7,** 433–440.

Voisey, P. W. 1977c. Some applications of the Ottawa Texture Measuring System. *Eng. Res. Serv., Agric. Can., Rep.* No. 7024–694.

Voisey, P. W., and D. J. Buckley. 1974. Apparatus for routine measurement of food deformation in texture tests at low deformation rates. *Can. Inst. Food Sci. Technol. J.* **7,** 163–165.

Voisey, P. W., and R. Crête. 1973. A technique for establishing instrumental conditions for measuring food firmness to simulate consumer evaluations. *J. Texture Stud.* **4,** 371–377.

Voisey, P. W., and J. M. deMan. 1970. An electronic recording viscometer for food products. *Can. Inst. Food Technol. J.* **3,** 130–135.

Voisey, P. W., and J. M. deMan. 1976. Applications of instruments for measuring food texture. *In* "Rheology and Texture in Food Quality" (J. M. deMan, P. W. Voisey, V. F. Rasper, and D. W. Stanley, eds.), pp. 142–242. Westport, Connecticut.

Voisey, P. W., and E. Larmond. 1974. Examination of factors affecting performance of the Warner–Bratzler meat shear test. *Can. Inst. Food Sci. Technol. J.* **7,** 243–249.

Voisey, P. W., and E. Larmond. 1977. The effect of deformation rate on the relationship between sensory and instrumental measurements of meat tenderness by the Warner–Bratzler method. *Can. J. Food Sci. Technol.* **10,** 307–312.

Voisey, P. W., and I. L. Nonnecke. 1971. Measurement of pea tenderness. I. An appraisal of the FMC pea Tenderometer. *J. Texture Stud.* **2,** 348–364.

Voisey, P. W., and I. L. Nonnecke. 1972a. Measurement of pea tenderness. III. Field comparison of several methods of measurement. *J. Texture Stud.* **3,** 329–358.

Voisey, P. W., and I. L. Nonnecke. 1972b. Measurement of pea tenderness. IV. Development and evaluation of the test cell. *J. Texture Stud.* **3,** 459–477.

Voisey, P. W., and I. L. Nonnecke. 1973a. Some observations regarding pea Tenderometer standardization. *Agric. Can., Eng. Res. Serv., Rep.* No. 6820–5.

Voisey, P. W., and I. L. Nonnecke. 1973b. Summary of results. Pea Tenderometer tests. *Agric. Can., Eng. Res. Serv., Rep.* No. 6820–7.

Voisey, P. W., and I. L. Nonnecke. 1973c. Measurement of pea tenderness. V. The Ottawa Pea Tenderometer and its performance in relation to the Pea Tenderometer and the FTC Texture Test System. *J. Texture Stud.* **4,** 323–343.

Voisey, P. W., D. C. MacDonald, M. Kloek, and W. Foster. 1972. The Ottawa Texture Measuring System. An operational manual. *Agric. Can., Eng. Spec.* No. 7024.

Voisey, P. W., D. J. Buckley, and R. Crête. 1974. A system for recording deformations in texture tests. *J. Texture Stud.* **5,** 61–75.

Voisey, P. W., M. Kloek, K. Summers, and M. Gillette. 1979. A method for testing the ease of extrusion of icing marketed in plastic containers. *J. Texture Stud.* **10,** 435–448.

Volodkevich, N. N. 1938. Apparatus for measurement of chewing resistance or tenderness of foodstuffs. *Food Res.* **3,** 221–225.

Wade, P. 1968. A texture meter for the measurement of biscuit hardness. *In* "Rheology and Texture of Foodstuffs," Monograph No. 27, pp. 225–234. Soc. Chem. Ind., London.

Warner, K. F. 1928. Progress report of the mechanical test for tenderness of meat. *Proc. Am. Soc. Anim. Prod.* 114–116.

Washburn, R. M. 1910. Principles and practice of ice cream making. *Bull.—Vt. Agric. Exp. Stn.* No. 155.

Waugh, F. A. 1901. "Fruit Harvesting, Storing, Marketing," p. 44. Orange Judd Co., New York.

Waugh, L. M. 1937. Dental observations among Eskimos. *J. Dent. Res.* **16,** 356.
Weissenberg, K. 1949. Abnormal substances and abnormal phenomena of flow. *Proc. Int. Rheol. Cong., 1st, Scheveningen, Neth., 1948* pp. I-29–I-46.
Wender, N. 1895. The viscometer examination of butter for foreign fats. *J. Am. Chem. Soc.* **17,** 719–723.
Whitcombe, P. J., and C. W. Macosko. 1978. Rheology of xanthan gum. *J. Rheol.* **22,** 493–505.
White, G. W. 1970. Rheology in food research. *J. Food Technol.* **5,** 1–32.
White, K., and N. N. Mohsenin. 1967. Apparatus for determination of bulk modulus and compressibility of materials. *Trans. ASAE* **10,** 670–671.
Whittle, K. J. 1973. A multiple sampling technique for use with the cell-fragility method of determining deteriorative changes in frozen fish. *J. Sci. Food Agric.* **24,** 1383–1389.
Whittle, K. J. 1975. Improvement of the Torry–Brown homogenizer for the cell fragility method. *J. Food Technol.* **10,** 215–220.
Whorlow, R. W. 1980. "Rheological Techniques." Ellis Harwood, Chichester, England.
Wilder, H. K. 1947. Measurement of fiber content in canned asparagus. *Natl. Canners' Assoc., Res. Lab Rep.* No. 12313B.
Wiley, R. C., N. Elehwany, A. Kramer, and F. J. Hager. 1956. The shear press—an instrument for measuring the quality of foods. IV. Application to asparagus. *Food Technol.* **10,** 439–443.
Willard, J. T., and R. H. Shaw. 1909. Analysis of eggs. *Kans. Agric. Exp. Stn., Bull.* No. 159, pp. 143–177.
Willhoft, E. M. A. 1970. An empirical equation describing the firming of the crumb of bread. *Chem. Ind. (London)* 1017, 1018.
Willhoft, E. M. A. 1973. Mechanism and theory of staling of bread and baked goods, and associated changes in textural properties. *J. Texture Stud.* **4,** 292–322.
Williamson, R. V. 1929. The flow of pseudoplastic materials. *Ind. Eng. Chem.* **21,** 1108–1111.
Wood, A. H., and C. L. Parsons. 1891. Hardness of butter. *N.H. Agric. Exp. Stn., Bull.* No. 13, p. 8.
Worthington, J. T., and J. H. Yeatman. 1968. A statistical evaluation of objective measurement of apple firmness. *Proc. Am. Soc. Hortic. Sci.* **92,** 739–747.
Yang, Y. M., and N. Mohsenin. 1974. Analysis of the mechanics of the fruit pressure tester. *J. Texture Stud.* **5,** 213–238.
Yoshikawa, S., S. Nishimaru, T. Tashiro, and M. Yoshida. 1970a. Collection and classification of words for description and food texture. I. Collection of words. *J. Texture Stud.* **1,** 437–442.
Yoshikawa, S., S. Nishimaru, T. Tashiro, and M. Yoshida. 1970b. Collection and classification of words for description of food texture. II. Texture profiles. *J. Texture Stud.* **1,** 443–451.
Yoshikawa, S., S. Nishimaru, T. Tashiro, and M. Yoshida. 1970c. Collection and classification of words for description of food texture. III. Classification of multivariate analysis. *J. Texture Stud.* **1,** 452–463.
Yurkstas, A. A. 1965. The masticatory act. A review. *J. Prosthet. Dent.* **15,** 248–260.
Yurkstas, A., and W. A. Curby. 1953. Force analyses of prosthetic appliances during function. *J. Prosthet. Dent.* **3,** 82–87.

Index

A

B

C

I

J

K

L

M

N

O

P

Q

R

S

T

U

V

W

Y

Z

FOOD SCIENCE AND TECHNOLOGY

A SERIES OF MONOGRAPHS

Maynard A. Amerine, Rose Marie Pangborn, and Edward B. Roessler, PRINCIPLES OF SENSORY EVALUATION OF FOOD. 1965.

S. M. Herschdoerfer, QUALITY CONTROL IN THE FOOD INDUSTRY. Volume I — 1967. Volume II — 1968. Volume III — 1972.

Hans Reimann, FOOD-BORNE INFECTIONS AND INTOXICATIONS. 1969.

Irvin E. Leiner, TOXIC CONSTITUENTS OF PLANT FOODSTUFFS. 1969.

Martin Glicksman, GUM TECHNOLOGY IN THE FOOD INDUSTRY. 1970.

L. A. Goldblatt, AFLATOXIN. 1970.

Maynard A. Joslyn, METHODS IN FOOD ANALYSIS, second edition. 1970.

A. C. Hulme (ed.), THE BIOCHEMISTRY OF FRUITS AND THEIR PRODUCTS. Volume 1 — 1970. Volume 2 — 1971.

G. Ohloff and A. F. Thomas, GUSTATION AND OLFACTION. 1971.

George F. Stewart and Maynard A. Amerine, INTRODUCTION TO FOOD SCIENCE AND TECHNOLOGY. 1973.

C. R. Stumbo, THERMOBACTERIOLOGY IN FOOD PROCESSING, second edition. 1973.

Irvin E. Liener (ed.), TOXIC CONSTITUENTS OF ANIMAL FOODSTUFFS. 1974.

Aaron M. Altschul (ed.), NEW PROTEIN FOODS: Volume 1, TECHNOLOGY, PART A — 1974. Volume 2, TECHNOLOGY, PART B — 1976. Volume 3, ANIMAL PROTEIN SUPPLIES, PART A — 1978. Volume 4, ANIMAL PROTEIN SUPPLIES, PART B — 1981.

S. A. Goldblith, L. Rey, and W. W. Rothmayr, FREEZE DRYING AND ADVANCED FOOD TECHNOLOGY. 1975.

R. B. Duckworth (ed.), WATER RELATIONS OF FOOD. 1975.

Gerald Reed (ed.), ENZYMES IN FOOD PROCESSING, second edition. 1975.

A. G. Ward and A. Courts (eds.), THE SCIENCE AND TECHNOLOGY OF GELATIN. 1976.

John A. Troller and J. H. B. Christian, WATER ACTIVITY AND FOOD. 1978.

A. E. Bender, FOOD PROCESSING AND NUTRITION. 1978.

D. R. Osborne and P. Voogt, THE ANALYSIS OF NUTRIENTS IN FOODS. 1978.

Marcel Loncin and R. L. Merson, FOOD ENGINEERING: PRINCIPLES AND SELECTED APPLICATIONS. 1979.

Hans Riemann and Frank L. Bryan (eds.), FOOD-BORNE INFECTIONS AND INTOXICATIONS, second edition. 1979.

N. A. Michael Eskin, PLANT PIGMENTS, FLAVORS AND TEXTURES: THE CHEMISTRY AND BIOCHEMISTRY OF SELECTED COMPOUNDS. 1979.

J. G. Vaughan (ed.), FOOD MICROSCOPY. 1979.

J. R. A. Pollock (ed.), BREWING SCIENCE, Volume 1 — 1979. Volume 2 — 1980.

Irvin E. Liener (ed.), TOXIC CONSTITUENTS OF PLANT FOODSTUFFS, second edition. 1980.

J. Christopher Bauernfeind (ed.), CAROTENOIDS AS COLORANTS AND VITAMIN A PRECURSORS:
TECHNOLOGICAL AND NUTRITIONAL APPLICATIONS. 1981.

Pericles Markakis (ed.), ANTHOCYANINS AS FOOD COLORS. 1982.

Vernal S. Packard, HUMAN MILK AND INFANT FORMULA. 1982.

George F. Stewart and Maynard A. Amerine, INTRODUCTION TO FOOD SCIENCE AND TECHNOLOGY, SECOND EDITION. 1982.

Malcolm C. Bourne, FOOD TEXTURE AND VISCOSITY: CONCEPT AND MEASUREMENT. 1982.

R. Macrae (ed.), HPLC IN FOOD ANALYSIS. 1982.

In preparation

Héctor A. Iglesias and Jorge Chirife, HANDBOOK OF FOOD ISOTHERMS: WATER SORPTION PARAMETERS FOR FOOD AND FOOD COMPONENTS. 1982.

John A. Troller, SANITATION IN FOOD PROCESSING. 1983.